Tuco-Tucos

Thales Renato Ochotorena de Freitas
Gislene Lopes Gonçalves • Renan Maestri
Editors

Tuco-Tucos

An Evolutionary Approach to the Diversity
of a Neotropical Subterranean Rodent

 Springer

Editors
Thales Renato Ochotorena de Freitas
Department of Genetics
Federal University of Rio Grande do Sul
Porto Alegre, Rio Grande do Sul, Brazil

Gislene Lopes Gonçalves
Department of Genetics
Federal University of Rio Grande do Sul
Porto Alegre, Rio Grande do Sul, Brazil

Renan Maestri
Department of Ecology
Federal University of Rio Grande do Sul
Porto Alegre, Rio Grande do Sul, Brazil

ISBN 978-3-030-61681-6 ISBN 978-3-030-61679-3 (eBook)
https://doi.org/10.1007/978-3-030-61679-3

This Springer imprint is published by the registered company Springer Nature Switzerland AG
The registered company address is: Gewerbestrasse 11, 6330 Cham, Switzerland

Foreword

The tucotuco (Ctenomys brasiliensis) *is a curious small animal, which may be briefly described as a Gnawer, with the habits of a mole.*
Charles Darwin
The Voyage of the Beagle
Chapter III, Maldonado

Such begin Darwin's comments on tuco-tucos, based on his encounters with these animals along the southern coast of Uruguay. Now, 175 years after the original publication of Darwin's observations, biologists continue to be fascinated by these charismatic rodents. In part, this interest reflects the somewhat mysterious nature of tuco-tucos—even in areas where the animals are locally abundant and can at times be heard calling from all directions, it is often challenging to catch a glimpse of these largely subterranean mammals. Increasingly, though, interest in tuco-tucos reflects our growing understanding of the biology of these animals and the unexpected wealth of phenotypic and genotypic variation that they represent. Indeed, as evident from the contents of this volume, studies of tuco-tucos now encompass analyses of systematic, phylogenetic, morphological, physiological, ecological, and behavioral diversity, providing important opportunities to examine the evolutionary processes underlying divergence within *Ctenomys* and, concomitantly, convergence between these and other lineages of subterranean rodents.

One attribute of tuco-tucos that makes them both intriguing and at times challenging to study is the apparently rapid divergence of species within the genus *Ctenomys*. As Diego Verzi and colleagues indicate (Chap. 1), while the family Ctenomyidae appears to date to the Oligocene, the single extant genus *Ctenomys* has arisen much more recently, with current species-level diversification dating only to the early Pleistocene. This rapid burst of speciation has made it difficult to evaluate the evolutionary relationships among species, as molecular genetic analyses have often failed to reveal much phylogenetic structure, particularly for deeper nodes within the genus. Indeed, as Guillermo D'Elía and co-authors state (Chap. 2), phylogenetic analyses of tuco-tucos are in many ways still in their infancy and will benefit substantially from inclusion of additional data sets (e.g., genomic-level sequencing) as well as more comprehensive sampling of putative species and species groups. At the same time, the rapid diversification within *Ctenomys* creates exciting opportunities to explore the evolutionary mechanisms underlying speciation in this lineage. The role of chromosomal rearrangements in promoting speciation has been a particular focus for studies of tuco-tucos due to the often pronounced karyotypic differences among species, including those for which molecular genetic analyses fail to detect marked differentiation. This theme is examined by Thales de Freitas (Chap. 3), who concludes that evidence for distinct processes of speciation (e.g., allopatric, sympatric, chromosomal) varies in relation to the time since divergence among different members of the genus *Ctenomys*.

Geographically, tuco-tucos are widespread, occurring throughout much of sub-Amazonian South America. At the level of individual taxa, however, it has long been thought that allopatry dominates, with only a few examples of sympatry having been identified within *Ctenomys*. To explore how local spatial relationships among species translate into the genus-level distribution of these animals, Renan Maestri and Bruce Patterson (Chap. 4) characterize geographic variation in several attributes of *Ctenomys*, including patterns of species richness and range size. These authors report that although species ranges tend to be smaller in *Ctenomys*, the exclusivity of these ranges does not differ from that observed in other lineages of caviomorph rodents, providing no evidence that allopatry is particularly pronounced among tuco-tucos. Fernando Mapelli and colleagues (Chap. 5) add a genetic component to analyses of geographic variation, arguing that landscape features may impact the demographic processes that shape patterns of genetic differentiation within and among species of *Ctenomys*. Their review suggests that landscape-level genetic variation reflects a baseline pattern of isolation by distance that is modified by a complex, species-specific interplay between geographic features, environmental conditions, and demographic parameters.

In terms of their gestalt, it has been suggested that if you have seen one tuco-tuco, you have seen them all. This quip reflects the general expectation that the challenges associated with life in underground burrows have acted to constrain morphological and other forms of phenotypic diversification within *Ctenomys*. As knowledge of these animals has increased, it has become increasingly apparent that they are more phenotypically diverse than has been appreciated. Morphologically, variation is evident for multiple cranial traits, and, as reported by Rodrigo Fornel and co-authors (Chap. 6), this variation displays geographic but not phylogenetic signal, suggesting that environmental conditions may play a critical role in shaping skull structure in these animals. One obvious environmental factor that may contribute to this variation is the difference in the soils in which the animals live. As described by Aldo Vassallo and colleagues (Chap. 7), although tuco-tucos rely primarily on their forepaws to dig, they also routinely use their incisors to chew through obstructions or loosen hard chunks of soil. Accordingly, the structure of both the forelimbs and the skull may vary with soil type, and, conversely, constraints on the biomechanics of digging may preclude the animals from occupying particular soils. Soil may also be an important determinant of the underappreciated variation in pelage coloration that occurs within *Ctenomys*. Using comparisons of overall pelage color as well as the structure of individual hairs, Gislene Goncalves (Chap. 8) argues that differences in coloration among species of tuco-tucos from the Atlantic coasts of Brazil and Argentina reflect selection imposed by differences in soil color, with color matching to local substrates serving to protect animals from predation while active on the surface.

Interactions between individuals and their environments are also central to studies of the ecology and physiology of *Ctenomys*. The role of tuco-tucos as ecosystem engineers is examined by Bruno Kubiak and Daniel Gailano (Chap. 9), who also consider the effects of habitat parameters on species' distributions as well as spatial and social relationships among conspecifics. Although relevant data are lacking for many species, the emerging picture is one of greater than expected ecological and behavioral variation within the genus. One critical aspect of a species' ecology is its diet, which can affect not only where animals occur on the landscape but also how they acquire energy and nutrients, thereby providing a particularly direct link between external conditions and intrinsic processes. Although all tuco-tucos are herbivorous, surprisingly few detailed studies of the animals' diets have been conducted. As Carla Lopes (Chap. 10) reports, the growing use of DNA sequencing of fecal samples to characterize diets is creating new opportunities to examine dietary variation within and among members of the genus *Ctenomys*, including the role of diet partitioning in shaping the few examples of sympatry that have been reported for these animals. Maria Sol Fanjul and colleagues (Chap. 11) explore the inner workings of tuco-tucos in greater detail, revealing how differences in habitat conditions as well as differences in how individuals use their habitats contribute to adaptively important variation in multiple physiological systems, including processing of sensory information, response to external stressors, and regulation of both water and energy balance. Extrinsically generated differences in physiology may be mediated by variation in individual phenotypes (e.g., sex, reproductive status), thereby adding an additional layer of complexity to efforts in understanding how external conditions shape the internal biology of tuco-tucos. In the final chapter of the volume, Cristina Matzenbacher and Juliana da Silva (Chap. 12) take a more applied approach to interactions between tuco-tucos and their environments by examining the role of these animals as bioindicators of environmental change, specifically the introduction of heavy metals and other toxic compounds as a result of human activity. More generally, this discussion raises the issue of conservation of the genus *Ctenomys*, thereby serving to connect the previous chapters to the increasingly important need to ensure that members of this lineage are protected from an ever-growing list of threats.

In closing, one theme that resonates throughout this volume is diversity. From systematic and phylogenetic revisions of *Ctenomys* to analyses of interactions between the environment and specific physiological processes, it is clear that studies of tuco-tucos are revealing new and sometimes unexpected patterns of diversification in this relatively young clade of rodents. Coupled with an ever-growing suite of analytical tools, this diversity creates novel opportunities to examine long-standing questions regarding the biology of tuco-tucos. For example, efforts to understand the often marked karyotypic differences among otherwise closely related species should benefit from the use of genomic tools to identify the specific portions of the genome that are impacted by chromosomal rearrangements. Similarly, as our ability to characterize the genomic architecture of specific phenotypic traits increases, we will be better able to examine the genetic bases for adaptive traits such as the specialized morphological features associated with digging. More generally, the expanding catalog of diversity within *Ctenomys* means that members of this genus are increasingly recognized as important models for research on a wide range of evolutionary topics, including studies that explore the effects of pathogen communities on immunological function or the role of ecological factors in generating interspecific differences in social systems. All told, our growing understanding of the biology of tuco-tucos suggests that the "curious small animals" that intrigued Darwin will continue to play a central role in biological research for years to come.

Berkeley, USA   Eileen Lacey

Introduction

This book examines the biology of tuco-tucos (*Ctenomys*) from an evolutionary perspective. *Ctenomys* is a remarkable lineage of subterranean rodents widely distributed over the southern half of South America. It exhibits various adaptations for living underground—mostly solitarily, but in some species also in social groups. Such a peculiar lifestyle has long attracted the attention of scientists, including Charles Darwin. In 1832, during the voyage of the Beagle, Darwin had a memorable experience with tuco-tucos when he stayed in Maldonado, Uruguay, which he registered in his diary (published in 1839).

The next century of scientific studies of *Ctenomys* was mostly limited to species descriptions obtained on European expeditions to South America. Beginning in 1950, a wealth of knowledge on physiology, ecology, genetics, morphology, paleontology, and taxonomy has been documented in scientific journals, as well as in many theses and dissertations dedicated to this intriguing group. Most studies have documented local or regional patterns shown by tuco-tucos; however, global or comprehensive synopses are still needed. We seek to partly fill this gap by inviting investigators that have worked for years both in field and laboratory with extinct and extant tuco-tuco species to review major evolutionary topics and frame these essays with the breadth of current understanding. We hope that the combination of extensive reviews and original information on tuco-tucos, produced by numerous authors, will stimulate future studies.

Among the subterranean rodents, the tuco-tucos (*Ctenomys*) stand along with the Mediterranean mole rats (*Spalax*), North American pocket gophers (*Thomomys* and *Geomys*), and species of Bathyergidae in Africa as the major lineages well-known from long-standing studies. If one considers that tuco-tucos are endemic to the Neotropics—where funding for basic research is limited—the status of knowledge reached for *Ctenomys* is remarkable, resulting from the efforts and passion of many individuals.

With some frequency—before we decide for this book project—we used to ask colleagues during annual scientific meetings if they wonder what maintains the passion for tuco-tucos across distinct generations of scientists (since Darwin) and, particularly, what makes *Ctenomys* an exciting group to be studied. In general, people mentioned that when tuco-tucos are carefully observed in the field, enormous variation is found, such as in their behavior, pelage, skull, and digging ability. The facilities of working with this group, as in the capture of specimens in the field and the possibility of keeping them in captivity, and locating their populations and tracking them in time and space for long years, together with the underlying variation in a rodent that seems uniform, were considered, in a large degree, the triggers for maintaining the passion for tuco-tucos. While compiling background and putting them into context, it became clear—at least explicit—why tuco-tucos are fascinating from an evolutionary perspective, and we felt motivated to organize this volume.

Considering that the articles on *Ctenomys* started from the 1950s of the last century, we believe that there is currently sufficient data spread over various disciplines and well-established lines of research that, after 70 years, should be put in a book trying to make a synthesis of what already exists. We should mention previous books that included *Ctenomys*, starting with *Evolution of Subterranean Mammals at the Organismal and Molecular Levels*, edited by Eviatar Nevo and Oswaldo A. Reig, which included a whole chapter about the genus from an evolutionary point of view (Reig et al. 1990). The genus also appeared on books that featured subterranean rodents such as *Life Underground: The biology of subterranean rodents* (2000), authored by Eileen Lacey, James L. Patton, and Guy N. Cameron, and more recently, in 2010, the book *Subterranean Rodents: News from Underground*, by Sabine Gegall, Hynek Burda, and Cristian E. Schleich. Given how much the scientific community has learned about tuco-tucos since Reig et al. (1990), we believe it is time for *Ctenomys* to have their own book.

Species and local populations of tuco-tucos are the most interesting South American mammals for studying mechanisms underlying speciation. Basically, two principal interconnected aspects drive interest in tuco-tucos: chromosomes and species diversity. *Ctenomys* form one of the most karyotypically diverse clades known in mammals, with chromosomal diploid numbers ranging from 10 to 70 (Cook et al. 1990; Gallardo 1991; Reig et al. 1992; Ortells 1995). In addition, 65 species are recognized for the genus (Teta and D'Elia 2020, Chap. 2, this volume), more than any other group of subterranean rodents (Reig et al. 1990; Lessa and Cook 1998; Castillo et al. 2005; Woods and Kilpatrick 2005). Since *Ctenomys* appeared in the late Pliocene, their extant diversity was achieved by remarkable flurry of speciation events (Verzi et al. 2010; Parada et al. 2011). The age of the genus is quite recent, estimated at ca. 5 Ma according to molecular evidence (Parada et al. 2011; Upham and Patterson 2015), which agrees with the paleontological records (Verzi 1999, 2002; Verzi et al. 2010; Chap. 1, this volume).

Reig and Kiblisky (1969) were the first to propose that tuco-tucos are a prime example of chromosomal speciation. Reig et al. (1990) raised the idea, still accepted, that diversification may have been facilitated by the isolation of small demes that characterize population structure in most species and extensive chromosomal rearrangements (Reig and Kiblisky 1969; Cook et al. 1990; Gallardo 1991; Ortells 1995). In fact, the high intra- and interspecific chromosomal polymorphisms—once suggested as the main factor responsible for fast speciation of *Ctenomys* (Ortells 1995)—do not seem to be directly responsible for its species richness. Thus, rather than a triggering speciation in tuco-tucos, chromosomal rearrangements speed up the process by modulating gene flow rate in certain genome regions (Torgasheva et al. 2017), suggesting an even more complex scenario for cladogenesis.

The collection of chapters in this book articulates research views that are disseminated across major subjects as paleontology, systematics, evolutionary ecology, and genetics. To address such a broad range of topics from the perspective of a single mammal genus is unusual and transcends good-study-model reasoning. Such excess might be based on subjective aspects of tuco-tucos since it is a charismatic animal, which typically touches human feelings, including those from contributors and readers.

We are most grateful to the authors for their willingness to join us and make this book happen. Many thanks are also due to Eileen Lacey for writing the foreword and offering editorial suggestions, as well as to Bruce D. Patterson for helpful comments on our Introduction. We thank Springer Nature Publisher for bringing our project in the current form, particularly Luciana Christante de Mello, Vignesh Viswanathan, and Nolan Mallaigh, for their editorial assistance.

Porto Alegre, Rio Grande do Sul, Brazil   Thales Renato Ochotorena de Freitas
                                                Gislene Lopes Gonçalves
                                                           Renan Maestri

Literature Cited

Castillo AH, Cortinas MN, Lessa EP (2005) Rapid diversification of South American tuco-tucos (*Ctenomys*; Rodentia, Ctenomyidae): Contrasting mitochondrial and nuclear intron sequences. J Mammal 86:170–179

Cook JA, Anderson S, Yates TL (1990) Notes on Bolivian Mammals 6: the Genus *Ctenomys* (Rodentia, Ctenomyidae) in the Highlands. Am Mus Novit 2980:1–27

Gallardo MH (1991) Karyotypic evolution in *Ctenomys* (Rodentia, Ctenomyidae). J Mammal 72:1–21

Lessa EP, Cook JA (1998) The molecular phylogenetics of tuco-tucos (genus *Ctenomys*, Rodentia: Octodontidae) suggests an early burst of speciation. Mol Phylogenet Evol 9:88–99

Ortells MO (1995) Phylogenetic analysis of G-banded karyotypes among the South American subterranean rodents of the genus *Ctenomys* (Caviomorpha: Octodontidae), with special reference to chromosomal evolution and speciation. Biol J Linnean Soc 54:43–70

Parada A, D'Elia G, Bidau CJ, Lessa EP (2011) Species groups and the evolutionary diversification of tuco-tucos, genus *Ctenomys* (Rodentia: Ctenomyidae). J Mammal 92: 671–682

Reig OA, Kiblisky P (1969) Chromosome multiformity in the genus Ctenomys (Rodentia, Octodontidae). Chromosoma 28:211–244

Reig O, Busch C, Contreras J, Ortells M (1990) An overview of evolution, systematic, population biology and molecular biology. In: Nevo E, Reig OA (eds) Biology of Subterranean mammals. Wiley-Liss, New York, pp 71–96

Reig O, Massarini A, Ortells M, Barros M, Tiranti S, Dyzenchauz F (1992) New karyotypes and C-banding patterns of the subterranean rodents of the genus *Ctenomys* (Caviomorpha, Octodontidae) from Argentina. Mammalia 56:603–624

Teta P, D'Elia G (2020) Uncovering the species diversity of subterranean rodents at the end of the World: three new species of Patagonian tuco-tucos (Rodentia, Hystricomorpha, *Ctenomys*). PeerJ 8:e9259

Torgasheva AA, Basheva EA, Gómez Fernández MJ, et al (2017) Chromosomes and speciation in tuco-tuco (*Ctenomys*, Hystricognathi, Rodentia). Russ J Genet Appl Res 7:350–357

Upham NS, Patterson, BD (2015) Evolution of caviomorph rodents: a complete phylogeny and timetree for living genera. Biology of Caviomorph Rodents: Diversity and Evolution 1:63–120

Verzi DH, Olivares AI, Morgan CC (2010) The oldest South American tuco-tuco (late Pliocene, northwestern Argentina) and the boundaries of the genus Ctenomys (Rodentia, Ctenomyidae). Mamm Biol 75:243–252

Verzi DH (2002) Patrones de evolución morfológica en Ctenomyinae (Rodentia, Octodontidae). Mastozool Neotrop 9:309–328

Verzi DH (1999) The dental evidence on the differentiation of the ctenomyine rodents (Caviomorpha, Octodontidae, Ctenomyinae). Acta Theriologica 44:263–282

Woods C Kilpatrick C (2005) Infraorder Hystricognathi Brandt, 1855. In: Mammal species of the world: a taxonomic and geographic reference, vol 2, 3rd edn. pp 1538–1600

Contents

Contributors

Alicia Álvarez Instituto de Ecoregiones Andinas, CONICET, Jujuy, Argentina

C. Daniel Antenucci Grupo 'Ecología Fisiológica y del Comportamiento', Instituto de Investigaciones Marinas y Costeras, Universidad Nacional de Mar del Plata, Consejo Nacional de Investigaciones Científicas y Técnicas, Mar del Plata, Argentina

Ailin Austrich Departamento de Biología, Facultad de Ciencias Exactas y Naturales, Universidad Nacional de Mar del Plata, Instituto de Investigaciones Marinas y Costeras (IIMyC), CONICET – UNMdP, Mar del Plata, Buenos Aires, Argentina

Federico Becerra Grupo Morfología Funcional y Comportamiento, Instituto de Investigaciones Marinas y Costeras (IIMyC, UNMDP-CONICET), Universidad Nacional de Mar del Plata (UNMdP), Consejo Nacional de Investigaciones Científicas y Técnicas (CONICET), Mar del Plata, Buenos Aires, Argentina

Valentina Brachetta Grupo 'Ecología Fisiológica y del Comportamiento', Instituto de Investigaciones Marinas y Costeras, Universidad Nacional de Mar del Plata, Consejo Nacional de Investigaciones Científicas y Técnicas, Mar del Plata, Argentina

Guido N. Buezas Grupo Morfología Funcional y Comportamiento, Instituto de Investigaciones Marinas y Costeras (IIMyC, UNMDP-CONICET), Universidad Nacional de Mar del Plata (UNMdP), Consejo Nacional de Investigaciones Científicas y Técnicas (CONICET), Mar del Plata, Buenos Aires, Argentina

Mariana Cohen Grupo Histología e Histoquímica, Instituto de Investigaciones Marinas y Costeras (IIMyC, UNMDP-CONICET), Universidad Nacional de Mar del Plata (UNMdP), Consejo Nacional de Investigaciones Científicas y Técnicas (CONICET), Mar del Plata, Buenos Aires, Argentina

Pedro Cordeiro-Estrela Departamento de Sistemática e Ecologia, Centro de Ciências Exatas e da Natureza – Campus I, Universidade Federal da Paraíba, Jardim Universitário s/n, João Pessoa, PB, Brazil

Ana Paula Cutrera Grupo 'Ecología Fisiológica y del Comportamiento', Instituto de Investigaciones Marinas y Costeras, Universidad Nacional de Mar del Plata, Consejo Nacional de Investigaciones Científicas y Técnicas, Mar del Plata, Argentina

Juliana da Silva Universidade LaSalle – UNILASALLE and Universidade Luterana do Brasil - ULBRA, Canoas, RS, Brazil

Thales Renato Ochotorena de Freitas Department of Genetics, Federal University of Rio Grande do Sul, Porto Alegre, Rio Grande do Sul, Brazil

Nahuel A. De Santi Sección Mastozoología, CONICET, Museo de La Plata, La Plata, Argentina

Guillermo D'Elía Instituto de Ciencias Ambientales y Evolutivas, Universidad Austral de Chile, Valdivia, Chile

Alcira O. Díaz Grupo Histología e Histoquímica, Instituto de Investigaciones Marinas y Costeras (IIMyC, UNMDP-CONICET), Universidad Nacional de Mar del Plata (UNMdP), Consejo Nacional de Investigaciones Científicas y Técnicas (CONICET), Mar del Plata, Buenos Aires, Argentina

Alejandra I. Echeverría Grupo Morfología Funcional y Comportamiento, Instituto de Investigaciones Marinas y Costeras (IIMyC, UNMDP-CONICET), Universidad Nacional de Mar del Plata (UNMdP), Consejo Nacional de Investigaciones Científicas y Técnicas (CONICET), Mar del Plata, Buenos Aires, Argentina

María Sol Fanjul Grupo 'Ecología Fisiológica y del Comportamiento', Instituto de Investigaciones Marinas y Costeras, Universidad Nacional de Mar del Plata, Consejo Nacional de Investigaciones Científicas y Técnicas, Mar del Plata, Argentina

Rodrigo Fornel Programa de Pós-Graduação em Ecologia, Departamento de Ciências Biológicas, Universidade Regional Integrada do Alto Uruguai e das Missões – Campus de Erechim, Erechim, RS, Brazil

Daniel Galiano Laboratório de Zoologia, Universidade Federal da Fronteira Sul, Realeza, Brazil

Gislene Lopes Gonçalves Departamento de Genética, Universidade Federal do Rio Grande do Sul, Porto Alegre, RS, Brazil

Departamento de Recursos Ambientales, Facultad de Ciencias Agronómicas, Universidad de Tarapacá, Arica, Chile

Marcelo Javier Kittlein Departamento de Biología, Facultad de Ciencias Exactas y Naturales, Universidad Nacional de Mar del Plata, Instituto de Investigaciones Marinas y Costeras (IIMyC), CONICET – UNMdP, Mar del Plata, Buenos Aires, Argentina

Bruno Busnello Kubiak Programa de Pós-Graduação em Genética e Biologia Molecular, Universidade Federal do Rio Grande do Sul, Porto Alegre, Brazil

Enrique P. Lessa Departamento de Ecología y Evolución, Facultad de Ciencias, Universidad de la República, Montevideo, Uruguay

M. Victoria Longo Grupo Histología e Histoquímica, Instituto de Investigaciones Marinas y Costeras (IIMyC, UNMDP-CONICET), Universidad Nacional de Mar del Plata (UNMdP), Consejo Nacional de Investigaciones Científicas y Técnicas (CONICET), Mar del Plata, Buenos Aires, Argentina

Carla Martins Lopes Departamento de Biodiversidade e Centro de Aquicultura, Instituto de Biociências, Universidade Estadual Paulista (UNESP), Rio Claro, SP, Brazil

Facundo Luna Grupo 'Ecología Fisiológica y del Comportamiento', Instituto de Investigaciones Marinas y Costeras, Universidad Nacional de Mar del Plata, Consejo Nacional de Investigaciones Científicas y Técnicas, Mar del Plata, Argentina

Renan Maestri Departamento de Ecologia, Universidade Federal do Rio Grande do Sul, Porto Alegre, RS, Brazil

Fernando Javier Mapelli División Mastozoología, Museo Argentino de Ciencias Naturales "Bernardino Rivadavia", Buenos Aires, Ciudad Autónoma de Buenos Aires, Argentina

Cristina A. Matzenbacher Programa de Pós-Graduação em Genética e Biologia Molecular, Universidade Federal do Rio Grande do Sul - UFRGS, Porto Alegre, Brazil

Matías Sebastián Mora Departamento de Biología, Facultad de Ciencias Exactas y Naturales, Universidad Nacional de Mar del Plata, Instituto de Investigaciones Marinas y Costeras (IIMyC), CONICET – UNMdP, Mar del Plata, Buenos Aires, Argentina

Cecilia C. Morgan Sección Mastozoología, CONICET, Museo de La Plata, La Plata, Argentina

A. Itatí Olivares Sección Mastozoología, CONICET, Museo de La Plata, La Plata, Argentina

Bruce D. Patterson Negaunee Integrative Research Center, Field Museum of Natural History, Chicago, IL, USA

Cristian E. Schleich Grupo 'Ecología Fisiológica y del Comportamiento', Instituto de Investigaciones Marinas y Costeras, Universidad Nacional de Mar del Plata, Consejo Nacional de Investigaciones Científicas y Técnicas, Mar del Plata, Argentina

Pablo Teta División Mastozoología, Museo Argentino de Ciencias Naturales "Bernardino Rivadavia", Buenos Aires, Argentina

Aldo I. Vassallo Grupo Morfología Funcional y Comportamiento, Instituto de Investigaciones Marinas y Costeras (IIMyC, UNMDP-CONICET), Universidad Nacional de Mar del Plata (UNMdP), Consejo Nacional de Investigaciones Científicas y Técnicas (CONICET), Mar del Plata, Buenos Aires, Argentina

Diego H. Verzi Sección Mastozoología, CONICET, Museo de La Plata, La Plata, Argentina

Roxana R. Zenuto Grupo 'Ecología Fisiológica y del Comportamiento', Instituto de Investigaciones Marinas y Costeras, Universidad Nacional de Mar del Plata, Consejo Nacional de Investigaciones Científicas y Técnicas, Mar del Plata, Argentina

About the Editors

Thales Renato Ochotorena de Freitas has been a full professor in the Department of Genetics, Institute of Biosciences, Federal University of Rio Grande do Sul (UFRGS), since 1985. He earned his Ph.D. in genetics and molecular biology (1990) from UFRGS. He has been a visiting scholar at the Museum of Vertebrate Zoology, University of California-Berkeley (1996–1997), and visiting researcher at Laboratoire d'Écologie Alpine, University of Grenoble (2013–2014), where he worked on cutting-edge genomic tools. Dr. Freitas is currently investigating variation at population and species levels in *Ctenomys*, particularly from groups on the border of the Amazon Forest and coastal plain, using cytogenetic and molecular genetics. Dr. Freitas serves as director of the Graduate Program in Genetics and Molecular Biology at UFRGS and was the founder president of the Brazilian Society of Mastozoology (SBMZ). He teaches classical genetics and conservation genetics to undergraduate and graduate students.

Gislene Lopes Gonçalves has been a research collaborator in the Department of Genetics, Federal University of Rio Grande do Sul (UFRGS), Brazil, and at the Universidad de Tarapacá, Chile, since 2014. She earned her Ph.D. in genetics and molecular biology (2011) from UFRGS, where she completed her postdoctoral studies in zoology (2012–2016). She has been a visiting researcher at the Museum of Comparative Zoology, Harvard University (2009–2010), where she worked on the genetic basis of hair pigmentation in rodents with Professor Hopi Hoekstra. Dr. Gonçalves is currently interested in editorial works and collaborates with numerous research groups in the scope of evolutionary genetics.

Renan Maestri has been an assistant professor in the Department of Ecology, Institute of Biosciences, Federal University of Rio Grande do Sul (UFRGS), since 2018. He is also a research associate at The Field Museum of Natural History, Chicago, IL, USA, since 2020. He earned his Ph.D. in ecology (2017) from UFRGS, with a period in The Field Museum, and completed his postdoctoral studies in the Graduate Program in Animal Biology (2017–2018), UFRGS. Dr. Maestri teaches and advises students in the field of ecology and evolution, and coordinates the

Laboratory of Ecomorphology and Macroevolution at the Department of Ecology, UFRGS. He is interested in the areas of evolutionary biology, ecology, and biogeography, with an emphasis on morphological evolution, macroevolution, and macroecology of complex phenotypes.

Part I
Evolution of *Ctenomys*

Chapter 1
The History of *Ctenomys* in the Fossil Record: A Young Radiation of an Ancient Family

Diego H. Verzi, Nahuel A. De Santi, A. Itatí Olivares, Cecilia C. Morgan, and Alicia Álvarez

1.1 Introduction

Ctenomyidae is a clade of South American hystricomorph rodents with a peculiar evolutionary history characterized by: strong morphological differentiation, i.e., modernization that took place in the late Miocene; extinction of lineages during the Plio-Pleistocene, which led to *Ctenomys* being the only representative of the clade in the living fauna; and an extremely high rate of speciation of the latter genus, which is unmatched among caviomorphs (Reig et al. 1990; Lessa et al. 2008; Verzi 2008; Verzi et al. 2014, 2016; Álvarez et al. 2017, 2020). The stage of morphological differentiation is defined by the acquisition of a unique dental morphology, which persists in living species (Reig 1970). Because of its uniqueness, the appearance of this dental morphology has dictated the recognition of ctenomyids in the fossil record (Wood 1955; Reig et al. 1990; Arnal and Vucetich 2015). In addition, the skeletal morphology of modern ctenomyids diversified in adjustment to life underground. Because of their unequivocal recognition, as well as their appealing adaptive diversification, these modern representatives have attracted the attention of paleontologists almost exclusively; the corpus of information produced, primarily systematic and paleobiological, has provided knowledge on the boundaries of specialization explored by at least part of the clade throughout its history (Reig and Quintana 1992; Casinos et al. 1993; Quintana 1994; Fernández et al. 2000; Vieytes et al. 2007; Lessa et al. 2008; Verzi 2008; Morgan and Verzi 2011).

With regard to the other major contribution of fossils, i.e., the estimation of the time of origin and extinction of lineages and clades, this is an issue that remains still

D. H. Verzi (✉) · N. A. De Santi · A. I. Olivares · C. C. Morgan
Sección Mastozoología, CONICET, Museo de La Plata, La Plata, Argentina
e-mail: dverzi@fcnym.unlp.edu.ar; ndesanti@fcnym.unlp.edu.ar; iolivares@fcnym.unlp.edu.ar; cmorgan@fcnym.unlp.edu.ar

A. Álvarez
Instituto de Ecoregiones Andinas, CONICET, Jujuy, Argentina

© Springer Nature Switzerland AG 2021
T. R. O. de Freitas et al. (eds.), *Tuco-Tucos*,
https://doi.org/10.1007/978-3-030-61679-3_1

3

partially unresolved and unclear for ctenomyids. This stems not only from the need for more, and more exhaustive, phylogenetic analyses that include extinct species but also from the dissimilar current interpretations regarding the evolutionary meaning of these extinct taxa. Fossils provide estimations of the divergence times of clades as raw information, and subsequently through the calibration of molecular clocks (Benton and Donoghue 2007; Ronquist et al. 2016). Consequently, they play a central role in the analysis of evolutionary patterns, models, and rates. Nevertheless, reliable age estimations require hypotheses regarding the correspondence of fossils to the different evolutionary stages of a clade: origin, modernization, and establishment of the crown group (Hennig 1965). As previously mentioned, the paleontological studies of ctenomyids have been essentially focused on the modern species; however, in this context of analysis, these have little to contribute to the knowledge of the origin of the clade, or even, depending on their phylogenetic position, of the origin and diversification of the crown group. Thus, achieving an understanding of the evolutionary pattern of this family, including the times and rates of taxonomic and morphological diversification of living species (e.g., Álvarez et al. 2017; Caraballo and Rossi 2018), still requires a more accurate interpretation of the fossil record.

In this chapter, we offer a brief review of the history of the family Ctenomyidae such as it can be interpreted through its fossil record. In addition to describing major characteristics of this history, we provide a critical assessment of the potential contribution of available information to the estimation of divergence times through the phylogeny of the family, with emphasis on its single living representative, *Ctenomys*.

1.2 Stem Ctenomyids and the Understanding of Ctenomyid Origin

The family Ctenomyidae has been traditionally recognized by the rootless molars with exceptionally simplified occlusal surfaces that characterize its late Miocene to Recent representatives (Simpson 1945; Wood 1955; Reig et al. 1990; Vucetich et al. 1999; Arnal and Vucetich 2015). Alternatively, Verzi (1999) proposed an octodontoid with conservative rootless molars with lophids and flexids, the lower late Miocene †*Chasichimys*, as potential ancestor of the modern ctenomyids (see also Verzi et al. 2004a). Later phylogenetic analyses supported the position of †*Chasichimys*, the related †*Chasicomys* (late Miocene), and the older †*Sallamys* (late Oligocene), †*Willidewu*, and †*Protadelphomys* (early Miocene) as stem ctenomyids (Fig. 1.1; Verzi et al. 2014, 2016). However, this unorthodox phylogenetic hypothesis is far from consensus. With the exception of †*Chasicomys* (see Pascual 1967), these genera were initially assigned to Echimyidae (e.g., Simpson 1945; Wood 1955; Wood and Patterson 1959; Patterson and Pascual 1968; Patterson and Wood 1982; Vucetich and Verzi 1991), a family whose living representatives maintain rooted molars with conservative morphologies (Verzi et al. 2016, Fig. 1.1). Phylogenetic analyses based essentially on dental characters have supported the

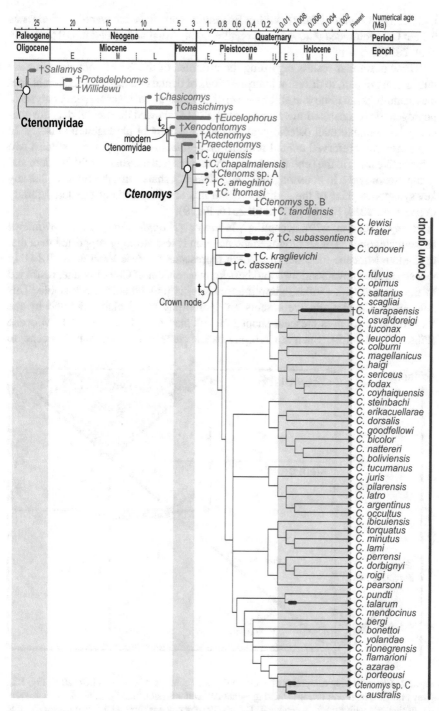

Fig. 1.1 Strict consensus of eight most parsimonious trees resulting from parsimony analysis of combined morphological and molecular data. Divergence times for species of the crown group is according to a Bayesian tip-dating analysis by De Santi et al. (unpubl. results)

inclusion of †*Protadelphomys* within Echimyidae and have recovered †*Sallamys* as a stem Octodontoidea (Carvalho and Salles 2004; Arnal and Vucetich 2015; Boivin et al. 2019).

These dissimilar results regarding the affinities of these early octodontoids are due, at least in part, to different interpretations of dental characters for different species samplings, an issue that still needs critical revision. In a recent meta-analysis of phylogenetic reconstructions, Sansom et al. (2017) showed that dental data are generally less reliable than osteological data as indicators of phylogenetic history. In this sense, the preserved cranial remains of †*Protadelphomys* possess at least two informative traits in the orbital and auditory regions that are shared with modern and living ctenomyids; in addition, this genus does not share with the Echimyidae any key synapomorphies of the auditory region that are diagnostic of this latter family (Verzi et al. 2014, Fig. 6; Verzi et al. 2016, Fig. 9).

In any case, even when accepting †*Sallamys*, †*Protadelphomys*, and †*Willidewu* as stem ctenomyids, their phylogenetic position is less strongly supported than that of the late Miocene-Pleistocene modern representatives (see Verzi et al. 2014). In this sense, the abovementioned and best-known concept of Ctenomyidae, restricted to the species with rootless simplified molars, is undoubtedly more stable. This concept of Ctenomyidae represents an apomorphy-based clade, defined by the acquisition of rootless crescent-shaped molars that took place in the late Miocene (Figs. 1.1 and 1.2). Such apomorphy-based clade comprises late Miocene to

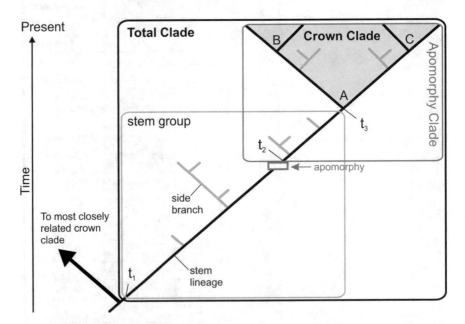

Fig. 1.2 Chart showing categories of clades and related concepts after de Queiroz (2007, Fig. 1.2). Grey branches represent lineages lacking extant descendants (side branches); black branches represent lineages with extant descendants. The apomorphy clade is represented as corresponding to stage t2 by assuming that the marked apomorphy is that which defines the beginning of this stage

Pleistocene stem representatives and the living *Ctenomys* (Fig. 1.1). Differentiation of lineages and genera among these modern ctenomyids would have resulted from the acquisition of disparate adaptations to digging and life underground (Reig and Quintana 1992). Four cohesive lineages are recognized: †*Eucelophorus* (early Pliocene-middle Pleistocene), †*Xenodontomys*-†*Actenomys* (late Miocene-late Pliocene), †*Praectenomys* (Pliocene), and †*Ctenomys* (late Pliocene-Recent). The †*Xenodontomys*-†*Actenomys* lineage would have had fossorial habits, while *Ctenomys* and †*Eucelophorus* independently acquired craniodental specializations for subterranean life (definition of fossorial and subterranean habits follows Lessa et al. 2008); †*Praectenomys* would have been at least fossorial (Verzi 2008; Verzi et al. 2010).

Beyond their different support or stability, the previously mentioned alternative definitions of Ctenomyidae are conceptually different and represent different times of the history of the clade. Three successive stages can be recognized in the evolutionary history of any clade with living representatives, referred to as t1, t2, and t3 by Hennig (1965: Fig. 1.4): t1, the time of its origin by divergence from the most closely related clade with living representatives; t2, its time of morphological differentiation or modernization by the acquisition of the apomorphy or apomorphies that characterize its extant members; and t3, the time of the origin of the last common ancestor of the living representatives. The nested clades that result from each of these points of origin are defined as a total clade, apomorphy clade, and crown clade, respectively (Fig. 1.2; de Queiroz 2007). A total clade comprises the crown clade and its corresponding stem group. The stem group is by definition paraphyletic and includes both extinct species that are directly ancestral to the crown, i.e., those belonging to the stem lineage, and those that are not directly ancestral, i.e., side branches.

In this context, modern ctenomyids with derived molars represent Hennig's stage t2 (Fig. 1.1). Hennig (1965: 114) pointed out that the delimitation of the stage of morphological differentiation, t2, depends on subjective criteria concerning the interpretation of the emergence of particular "types" or "Baupläne". We consider that this stage is related to change within lineages, and although its delimitation may imply subjectivity, it can yield important evolutionary information on environmentally-driven morphological changes (Verzi et al. 2014, 2015). Beyond this, even though many of the fossils at this stage of morphological differentiation may be stem representatives, as occurs in ctenomyids (Fig. 1.1), they do not provide relevant contributions to the interpretation of the origin of the total clade within which they are nested. The practice of interpreting the origin of clades from the first appearances of the main diagnostic characters shared with extant representatives should be assumed as an operational restriction. As pointed out by de Queiroz (2007: 968), the origins of total clades have to do with lineage splitting rather than with character state transformations. Consequently, for an apomorphy to be present in the earliest members of a total clade, that apomorphy would have to have arisen and become fixed simultaneously with the lineage-splitting event in which the clade originated. Because of the nature of evolutionary processes and hierarchies involved in that lineage-splitting event, the latter is not to be expected. As a result, early stem members share few "non-key" apomorphies with their corresponding crown-group (Steiper and Young 2008). In any case, the difficulty of separating the earliest representatives of two diverging extant clades does not negate the validity of the splitting point as the origin of the resulting clades (Briggs and Fortey 2005: 100). Thus, efforts focused on the recognition of plesiomorphic early stem lineages and side branches are indispensable to interpret the deep history of a surviving clade.

Here we lend more support to the idea of applying the name Ctenomyidae to the total clade (see below). †*Sallamys*, recorded in the late Oligocene of Bolivia and Peru, is its earliest stem representative (Fig. 1.1; Patterson and Wood 1982; Shockey et al. 2009). Recently, Pérez et al. (2019) transferred the species †*Sallamys quispea* from the late Oligocene of Peru to the genus †*Migraveramus*; we consider that the molar morphology of †*S. quispea* is comparable to that of the type species †*Sallamys pascuali* albeit less abbreviated (Shockey et al. 2009, Fig. 6), and therefore suggest that the former species should remain to be assigned to †*Sallamys*.

The proposals of paleontological ages younger than 10 My for this family (e.g., Reig 1989; Reig et al. 1990; Vucetich et al. 1999; Arnal and Vucetich 2015; Vucetich et al. 2015) should be reinterpreted as associated to the beginning of the modernization stage, t2 (Fig. 1.1). According to biochronological data, the earliest species corresponding to this stage, †*Xenodontomys simpsoni*, is approximately 6 My old.

1.3 The Genus *Ctenomys*

Information on the early history of the lineage that leads to *Ctenomys* is fragmentary and unclear. The available data hinder a temporal assignment more precise than the entire Pliocene for the divergence of this lineage from the sister genus †*Praectenomys* (see review of the age of Umala Formation in Cione and Tonni 1996). An unpublished mandibular fragment affine to *Ctenomys*, but with only a slight reduction of m3, was recently found in the early Pliocene of western Argentina (Verzi unpublished). Although no phylogenetic analyses have yet been made, this new fossil would represent an intermediate step in the acquisition of the apomorphies that characterize the *Ctenomys* lineage, being closer to the latter than to the one currently considered as sister genus, †*Praectenomys*.

Thus, while *Ctenomys* is a morphologically cohesive genus in the living fauna (Reig et al. 1990; Vassallo and Mora 2007), its boundaries become less evident when the variation of the oldest related extinct species is considered. By application of an adaptation-rooted criterion, which involves an assessment of both the monophyly and the adaptive profiles to delimit genera in the fossil record (Wood and Collard 1999; Cela-Conde and Ayala 2003), the species †*Ctenomys uquiensis* (late Pliocene) and †*Paractenomys chapalmalensis* (lower early Pleistocene) have been considered as the earliest members of the genus *Ctenomys* (Fig. 1.3; Verzi 2008; Verzi et al. 2010). Although part of their traits are undoubtedly plesiomorphic with respect to living species (Verzi 2002; Morgan and Verzi 2006, 2011), functionally significant specializations and the conserved allometry of their masseteric

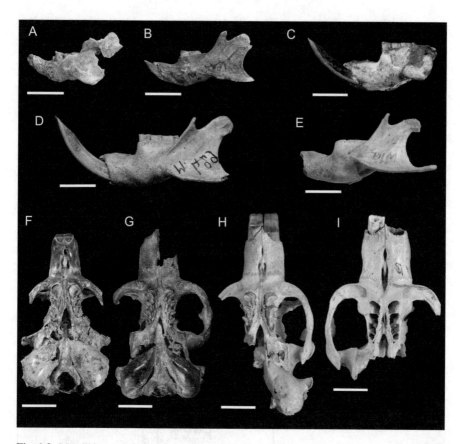

Fig. 1.3 Mandibles and skulls of some extinct species of *Ctenomys* mentioned in the text. Left mandible of: (**a**). †*C. uquiensis* MLP 96-II-29-1 (holotype); (**b**). †*C. chapalmalensis* MMP 1622-M (right reversed); (**c**). †*C. dasseni* PVL 739 (holotype); (**d**). †*C. kraglievichi* MMP M-429 (right reversed); (**e**). †*C. viarapaensis* MLP 2966. Ventral view of skull of: (**f**). †*C. chapalmalensis* MMP 481-S; (**g**). †*C. dasseni* (holotype of †*C. intermedius* MACN 1849); (**h**). †*C. kraglievichi* MSC MS 20–1; (**i**). †*C. viarapaensis* MLP 2935 (holotype). MACN, Museo Argentino de Ciencias Naturales, Buenos Aires, Argentina; MLP, Museo de La Plata, Argentina; MMP, Museo de Ciencias Naturales de Mar del Plata, Argentina; MSC, Museo de Ciencias Naturales de Santa Clara, Argentina; PVL: Colección Paleontología Vertebrados, Instituto Miguel Lillo, San Miguel de Tucumán, Argentina

morphology, in the comparative context of the modern ctenomyids, support the inclusion of these species within the genus (Verzi 2008; Verzi et al. 2010).

The genus *Ctenomys* thus delimited represents an apomorphy clade with a minimum age, given by †*C. uquiensis*, close to 3.5 My (Fig. 1.1). However, given that these species are ancestral to the crown clade (Verzi 2008; Verzi et al. 2010; De Santi et al. 2020), this estimation marks a maximum constraint (softbound) on the age of the crown (Benton and Donoghue 2007). Ongoing phylogenetic analyses (De Santi et al. 2020; unpublished results) that include the most complete fossil materials of extinct species accepted as valid suggest that the crown clade *Ctenomys* is

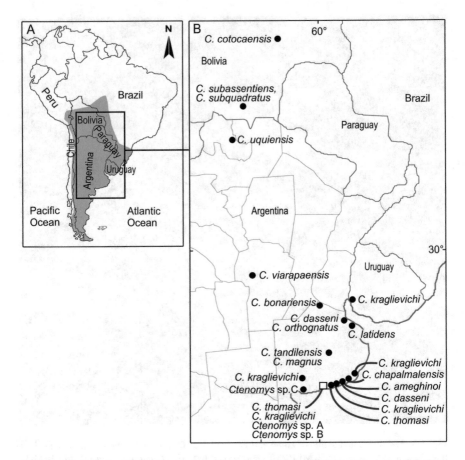

Fig. 1.4 Geographic distribution of the extinct species of *Ctenomys* accepted as valid. Open square, Necochea locality (see text)

younger than previously considered; this is supported both by evidence from the fossil record and by a calibrated tree obtained through Bayesian analysis (Fig. 1.1 and Table 1.1). Even species from the early and middle Pleistocene of central Argentina (Figs. 1.1, 1.3, and 1.4), markedly younger than †*C. uquiensis* and †*C. chapalmalensis*, are stem representatives according to the phylogenetic hypotheses obtained.

An alternative stance could be that the name *Ctenomys* be restricted to the crown clade (see de Queiroz 2007), in which case other genera should be erected for the stem species. If such a definition were adopted, the content of the clade (as of any other crown clade) would depend on extinction (see Budd and Mann 2020). In fact, the decoupling between the origin (t1) and the rise of the crown group (t3) in any clade results from extinction (Verzi et al. 2016). Although extinction within the variation of *Ctenomys* remains to be studied, species and populations of this genus are vulnerable to this phenomenon on account of some of their distinctive

Table 1.1 Estimated ages (in My) of origin of the crown clade *Ctenomys* and the total clade Ctenomyidae (i.e., Ctenomyidae/Octodontidae divergence). Values from this study are fossil-based estimates of minimum ages; value from De Santi et al. (unpubl.) is from a Bayesian analysis using fossil constraints after the tree in Fig. 1.1

	Ctenomys	Octodontidae/Ctenomyidae
This study	* ~ 0.78	** ~ 26.25
De Santi et al. (unpubl.)	1.3	–
Caraballo and Rossi (2018)	5.88 (4.32–7.54)	21.35 (15.59–26.69)
Álvarez et al. (2017)	11.13	25.96
Upham and Patterson (2015)	6.0 (4.6–7.6)	18.9 (15.7–22.1)
Upham and Patterson (2015)	4.3 (2.2–7.4)	19.1 (14.3–23.5)
Parada et al. (2011)	9.2 (6.4–12.6)	17.9 (13.5–23.0)
Castillo et al. (2005)	3.7, 1.3	–
Lessa and Cook (1998)	5.1 (3.3–6.9), 1.4, 1.1	–

*†*Ctenomys dasseni* (Soibelzon et al. 2009); **†*Sallamys quispea* (Shockey et al. 2009)

ecological features, such as patchy distribution, limited vagility, and small effective numbers (Reig et al. 1990). In support of the latter, a marked turnover in the diversity of *Ctenomys* that involved extinction is detected along a 500 kyr Pleistocene stratigraphic sequence (ca. 1 Ma to 0.5 Ma) in the locality of Necochea, in east-central Argentina (Fig. 1.4; Bidegain et al. 2005). In addition, the recently described †*C. viarapaensis* from the Holocene of central Argentina is abundant in the fossil record for ca. 7 kyr until its (at least local) extinction at 0.36 kyr BP (De Santi et al. 2020). This longevity pattern is quite different from that of species of other modern ctenomyid genera (and even other small mammals, see Prothero 2014, Table 1.1) whose duration is over a million years (Verzi et al. 2015, Fig. 5.3).

In addition, the application of a name to a crown clade that contains extinct species also entails a decision regarding the hierarchy of the clade being delimited. In the case of Fig. 1.2, two sister crown clades, B and C, could alternatively be recognized within the indicated crown clade A.

Accordingly, defining *Ctenomys* based on the apomorphy clade is more stable, given that it is not dependent on extinction.

Beyond which definition of *Ctenomys* is adopted, it is clear that this genus reached its unusually high current taxonomic and ecological diversification, as well as its widespread distribution and considerable morphofunctional disparity (Lacey and Wieczorek 2003; Bidau 2015; Freitas 2016; Borges et al. 2017; Álvarez et al. 2017; Morgan et al. 2017), in a surprisingly short lapse. The oldest representative of the crown clade *Ctenomys*, †*C. dasseni,* is about 0.78 Ma old according to biochronology and magnetostratigraphy (level D of Punta Hermengo locality, central Argentina; Soibelzon et al. 2009). Remarkably, the three Pleistocene species recovered as members of the crown group, i.e., †*C. dasseni*, †*C. subassentiens*, and †*C. kraglievichi*, were clustered into the earliest diverged clade known as *frater* group (Parada et al. 2011; Bidau 2015), which currently inhabits the Bolivian-Paraguayan Chaco and Andean zones. This supports the previous interpretation of the record of †*C. kraglievichi* as representing the irruption of a member of this clade into the pampean region (currently dominated by the *mendocinus* group), associated to an important middle Pleistocene warm pulse (near 0.5 Ma; Verzi et al. 2004b).

1.4   Final Considerations

Like the other families of South American hystricomorph rodents (Upham and Patterson 2015; Álvarez et al. 2017), Ctenomyidae has a deep evolutionary history, which is evidenced in the fossil record of this family since the late Oligocene. These earliest fossils, closer to the origin of the clade, and the first (middle Pleistocene) members of the crown clade are the ones that contribute relevant temporal information for the calibrations of molecular phylogenies. The temporally extended stem-group evidences strong extinction, and includes even middle Pleistocene species of the only surviving genus, *Ctenomys*. The history of the strikingly diverse crown clade which includes at least 69 extant species (Freitas 2016) as well as a few extinct species is, on the contrary, surprisingly short. Although this could be influenced by biases inherent to the fossil record, it is nevertheless significant that most Pleistocene species included in the analyses are recovered as members of the stem group. This entails a new perspective regarding the timing of taxonomic, ecological, and morphofunctional diversification of extant *Ctenomys*, which should be taken into account in future evolutionary analyses of the genus.

Acknowledgments   We are especially grateful to the editors, T Freitas, G Lopes Gonçalves, and R Maestri, for the invitation to contribute to this volume. For access to materials under their care, we thank P Teta, A Martinelli, M Ezcurra, L Chornogubsky (Museo Argentino de Ciencias Naturales, Argentina), P Ortiz, M Díaz (Facultad de Ciencias Naturales e Instituto Miguel Lillo, Argentina), S Bogan (Fundación de Historia Natural Azara, Argentina), M Pérez, E Ruigómez (Museo Paleontológico Egidio Feruglio), M Rosi (IADIZA, CONICET, Argentina), J Oliveira (Museu Nacional, Universidade Federal do Rio de Janeiro, Brasil), D Romero, M Taglioretti, F Scaglia, †A Dondas (Museo de Ciencias Naturales de Mar del Plata, Argentina), P Straccia (Museo de Ciencias Naturales de Santa Clara, Argentina), J Vargas Mattos (Colección Boliviana de Fauna, Bolivia), E Tonni, and M Reguero (Museo de La Plata, Argentina). This research was supported by Agencia Nacional de Promoción Científica y Tecnológica PICT 2016-2881.

Literature Cited

Álvarez A, Arévalo RLM, Verzi DH (2017) Diversification patterns and size evolution in caviomorph rodents. Biol J Linnean Soc 121:907–922
Álvarez A, Ercoli MD, Verzi DH (2020) Integration and diversity of the caviomorph mandible (Hystricomorpha, Rodentia): accessing to the evolutionary history through fossils and ancestral shape reconstructions. Zool J Linnean Soc 188:276–301
Arnal M, Vucetich MG (2015) Main radiation events in Pan-Octodontoidea (Rodentia, Caviomorpha). Zool J Linnean Soc 175:587–606
Benton MJ, Donoghue PCJ (2007) Paleontological evidence to date the tree of life. Mol Biol Evol 24:26–53

Bidau CJ (2015) Ctenomyidae. In: Patton JL, Pardiñas UFJ, D'Elía G (eds) Mammals of South America. 2. Rodents. University of Chicago Press, Chicago, pp 818–877

Bidegain JC, Soibelzon E, Prevosti FJ, Rico Y, Verzi DH, Tonni EP (2005) Magnetoestratigrafía y bioestratigrafía de las barrancas costeras de Necochea (provincia de Buenos Aires, Argentina). Actas XV Congreso Geológico Argentino 4:239–246

Boivin M, Marivaux L, Antoine PO (2019) L'apport du registre paléogène d'Amazonie sur la diversification initiale des Caviomorpha (Hystricognathi, Rodentia): implications phylogéné-tiques, macroévolutives et paléobiogéographiques. Geodiversitas 41:143–245

Borges LR, Maestri R, Kubiak BB, Galiano D, Fornel R, Freitas TRO (2017) The role of soil features in shaping the bite force and related skull and mandible morphology in the subterranean rodents of genus *Ctenomys* (Hystricognathi: Ctenomyidae). J Zool 301:108–117

Briggs DEG, Fortey RA (2005) Wonderful strife: systematics, stem groups, and the phylogenetic signal of the Cambrian radiation. Paleobiology 31:94–112

Budd GE, Mann RP (2020) The dynamics of stem and crown groups. Sci Adv 6:eaaz1626

Caraballo DA, Rossi MS (2018) Spatial and temporal divergence of the *torquatus* species group of the subterranean rodent *Ctenomys*. Contrib Zool 87:11–24

Carvalho GAS, Salles OL (2004) Relationships among extant and fossil echimyids (Rodentia: Hystricognathi). Zool J Linnean Soc 142:445–477

Casinos A, Quintana CA, Viladiu C (1993) Allometry and adaptation in the long bones of a digging group of rodents (Ctenomyinae). Zool J Linnean Soc 107:107–115

Castillo AH, Cortinas MN, Lessa EP (2005) Rapid diversification of South American tuco-tucos (*Ctenomys*; Rodentia, Ctenomyidae): Contrasting mitochondrial and nuclear intron sequences. J Mammal 86:170–179

Cela-Conde CJ, Ayala FJ (2003) Genera of the human lineage. Proc Natl Acad Sci 100:7684–7689

Cione AL, Tonni EP (1996) Reassessment of the Pliocene–Pleistocene continental time scale of southern South America. Correlation of the type Chapadmalal and with Bolivian sections. J South Am Earth Sci 9:221–236

De Queiroz K (2007) Toward an integrated system of clade names. Syst Biol 56:956–974

De Santi NA, Verzi DH, Olivares AI, Piñero P, Morgan CC, Medina ME, Rivero DE, Tonni EP (2020) A new peculiar species of the subterranean rodent *Ctenomys* (Rodentia, Ctenomyidae) from the Holocene of central Argentina. J South Am Earth Sci 100:102499

Fernández ME, Vassallo AI, Zárate M (2000) Functional morphology and palaeobiology of the Pliocene rodent *Actenomys* (Caviomorpha: Octodontidae): the evolution to a subterranean mode of life. Biol J Linnean Soc 71:71–90

Freitas TRO (2016) Family Ctenomyidae. In: Wilson DE, Lacher TE, Mittermeier RA (eds) The Handbook of mammals of the world. Lagomorphs and rodents I. Lynx Edicions, Barcelona, pp 498–534

Hennig W (1965) Phylogenetic systematics. Annu Rev Entomol 10:97–116

Lacey EA, Wieczorek JR (2003) Ecology of sociality in rodents: a ctenomyid perspective. J Mammal 84:1198–1211

Lessa EP, Cook JA (1998) The molecular phylogenetics of tuco-tucos (genus Ctenomys, Rodentia: Octodontidae) suggests an early burst of speciation. Mol Phylogenetics Evol 9:88–99

Lessa EP, Vassallo AI, Verzi DH, Mora M (2008) Evolution of morphological adaptations for digging in living and extinct ctenomyid and octodontid rodents (Caviomorpha). Biol J Linnean Soc 95:267–283

Morgan CC, Verzi DH (2006) Morphological diversity of the humerus of the south American subterranean rodent *Ctenomys* (Rodentia, Ctenomyidae). J Mammal 87:1252–1260

Morgan CC, Verzi DH (2011) Carpal- metacarpal specializations for burrowing in south American octodontoid rodents. J Anat 219:167–175

Morgan CC, Verzi DH, Olivares AI, Vieytes EC (2017) Craniodental and forelimb specializations for digging in the south American subterranean rodent *Ctenomys* (Hystricomorpha, Ctenomyidae). Mamm Biol 87:118–124

Parada A, D'Elía G, Bidau CJ, Lessa EP (2011) Species groups and the evolutionary diversification of tuco-tucos, genus *Ctenomys* (Rodentia: Ctenomyidae). J Mammal 92:671–682

Pascual R (1967) Los roedores Octodontoidea (Caviomorpha) de la Formación Arroyo Chasicó (Plioceno inferior) de la Provincia de Buenos Aires. Revista del Museo de La Plata Paleontología 5:259–282

Patterson B, Pascual R (1968) New echimyid rodents from the Oligocene of Patagonia, and a synopsis of the family. Bull Mus Comp Zool Breviora 301:1–14

Patterson B, Wood AE (1982) Rodents from the Deseadan Oligocene of Bolivia and the relationships of the Caviomorpha. Bull Mus Comp Zool 149:371–543

Pérez ME, Arnal M, Boivin M, Vucetich MG, Candela A, Busker F, Quispe BM (2019) New caviomorph rodents from the late Oligocene of Salla, Bolivia: taxonomic, chronological, and biogeographic implications for the Deseadan faunas of South America. J Syst Palaeontol (10):821–847

Prothero DR (2014) Species longevity in north American fossil mammals. Integr Zool 9:383–393

Quintana CA (1994) Sistemática y anatomía funcional del roedor Ctenomyinae *Praectenomys* (Caviomorpha: Octodontidae) del Plioceno de Bolivia. Revista Técnica de Yacimientos Petrolíferos Fiscales Bolivianos 15:175–185

Reig OA (1970) Ecological notes on the fossorial octodont rodent *Spalacopus cyanus* (Molina). J Mammal 51:592–601

Reig OA (1989) Karyotypic repatterning as one triggering factor in cases of explosive speciation. In: Fontdevila A (ed) Evolutionary biology of transient unstable populations. Springer-Verlag, Berlin, pp 246–289

Reig OA, Quintana CA (1992) Fossil ctenomyine rodents of the genus *Eucelophorus* (Caviomorpha: Octodontidae) from the Pliocene and early Pleistocene of Argentina. Ameghiniana 29:363–380

Reig OA, Busch C, Ortells MO, Contreras JR (1990) An overview of evolution, systematics, population biology, cytogenetics, molecular biology and speciation in *Ctenomys*. In: Nevo E, Reig OA (eds) Evolution of Subterranean Mammals at the Organismal and Molecular Levels. Wiley-Liss, New York, pp 71–96

Ronquist F, Lartillot N, Phillips MJ (2016) Closing the gap between rocks and clocks using total evidence dating. Philos Trans R Soc B 371:20150136

Sansom RS, Wills MA, Williams T (2017) Dental data perform relatively poorly in reconstructing mammal phylogenies: morphological partitions evaluated with molecular benchmarks. Syst Biol 66:813–822

Shockey BJ, Salas-Gismondi R, Gans P, Jeong A, Flynn JJ (2009) Paleontology and geochronology of the Deseadan (late Oligocene) of Moquegua, Perú. Am Mus Novit 3668:1–24

Simpson GG (1945) The principles of classification and a classification of mammals. Bull Am Mus Nat Hist 85:1–350

Soibelzon E, Prevosti F, Bidegain J, Rico Y, Verzi DH, Tonni EP (2009) Correlation of late Cenozoic sequences of southeastern Buenos Aires province: biostratigraphy and magnetostratigraphy. Quat Int 210:51–56

Steiper ME, Young NM (2008) Timing primate evolution: lessons from the discordance between molecular and paleontological estimates. Evol Anthropol 17:179–188

Upham NS, Patterson BD (2015) Phylogeny and evolution of caviomorph rodents: a complete time tree for living genera. In Vassallo AI, Antenucci D (eds.) Biology of caviomorph rodents: diversity and evolution. SAREM Series A, Mastozoological Research. Buenos Aires, pp 63–120

Vassallo AI, Mora MS (2007) Interspecific scaling and ontogenetic growth patterns of the skull in living and fossil ctenomyid and octodontid rodents (Caviomorpha: Octodontoidea). In Kelt DA, Lessa EP, Salazar-Bravo J, Patton JL (eds) The quintessential naturalist: honoring the life and legacy of Oliver P. Pearson. University of California Publications in Zoology. California, pp 945–968

Verzi DH (1999) The dental evidence on the differentiation of the ctenomyine rodents (Caviomorpha, Octodontidae, Ctenomyinae). Acta Theriol 44:263–282

Verzi DH (2002) Patrones de evolución morfológica en Ctenomyinae (Rodentia, Octodontidae). Mastozool Neotrop 9:309–328

Verzi DH (2008) Phylogeny and adaptive diversity of rodents of the family Ctenomyidae (Caviomorpha): delimiting lineages and genera in the fossil record. J Zool (Lond) 274:386–394

Verzi DH, Vieytes EC, Montalvo CI (2004a) Dental evolution in *Xenodontomys* and first notice on secondary acquisition of radial enamel in rodents (Rodentia, Caviomorpha, Octodontidae). Geobios 37:795–806

Verzi DH, Deschamps CM, Tonni EP (2004b) Biostratigraphic and palaeoclimatic meaning of the Middle Pleistocene South American rodent *Ctenomys kraglievichi* (Caviomorpha, Octodontidae). Palaeogeogr Palaeoclimatol Palaeoecol 212:315–329

Verzi DH, Olivares AI, Morgan CC (2010) The oldest South American tuco-tuco (Pliocene, Northwestern Argentina) and the boundaries of genus *Ctenomys* (Rodentia, Ctenomyidae). Mamm Biol 75:243–252

Verzi DH, Olivares AI, Morgan CC (2014) Phylogeny and evolutionary patterns of South American octodontoid rodents. Acta Palaeontol Pol 59:757–769

Verzi DH, Morgan CC, Olivares AI (2015) The history of South American octodontoid rodents and its contribution to evolutionary generalisations. In: Cox P, Hautier L (eds) Evolution of the Rodents, advances in phylogeny, functional morphology, and development. Cambridge studies in morphology and molecules: new paradigms in evolutionary biology. Cambridge University Press, Cambridge, pp 139–163

Verzi DH, Olivares AI, Morgan CC, Álvarez A (2016) Contrasting phylogenetic and diversity patterns in octodontoid rodents and a new definition of the family Abrocomidae. J Mamm Evol 23:93–115

Vieytes EC, Morgan CC, Verzi DH (2007) Adaptive diversity of incisor enamel microstructure in south American burrowing rodents (family Ctenomyidae, Caviomorpha). J Anat 211:296–302

Vucetich MG, Verzi DH (1991) Un nuevo Echimyidae (Rodentia, Hystricognathi) de la Edad Colhuehuapense de Patagonia y consideraciones sobre la sistemática de la familia. Ameghiniana 28:67–74

Vucetich MG, Verzi DH, Hartenberger JL (1999) Review and analysis of the radiation of the South American Hystricognathi (Mammalia, Rodentia). Comptes rendus de l'Académie des Sciences II A329:763–769

Vucetich MG, Arnal M, Deschamps CM, Pérez ME, Vieytes CE (2015) A brief history of caviomorph rodents as told by the fossil record. In Vassallo AI, Antenucci D (eds) Biology of caviomorph rodents: diversity and evolution. SAREM Series A, Mastozoological Research, Buenos Aires, pp. 11–62

Wood AE (1955) A revised classification of the rodents. J Mammal 36:165–187

Wood B, Collard M (1999) The human genus. Science 284:65–71

Wood AE, Patterson B (1959) Rodents of the Deseadan Oligocene of Patagonia and the beginnings of South American rodent evolution. Bull Mus Comp Zool 120:279–428

Chapter 2
A Short Overview of the Systematics of *Ctenomys*: Species Limits and Phylogenetic Relationships

Guillermo D'Elía, Pablo Teta, and Enrique P. Lessa

2.1 Introduction

Ctenomys Blainville, 1826 is one of the most species-rich genera of Mammalia; the current count indicates that the genus has 64 living species (see below). Tuco-tuco species have mostly allopatric distributions, with some of them having very restricted geographical ranges (Bidau 2015) and in some cases being only known from their type localities (Teta and D'Elía 2019). The genus displays one the broadest range of chromosomic variation (2n = 10 to 2n = 70; Cook et al. 1990; Novello and Lessa 1986) of any mammal genus. As such, *Ctenomys* has long attracted the attention of evolutionary biologists aimed to characterize its diversity and to understand the drivers of such impressive radiation. These studies have been hampered, however, by a poor understanding of species boundaries and the species phylogenetic relationships.

In recent year several studies have advanced our knowledge of distinct aspects of the evolutionary biology and ecology of *Ctenomys* (e.g., Martínez and Bidau 2016; Morgan et al. 2017; Tomasco et al. 2019; Kubiak et al. 2020); this book is indeed a more than welcome proof of these advances. Notwithstanding, regarding the evolutionary history of the genus several unclear areas remain. At the base of this scenario lays the fact that the taxonomic status of several populations and nominal forms is still dubious, either because they have not been evaluated (e.g., those of several unsampled geographic areas) or results are inconclusive. As such, the

G. D'Elía (✉)
Instituto de Ciencias Ambientales y Evolutivas, Universidad Austral de Chile, Valdivia, Chile

P. Teta
División Mastozoología, Museo Argentino de Ciencias Naturales "Bernardino Rivadavia", Buenos Aires, Argentina

E. P. Lessa
Departamento de Ecología y Evolución, Facultad de Ciencias, Universidad de la República, Montevideo, Uruguay

© Springer Nature Switzerland AG 2021
T. R. O. de. Freitas et al. (eds.), *Tuco-Tucos*,
https://doi.org/10.1007/978-3-030-61679-3_2

taxonomy in *Ctenomys* is still much unstable (see the species accounts in Bidau 2015). In turn, this scenario hampers the understanding of species phylogenetic relationships and secondarily all studies that need a phylogeny as input (e.g., studies of historical biogeography or morphological evolution).

Even though several fossil species are known (e.g., Rusconi 1931; Verzi et al. 2004; De Santi et al. 2020), we focused on living forms and structured the core of this review into two main sections. The first one deals with species boundaries while the second one focus on phylogenetic relationships among species of tuco-tucos. Both sections, after a short summary on the historical development of the knowledge, focus on its current state; emphasis is made on what we known, but also on gray areas that need to be the focus of future research. We close our literature review in June 2020.

2.2 Taxonomy

The taxonomic history of *Ctenomys* is long and complex. In a statement that has been cited several times, Sage et al. (1986) claimed that "The current taxonomy of *Ctenomys* is in a state of general chaos." As would be commented below several issues contribute to this view, including some taxonomic practices that should be avoided (see below), the general scarcity of study samples, as well as the intrinsic biological complexity of this genus. For instance, it has been widely acknowledged that the taxonomy of *Ctenomys* is challenged by the remarkable morphological homogeneity, both external and cranial, that exists among species. However, De Santi et al. (2020) shown that morphological differences among species and groups of species can be uncovered when cranial or dental characters are assessed in detail. Along the same lines, linear and geometric morphometric analyses have been able to differentiate species, although differences tend to blur if multiple species are considered simultaneously (Tiranti et al. 2005; Freitas 2005). Some species exhibit large chromosomal variation (e.g., 45 distinct karyotypes are known for *C. minutus* Nehring, 1887; see Lopes et al. 2013) that, if not considered altogether, may mislead species delimitation. Another fact that challenges the establishment of species boundaries is that, commonly, there is incongruence among the variation pattern of distinct data sets (e.g., the large and well-structured chromosomic variation of *C. pearsoni* Lessa and Langguth, 1983 contrasting with the patterns of variation of the mitochondrial genome and the skull morphology; Tomasco and Lessa 2007; D'Anatro and D'Elía 2011). These challenges are slowly starting to be overcome with work that incorporates distinct evidence lines and analytical tools.

What follows does not intend to be an exhaustive review of the taxonomy of *Ctenomys* (for that we refer the reader to the species accounts written by Bidau 2015; although below we pose some departures); rather, by commenting some study cases, we highlight distinct aspects of the taxonomic history and practice of *Ctenomys*. We aim to give our vision of the current taxonomic knowledge of *Ctenomys*, including the identification of some areas in need of additional research, as well as to advocate for the abandonment of some practices that have unnecessarily contributed to complicate tuco-tuco taxonomy.

2.2.1   Currently Considered Distinct Species

The first species of *Ctenomys* to be described was *C. brasiliensis* Blainville, 1826. At the same time, this was the first mention of *Ctenomys*; as such, this was one of the first named genera of Neotropical rodents. In the remaining of the nineteenth century, 10 other species of tuco-tucos were described (Table 2.1); of these, four were described in 1848. An important increase in species description occurred during the first three decades of the twentieth century. Between 1903 and 1926, 28 species, currently considered distinct, were described; Oldfield Thomas described 24 of these species (23 by himself and the other with Jane St. Leger). After that, there was a large period of over five decades (1927–1980) in which only five species were described. In addition, during this period, the number of recognized species suffered a marked reduction (e.g., Cabrera 1961 recognized only 26 species within this genus), with several taxa being uncritically included into the synonymy of others, and usually retained as subspecies (e.g., *azarae* Thomas, 1903; *bergi* Thomas, 1902; *fochi* Thomas, 1919; *haigi* Thomas, 1919; *juris* Thomas, 1920; *latro* Thomas, 1918; *lentulus*, Thomas, 1919; *occultus* Thomas, 1920; *pundtii* Nehring, 1900; *recessus* Thomas, 1912, and *tucumanus* Thomas, 1900, were regarded as synonyms of *mendocinus* Philippi, 1869). The imprint of this classification is still present in some species groups, such as is the case of *C. frater* Thomas, 1902 (see below). Over the next two decades (1981–2001), a new period of intense species description occurred, where 12 new species were named and described. Most of these species were authored by Julio R. Contreras. During the period 2002–2011, no new species was described. Finally, from 2012 up today (September 24, 2020) eight new species have been erected. As such, the cumulative count of species still shows a steady increase in known species richness (Fig. 2.1).

As expected from the dates of description, most species of *Ctenomys* were only described based on morphologic (mostly qualitative) differences (see a discussion of the general trend for rodents in D'Elía et al. 2019). The first species of *Ctenomys* whose description included chromosomic evidence was *C. coyahiquensis* by Kelt and Gallardo in 1994. As this form is now regarded as a synonym of *C. sericeus* J. A. Allen, 1903 (see Teta and D'Elía 2020), the first description incorporating chromosomic data of what we consider to be a distinct species is from one year later and corresponds to that of *C. osvaldoreigi* Contreras, 1995. Although several candidate species have been identified based on the analysis of DNA sequences (e.g., some Bolivian forms included in the analysis of Lessa and Cook 1998; some Patagonian forms in the study of Parada et al. 2011), the first species description incorporating analysis of DNA sequences dates from less than a decade ago, corresponding to that of *C. ibicuensis* Freitas, Fernandes, Fornel, and Roratto, 2012.

Table 2.1 List of the 64 living species of *Ctenomys* recognized in this study

	Species	Sperm type	Species group
1	*Ctenomys andersoni* Gardner, Salazar-Bravo and Cook, 2014		*boliviensis*
2	*Ctenomys argentinus* Contreras and Berry, 1982	symmetric	*tucumanus*
3	*Ctenomys australis* Rusconi, 1934	asymmetric	*mendocinus*
4	*Ctenomys bergi* Thomas, 1902	asymmetric	*mendocinus**
5	*Ctenomys bicolor* Miranda-Ribeiro, 1914		*boliviensis*
6	*Ctenomys bidaui* Teta and D'Elía, 2020	asymmetric	*magellanicus*
7	*Ctenomys boliviensis* Waterhouse, 1848	symmetric	*boliviensis*
8	*Ctenomys bonettoi* Contreras and Berry, 1982	asymmetric	*mendocinus**
9	*Ctenomys brasiliensis* Blainville, 1826		
10	*Ctenomys coludo* Thomas, 1920		
11	*Ctenomys conoveri* Osgood, 1946	asymmetric	*frater*
12	*Ctenomys contrerasi* Teta and D'Elía, 2020	asymmetric	*magellanicus*
13	*Ctenomys dorbigny* Contreras and Contreras, 1984		*torquatus*
14	*Ctenomys dorsalis* Thomas, 1900		*boliviensis*
15	*Ctenomys emilianus* Thomas and St. Leger, 1926		
16	*Ctenomys erikacuellarae* Gardner, Salazar-Bravo and Cook, 2014		*boliviensis*
17	*Ctenomys famosus* Thomas, 1920		*mendocinus*
18	*Ctenomys flamarioni* Travi, 1981	asymmetric	*mendocinus*
19	*Ctenomys fochi* Thomas, 1919		
20	*Ctenomys fodax* Thomas, 1910	asymmetric	*magellanicus*
21	*Ctenomys frater* Thomas, 1902	symmetric	*frater*
22	*Ctenomys fulvus* Phillippi, 1860	symmetric	*opimus*
23	*Ctenomys haigi* Thomas, 1919	asymmetric	*magellanicus*
24	*Ctenomys ibicuensis* Freitas, Fernandes, Fornel and Roratto, 2012		*torquatus*
25	*Ctenomys johannis* Thomas, 1921		
26	*Ctenomys juris* Thomas, 1920	asymmetric	*tucumanus**
27	*Ctenomys knighti* Thomas, 1919		
28	*Ctenomys lami* Freitas, 2001		*torquatus*
29	*Ctenomys latro* Thomas, 1918		*tucumanus*
30	*Ctenomys lessai* Gardner, Salazar-Bravo and Cook, 2014		*frater*
31	*Ctenomys leucodon* Waterhouse, 1848		*no group*
32	*Ctenomys lewisi* Thomas, 1926	symmetric	*frater*
33	*Ctenomys magellanicus* Bennet, 1836	asymmetric	*magellanicus*
34	*Ctenomys maulinus* Phillippi, 1872	asymmetric	*no group*
35	*Ctenomys mendocinus* Phillippi, 1869	asymmetric	*mendocinus*
36	*Ctenomys minutus* Nehring, 1887	symmetric	*torquatus*
37	*Ctenomys nattereri* Wagner, 1848		
38	*Ctenomys occultus* Thomas, 1920	asymmetric	*tucumanus*
39	*Ctenomys opimus* Wagner, 1848	symmetric	*opimus*

(continued)

Table 2.1 (continued)

	Species	Sperm type	Species group
40	*Ctenomys osvaldoreigi* Contreras, 1995	asymmetric	
41	*Ctenomys paraguayensis* Contreras, 2000		
42	*Ctenomys pearsoni* Lessa and Langguth, 1983	symmetric	*torquatus*
43	*Ctenomys perrensi* Thomas, 1896	symmetric	*torquatus*
44	*Ctenomys peruanus* Sanborn and Pearson, 1947		
45	*Ctenomys pilarensis* Contreras, 1993	asymmetric	*tucumanus**
46	*Ctenomys pontifex* Thomas, 1918		
47	*Ctenomys pundti* Nehring, 1900	symmetric	*talarum*
48	*Ctenomys rionegrensis* Langguth and Abella, 1970	asymmetric	*mendocinus*
49	*Ctenomys roigi* Contreras, 1988	symmetric	*torquatus*
50	*Ctenomys rondoni* Miranda-Ribeiro, 1914		
51	*Ctenomys saltarius* Thomas, 1912		*opimus*
52	*Ctenomys scagliai* Contreras, 1999	symmetric	*opimus*
53	*Ctenomys sericeus* Allen, 1903	asymmetric	*magellanicus*
54	*Ctenomys sociabilis* Peason and Christie, 1985	asymmetric	*no group*
55	*Ctenomys steinbachi* Thomas, 1907	symmetric	*boliviensis*
56	*Ctenomys talarum* Thomas, 1898	symmetric	*talarum*
57	*Ctenomys thalesi* Teta and D'Elía, 2020	asymmetric	*magellanicus*
58	*Ctenomys torquatus* Lichtenstein, 1830	symmetric	*torquatus*
59	*Ctenomys tuconax* Thomas, 1925	asymmetric	*tucumanus**
60	*Ctenomys tucumanus* Thomas, 1900	symmetric	*tucumanus*
61	*Ctenomys tulduco* Thomas, 1921		
62	*Ctenomys validus* Contreras, Roig and Suzarte, 1977		
63	*Ctenomys viperinus* Thomas, 1926		
64	*Ctenomys yatesi* Gardner, Salazar-Bravo and Cook, 2014		*boliviensis*

For each species, if information is available, we indicate sperm type and species group as defined by Parada et al. (2011). An asterick (*) indicates that species group allocation is inferred from the topology presented by Mascheratti et al. (2000)

After almost two centuries of taxonomic studies centered in *Ctenomys*, here we recognize 64 living species of tuco-tucos (Table 2.1). In the last review of *Ctenomys* presented by Claudio Bidau (2015) 65 (not 64 as stated in page 820) living species were listed for the genus. The list of Bidau and ours present several departures (Table 2.2). Bidau (2015) listed three informal forms that are not nomenclatorially available; wse omit them here for that reason. Bidau (2015) explicitly stated in the text (and by using quotation marks) that two of those forms were not available, but chose to list them as a form of recognizing their distinction at the species level as well as the need to formalize them. The first of these two is the Bolivian form *C. "mariafarelli"* described by Azurdy (2005); our assessment of the literature suggest us that *C. "mariafarelli"* seems to represent the same lineage of species level represented by the recently described taxon *C. erikacuellarae* Gardner, Salazar-Bravo and Cook, 2014; as such, if it would have been properly described, *C.*

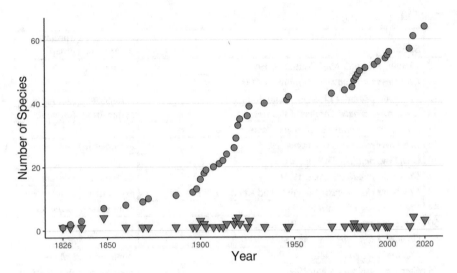

Fig. 2.1 Number of currently considered distinct living species of *Ctenomys* proposed between the years 1826 and May 2020 (triangles). Species accumulated in the same interval of time (circles)

"*mariafarelli*" would have priority over *C. erikacuellarae*. Bidau (2015) also included in his list *C.* "*yolandae*"; a form from Santa Fe Province in Argentina, which was enunciated by Contreras and Berry (1984) in a meeting abstract. Available evidence, including it being the unique known species with a complex asymmetric sperm type (Vitullo et al. 1988) and showing a distinctive karyotype of $2n = 50$, FN = 67, 70, 78 (Ortells et al. 1990; Bidau et al. 2005), indicates "*yolandae*" in fact represents a distinct species of tuco-tucos; as such, its formal description is needed. Lastly, Bidau (2015) listed *C.* "*rosendopascuali*" Contreras, 1995 (without quotation marks in Bidau 2015), as a distinct species, although raising doubts regarding its availability. Here we considered that this name is unavailable (see below) and as such, we exclude it from the species list. As a form to emphasize the need to formally describe species (see below), we have omitted from our list nonavailable binomens even if available evidence suggests they represent distinct lineages of species level.

Seven species included in our list (i.e., *C. andersoni* Gardner, Salazar-Bravo and Cook, 2014; *C. bidaui* Teta and D'Elía, 2020; *C. contrerasi* Teta and D'Elía, 2020; *C. erikacuellarae* Gardner, Salazar-Bravo, and Cook, 2014; *C. lessai* Gardner, Salazar-Bravo, and Cook, 2014; *C. thalesi* Teta and D'Elía, 2020; and *C. yatesi* Gardner, Salazar-Bravo, and Cook, 2014) were described after the review of Bidau (2015) entered the publication process. In addition, the forms *C. colburni* J. A. Allen, 1903, *C. coyahiquensis*, and *C. goodfellowi* Thomas, 1921 considered as distinct species by Bidau (2015) were later synonymized, respectively, under *C. magellanicus* Bennet, 1836, *C. sericeus*, and *C. boliviensis* Waterhouse, 1848 based on morphologic, karyotypic, and molecular data (Teta et al. 2020; Teta and D'Elía 2020; Gardner et al. 2014).

Table 2.2 Departures in the list of distinct living species of *Ctenomys* presented in this study (see Table 2.1) from the list of Bidau (2015), the last published taxonomic catalog of *Ctenomys*

Species	Bidau (2015)	This work
Ctenomys "mariafarelli"	Signaled as a distinct species with a nonavailable name.	Nnot considered because the name is not available (it probably represents the same species than *C. erikacuellarae*)
Ctenomys "yolandae"	Signaled as a distinct species with a non-available name.	Not considered because the name is not available even when it may be distinct
Ctenomys andersoni		Distinct (described after Bidau 2015 entered the press)
Ctenomys bidaui		Distinct (described after Bidau 2015)
Ctenomys colburni	Distinct	Synonym of *C. magellanicus*
Ctenomys contrerasi		Distinct (described after Bidau 2015)
Ctenomys coyahiquensis	Distinct	Synonym of *C. sericeus*
Ctenomys erikacuellarae		distinct (described after Bidau 2015 entered the press)
Ctenomys goodfellowi	Distinct	Synonym of *C. boliviensis*
Ctenomys lessai		Distinct (described after Bidau 2015 entered the press)
Ctenomys "rosendopascuali"	Signaled as a distinct species although expressing doubts on its availability	not considered because the name is not available
Ctenomys thalesi		Distinct (described after Bidau 2015)
Ctenomys yatesi		Distinct (described after Bidau 2015 entered the press)
Ctenomys azarae	Distinct	Synonym of *C. mendocinus*
Ctenomys porteousi	Distinct	Synonym of *C. mendocinus*

Finally, our list differs from that of Bidau (2015) in that two of the species listed by him as distinct are here synonymized in what constitutes the single nomenclatorial act of this contribution. Here we consider *C. azarae* and *C. porteuosi* Thomas, 1919 as subjective junior synonyms of *C. mendocinus*. Both *azarae* and *porteuosi* are forms of central Argentina that were described by Oldfield Thomas during the first quarter of the twentieth century. Both forms together with *C. australis* Rusconi, 1934, *C. flamarioni* Travi, 1981, *C. mendocinus*, and *C. rionegrensis* Langguth and Abella, 1970 are part of the *C. mendocinus* species group (Massarini et al. 1991; D'Elía et al. 1999; Parada et al. 2011). The distinction of *azarae* and *porteousi* from *C. mendocinus* has been largely questioned in the literature (e.g., Massarini et al. 1991: 141). The three taxa present the same karyotype polymorphism of 2n = 46, 47, and 48, as well as patterns of G-bands (Massarini et al. 1991, 1998); in addition, *C. mendocinus* also displays a cytotype of 2n = 50 (e.g., Parada et al. 2012). Multivariate analyses of skull measurements show that neither *azarae* nor *porteousi* differentiate from *C. mendocinus*, as the three nominal forms have completely overlapping morphospaces (Massarini and Freitas 2005; Fornel et al. 2018). Finally, genealogical analysis of mitochondrial DNA sequences, including sequences of specimens collected at the three type localities, shows that variants recovered from specimens of *azarae* and *porteousi* mix with variants of *C. mendocinus* (Mapelli et al. 2017). The lack of monophyly of the three nominal forms goes beyond what one expects for a gene tree not recovering the species tree (as it seems to be the case of *C. australis* in the same gene tree where its variants form a clade nested within the variation of *C. mendocinus*); variants accommodate in three main clades that are not allopatric, including the fact that mitochondrial clades overlap in different combinations at same localities (even when almost half of the sampled localities have only one or two sampled specimens and no locality has more than five sequenced specimens). In addition, the divergence observed between clade pairs is relatively low (up to 1.5% for a fragment of 899 bp of the control region plus 402 bp of the cyt b gene). As such, for the moment, no emerging property (e.g., morphological or karyotypic diagnosibility, monophyly) supports the distinction at the species level of *C. azarae* or *C. porteusi*. Then, as here delimited, *C. mendocinus* is a species with a large distribution in central Argentina including localities in the provinces of Buenos Aires, Córdoba, La Pampa, and San Luis.

We close this section noting that, so far, no study of species delimitation of *Ctenomys* has used a genomic approach neither the multispecies coalescent model. As data acquisition is getting easier and cheaper, we expect that in the next few years the current taxonomic scheme be further tested with these approaches (see comment in Lessa et al. 2014). In this regard, it would be of much interest to see how robust are current inferences that, regarding molecular data are mostly, if not exclusively, based on mitochondrial DNA sequences. Another relevant aspect is that large geographic areas still remain unexplored, a scenario that should be overcome with further specimen collection. To visualize the importance of this fact, it is relevant to note that several candidate species have been proposed (see below). Similarly, the distinction of some nominal forms has been called into question, while the distinction of a large fraction of currently considered distinct species has not been adequately evaluated (see below). As such, given these precedents and that currently several research groups in distinct countries are working on the taxonomy of *Ctenomys*, we anticipate that our list of living species would soon be outdated.

2.2.2 Candidate Species

As it happens when any group that is taxonomically assessed, distinct candidate species of *Ctenomys* have been identified. As the practice dictates, the distinction of these forms should be properly evaluated with additional studies, and if proven to be distinct, we advocate for formally describe them. For example, Teta and D'Elía (2020) assessed the distinction of some candidate species previously identified on the base on DNA sequences and karyology by Bidau et al. (2003) and Parada et al. (2011), and after corroborating their distinction using an integrative taxonomic approach, named and described *C. bidaui*, *C. contrerasi*, and *C. thalesi*.

Candidate species of *Ctenomys* have been identified based on distinct sources of evidence. Next, we illustrate some cases of candidate species involving Argentinean, Brazilian, and Chilean tuco-tucos, which differ in their complexity and in the evidence used to identify them.

A series of populations from pre-Andean areas of the Araucania Región, Chile, display a distinct cytotype ($2n = 28$; FN = 48) from nearby populations of *C. maulinus brunneus* Osgood, 1943 ($2n = 26$, FN = 50; Gallardo 1979). This difference prompted Gallardo (1979) to suggest $2n = 28$ populations represent an undescribed species and referred to them as *C.* sp. No published study has further assessed the taxonomic status of these populations (but see Aguilar 1988). Similarly, Leipnitz et al. 2020) identified, on the basis of mitochondrial DNA sequences, two lineages in the Brazilian states of Mato Grosso and Rondônia, hypothesizing that they are lineages of species level. Authors referred to these lineages as *C.* sp. "xingu" and *C.* sp. "central" awaiting new analyses to test their distinction.

Another example where candidate species have been identified pertains to populations of the Corrientes group, an assemblage of populations from Corrientes province, Argentina, which is part of the *C. torquatus* species group (Parada et al. 2011) and exhibit large chromosomic variation, ranging from $2n = 41$ to $2n = 70$ (e.g., Ortells et al. 1990; Giménez et al. 2002). Three nominal forms, *C. dorbignyi* Contreras and Contreras, 1984; *C. perrensi* Thomas, 1896, and *C. roigi* Contreras, 1988 are associated with the Corrientes group and are considered as representing distinct species (e.g., Bidau 2015; Teta et al. 2018; Table 2.1); however, the biological boundaries, and then also the geographic distribution, of these species, in particular of *C. dorbignyi* and *C. perrensi*, have been long discussed. The taxonomic literature is relatively vast, at least when compared to that available for tuco-tucos from most of the other areas and complexes (the exception would be those well-studied taxa from southern Uruguay and Rio Grande do Sul, Brazil). Mitochondrial DNA sequences analyses fail to recover *C. dorbignyi* and *C. perrensi* as monophyletic, while *roigi* was found as monophyletic and nested within *perrensi* (Giménez et al. 2002; Mirol et al. 2010; Buschiazzo et al. 2018; Caraballo and Rossi 2018a; see also Gómez Fernández et al. 2012). In addition, there is a lack of correspondence between cytotypes and haplogroups (e.g., Giménez et al. 2002; Caraballo and Rossi 2018a), blurring the delineation of lineages of species level. This situation is accentuated by the fact that, surprisingly given that these populations have been extensively studied since the decade of 1980, there is not a comprehensive assessment of the morphological variation of populations of the Corrientes group. Available morphologic information is limited to that presented on the species descriptions, which were based on quantitative assessments of small sample sizes, and to a single quantitative analysis conducted by Contreras and Scolaro (1986) centered on four groups of populations then assigned to *C. dorbignyi* (one of these groups is more related to *C. perrensi* than to *C. dorbignyi*; Gimenez et al. 2002). Notwithstanding, Giménez et al. (2002) suggested the existence of two candidate species, which were labeled as *Ctenomys* sp. α and *Ctenomys* sp. β; both candidate species encompass populations assigned to *C. perrensi* (those of *C.* sp. β were the ones earlier assigned to *C. dorbignyi* by Contreras and Scolaro 1986). This suggestion was reinforced by the results of Caraballo et al. (2012). In addition, Caraballo and Rossi (2018a) identified another candidate species, termed "Sarandicito," encompassed by an isolated population in southern Corrientes that has been previously considered as part of *C. dorbignyi*. We expect that this changing and complex scenario starts to settle down when patterns of morphological variation are thoroughly assessed and nuclear gene DNA sequences are incorporated in the analysis of this complex group of tuco-tucos.

The proposition of candidate species is a natural outcome of taxonomic work. We advocate conducting species validation analysis to test the distinction of the candidate species here reviewed as well as the other presented in the literature (including some for which if proven distinct, names may be available; e.g., *C. lentulus*; see Teta et al. 2020). Then, if results of the validation analyses show that they in fact constitute distinct species, we advocate to formally describe them to, among other reasons, avoid situations as the ones commented below. Informal names have no regulation; as such, the literature shows that they vary from study to study (e.g., a lineage first referred as *C.* sp. α and later as Iberá), at the time that the same denotation (e.g., *C.* sp. 1) can be applied to distinct candidate species. These facts hamper effective communication and should be avoided.

2.2.3 Unavailable Names

A common issue in the literature of *Ctenomys* is the extensive usage of nonavailable names to refer to putative lineages of species level. These cases, which includes that of the already commented "*yolandae*," depart from the traditional way of labeling candidates species (e.g., sp. α; sp. "xingu") in that putative distinct species are identified with binomials (C. "*yolandae*"), which did not accomplish the provisions of The International Code of Zoological Nomenclature (ICZN 1999) and as such, are unavailable. Most commonly, these names have been introduced in papers without selecting a holotype and providing a diagnosis or in meeting abstracts. One of these names is "*eremofilus*," (also referred as *C. eremophilus* or *C. eremicus*), which was advanced by Contreras and Roig (1975) for populations from the Ñacuñán Biosphere Reserve, Mendoza, Argentina; this name was used regularly in the literature until Parada et al. (2012) showed that those populations belong to *C. mendocinus*. As such, "*eremofilus*," besides not being available, does not represent a distinct species, differing this way from the case of "*yolandae*" (see above), at the time that resembling that of *C.* "*chasiquensis*." Perhaps the most extensively used informal name in the literature of *Ctenomys* is *C.* "*chasiquensis*," mentioned for the first time in a meeting abstract by Contreras, Manceñido and Ripa Alsina (1970) and mentioned by Contreras and Maceiras (1970) as a subspecies of *C. azarae* (*C. a.* "*chasiquensis*") to refer to tuco-tuco populations from the area of the Chasicó Lagoon, southwestern Buenos Aires Province, Argentina. The name, now at the species level, keeps been used up today (e.g., Mora et al. 2016; Mapelli et al. 2019). However, in addition to have never been formalized, to be best of our knowledge no evidence supports the distinction of *"chasiquensis"* as a distinct species. On the contrary, karyotypic data (Massarini et al. 1991) and analysis of mitochondrial DNA sequences (Mapelli et al. 2017; see also Mora et al. 2016) strongly indicates that populations referred to as *"chasiquensis"* belong to *C. mendocinus*. It seems that the still ongoing mentions to *C. "chasiquensis"* are caused by some sort of taxonomic inertia. Here we made a call to no longer use this name because it does not satisfy nomenclatorial rules and there is no need to do it since populations referred to it belong to *C. mendocinus*.

There are other unavailable names, all of them coined by Contreras and collaborators (e.g., *C. "alvaromonesi," C. "avellanedae," C. "felixi," C. "nevoi," C. "paramilloensis"*; see Contreras and Bidau 1999: Bidau et al. 2003), that fortunately have not gained momentum on the literature. Other names, such as *beltzeri, marthae, monesi*, and *reigi*, described as subspecies of *C. minutus* by Contreras and Contreras (1984) in a meeting abstract are also unavailable; however, these names were considered as synonyms of *C. minutus* by Woods and Kilpatrick (2005). We highlight the relevance of complying with the provisions of International Code of Zoological Nomenclature (ICZN 1999), in particular, with the designating of a holotype. Name bearing type specimens constitute the only way to unambiguously link taxonomic names with the lineages to which they apply (see D'Elía 2012 for additional discussion).

Regarding type specimens, we call attention to an issue that seems to have passed unnoticed in the taxonomic literature of *Ctenomys*. It is the fact that Contreras, when describing *C. argentinus* Contreras and Berry, 1982, *C. bonettoi* Contreras and Berry, 1982, and *C. d'orbigny* (also the nonavailable *C. "yolandae"*) designed two specimens as "holotypes" for each of these species. The Code establishes that the holotype is a single specimen; as such, the two specimens listed as holotypes of a given species should be considered as syntypes forming part of a type series (Art. 72), for which one can be selected as a lectotype to serve as the name-bearing type (Art. 74; see also Art. 72.2).

We close this section mentioning an issue raised by Bidau (2015: 866) regarding the possibility that the name *C. "rosendopascuali"* be unavailable, given that it is not clear if the 1995 issue of Nótulas Faunísticas, were the species description appears, was in fact published or not. Yolanda Davis communicated one of us (PT) that that issue was probably not published. In addition, we note that in Contreras and Bidau (1999), *C. "rosendopascuali"* is indicated on page 4 without authorship and as in press 1999, while on page 11 is mentioned with the authorship of Contreras 1999. Recently, Agnolin et al. (2020) stated that in a personal communication to the first author, J.R. Contreras mentioned that in fact the description of *C. "rosendopascuali"* was never published. As such, we have not included *"rosendopascuali"* in the list of currently distinct and available species.

2.2.4 Synonyms

About nine species of *Ctenomys* have taxonomic names in their synonymy. Some of these names are traditionally recognized as subspecies (e.g., *C. maulinus brunneus*), although in most, if not all, of these cases, the distinction of these putative distinct lineages of subspecies level has not been assessed with contemporaneous approaches. For example, the forms *barbarus* Thomas, 1921; *budini* Thomas, 1913; *mordosus* Thomas, 1926; *sylvanus* Thomas, 1919; and *utibilis* Thomas, 1921, together with the nominotypical *frater* Thomas, 1902 are recognized as subspecies of *C. frater,* an arrangement that goes back to Cabrera (1961; see also Bidau 2015) and that since has not been assessed. However, without providing new data, Galliari et al. (1996) treated *barbarus*, *budini*, and *sylvanus* as distinct species; a similar scheme was presented by Wood and Kilpatrick (2005) who listed three distinct species within this complex: *budini* (including *barbarus* as a subspecies), *frater* (including *mordosus* as a subspecies), and *sylvanus* (including *utibilis* as subspecies). The degree of understanding of the variation within and among *frater* and associated names is a good example of the taxonomic knowledge of most species of *Ctenomys* with names associated in their synonymy: the distinction of the nominal forms has been, if so, seldom evaluated and their ranking has shifted over taxonomic history without necessarily providing evidence supporting changes. We expect this review to call the attention over these cases and prompted colleagues to study them.

In addition, the status of several currently considered distinct species is dubious, either because their description was based on minor differences shown by a small series of specimens and their distinction was not later evaluated or because there is evidence questioning their distinctiveness. Several species fall in the first category. For instance, *C. tulduco*, which was described by Thomas (1921) and is known from the surroundings of its type locality in San Juan Province, Argentina remains with unknown karyotype and sperm type (Bidau 2015) and has not been included in any taxonomic nor phylogenetic study. However, Cabrera (1961) regarded it as a subspecies of *C. fulvus* Philippi, 1860. Similarly, the status of *C. validus* Contreras, Roig, and Suzarte, 1977, a species from central-northern Mendoza with unknown karyotype and sperm type, is dubious. When it was described, Contreras et al. (1997) did not compared it with the nearby and widely distributed *C. mendocinus*. However, Rossi et al. (2002), on the basis of a morphometric analysis and unpublished allozimic data (page 281) of Gallardo and collaborators, questioned the distinction of *C. validus* with respect to *C. mendocinus*.

As an example of the species whose distinction has been questioned in the literature, we mention one involving the type species of the genus, *C. brasiliensis*. Fernandes et al. (2012) raised the possibility that *C. brasiliensis* traditionally regarded as from Minas Gerais, Brazil, an area where no tuco-tuco is known, be a senior synonym of *C. pearsoni*, a form known from southern Uruguay and nearby areas of Argentina. This hypothesis has its base on a reinterpretation of the placement of the type locality of *C. brasiliensis* as to Minas, Uruguay. Morphometrically the type specimen of *C. brasiliensis* falls in the morphospace of *C. pearsoni*

(Fernandes et al. 2012) and is close to that a *C. torquatus*, a species closely related to *C. pearsoni*; the known distribution of the later does not reach Minas, but it is no far from there. As in similar cases pending of resolution, comparisons of DNA sequences from putative topotypes and the type specimen of *C. brasiliensis* would allow clarify this issue. As the area of Minas is quite accessible and the type of *C. brasiliensis* is preserved as skin and skull, such a study seems feasible.

2.2.5 Final Remarks of the Taxonomy Section

As it is clear from the cases reviewed above the taxonomy of *Ctenomys* is far from been stable; Cabrera (1961) recognized 26 tuco-tuco species, Honacki et al. (1982) recognized 33, Woods (1993) recognized 38 species, Woods and Kirlpatrick (2005) listed 60 species, Bidau (2015) listed 65, while here we recognized 64 (with several departures from those 65 recognized by Bidau). This variation in numbers and contents not only reflects the discovery of new forms of species-level; it is also due to distinct considerations on the distinction of some populations, as well as distinct operational criteria adopted on a given list (e.g., to include or not names that are nomenclatorially unavailable, even if the evidence supports the hypothesis of them representing distinct species-level lineages).

We close this section on Taxonomy remarking that studies based on the integration of chromosomal, molecular, and morphological evidence are still scarce (e.g., Freitas et al. 2012; Gardner et al. 2014; Teta and D'Elía 2020) and that the sampling density for data types is, in general, uneven (e.g., species well characterized at the chromosomic level limits to those of Rio Grande do Sul, Brazil). Similarly, molecular evidence, even when several specimens have been analyzed in some studies, limits for the most part to mitochondrial DNA sequences. Moreover, in general, the geographic coverage of the taxonomic studies is narrow (e.g., the species that we just described are known from a handful of localities; Teta and D'Elía 2020). Given these facts, at the time that the taxonomic status of several populations and nominal forms is dubious, we anticipate that several departures from our list will start to accumulate in the near future.

2.3 Phylogenetic Relationships

Even though several studies have been published since the early 1990s, it can be said that tuco-tuco phylogenetic studies are still in their infancy. For instance, only one study, with a reduced taxonomic sampling, has incorporated nuclear DNA sequences; all other molecular studies analyze mitochondrial DNA sequences. The two analyses incorporating morphological evidence have also incomplete taxonomic sampling. In fact, none of the published studies have nearly complete species sampling.

2.3.1 Early Phylogenetic Analyses

Three strictly phylogenetic analyses of species of *Ctenomys* were published before
the incorporation of DNA sequences; they included only Bolivian or Argentinean
species. The first one was part of a coevolution study and analyzed qualitative morphological features and karyotypic characters of six Bolivian species (Gardner
1991). Cook and Yates (1994), based on allelic variation at 21 allozyme loci,
assessed relationships among seven species and an at the time undescribed form
from Bolivia (currently recognized as *C. yatesi*). The third study is that of Ortells
(1995) where, based on G-band patterns, relationships among 11 species and five
karyomorphs from Argentina were assessed.

2.3.2 DNA Sequence-Based Studies

During the last years of the decade of the 1990s, phylogenetic analyses of species of
Ctenomys incorporated DNA sequences as evidence. With a single exception, these
studies are based on sequences of the mitochondrial DNA cytochrome b (cytb)
gene. The single study incorporating nuclear DNA sequences is the one conducted
by Castillo and collaborators (2005) that, in addition to cytb sequences, included
sequences of two nuclear genes, namely the 4th intron of the rhodopsin gene and the
2nd intron of vimentin. This study, which included ca. 20 species, corroborated
some groups found in previous mitochondrial topologies but did not provide resolution at the base of the tree. It was not expanded in further studies; therefore what we
know about the phylogenetic relationships of species of *Ctenomys* is basically based
on cytb variation. This is certainly a main limitation of our current knowledge; we
expect that this scenario improves in the next few years with the adoption of a multilocus or even genomic approach and the use of an exhaustive taxonomic coverage.

Five cytb-based phylogenetics studies of tuco-tucos were published two decades
ago (Cook and Lessa 1998; Lessa and Cook 1998; D'Elía et al. 1999; Mascheratti
et al. 2000; Slamovits et al. 2001), starting the current era of phylogenetic studies of
Ctenomys. The first and last studies present phylogenies that were inferred to assess
the diversification rate of *Ctenomys* and the evolution of satellite DNA, respectively.
The other three studies were motivated with a more traditional systematic aim, the
clarification of species relationships. The study of Lessa and Cook (1998) includes
Argentinean and Bolivian species; that of D'Elía et al. (1999) builds over the taxonomic sampling of Lessa and Cook (1998) adding species from Brazil, Chile, and
Uruguay. Still, both studies covered a relatively small fraction of the species diversity. The study of Mascheratti et al. (2000) has a broader taxonomic sampling (ca.
28 species), including for the first time a Paraguayan species and additional species
from Argentina, but has a much-reduced character sampling (ca. less than 400 bp vs.
the complete gene in the other studies). This may be the reason behind why subsequent phylogenetic studies did not include sequences gathered by Mascheratti et al.

(2000); as such, the phylogenetic position of some species (e.g., *C. bergi, C. bon-netoi, C. pilarensis* Contreras, 1993) has not been further assessed after Mascheratti et al. (2000).

Early in this decade, Parada et al. (2011) published the results of a phylogenetic analysis that remains the reference topology of the phylogenetic relationships among species of *Ctenomys*. Considering that this study was published 10 years ago, it is clear that not much progress has been achieved since then. Subsequent studies mostly built on the matrix of Parada et al. (2011) adding sequences of previously not included species (i.e., *ibicuensis*, Freitas et al. 2012; *bicolor* Miranda-Ribeiro, 1914: Leipnitz et al. 2020, see also Stolz et al. 2013; *dorsalis* Thomas, 1900: Londoño-Gaviria et al. 2018; *famosus* Thomas, 1920: Sánchez et al. 2018). Notwithstanding, some species (e.g., *johanis* Thomas, 1921; *juris* Thomas, 1920; *osvaldoreigi*; *pilarensis*; *paraguayensis* Contreras, 2000; *tulduco*; *validus*) remain without being incorporated in any phylogenetic analysis. As follows from what has been said, the comments we present below are mostly built on the results of the study of Parada et al. (2011).

2.3.3 Species Relationships and Species Groups

Most species of *Ctenomys* fall in eight main clades that Parada et al. (2011) referred as to species groups and are named using the oldest species name of each group. These groups are the *boliviensis, frater, magellanicus, mendocinus, opimus, tala-rum, torquatus,* and *tucumanus* species groups. In Table 2.1, we provide the group affiliation of the species here recognized. The richness of the species groups is relatively homogeneous, ranging between 2 and 8 species. Except for the *magellanicus* and *torquatus* groups, the other species groups were previously identified, with some differences in composition, and referred with other names (mostly referring to their geographic distribution), by Contreras and Bidau (1999; see Table 2.1 in Parada et al. 2011). As after Parada et al. (2011), no gene other than cytb has been used to assess the phylogenetic relationships of *Ctenomys*, and the contents of species groups have been stable. Modifications only pertain to the inclusion of new species into already defined groups, as new sequences become available. For instance, *C. famosus* was shown to be a member of the *mendocinus* species group (Sánchez et al. 2018), and the *magellanicus* species group was expanded to include the recently described species *C. bidaui, C. contrerasi,* and *C. thalesi* (Teta and D'Elía 2020).

Four species, *C. leucodon* Waterhouse, 1848; *C. maulinus* Philippi, 1872; *C. sociabilis* Pearson and Christie, 1985; and *C. tuconax* Thomas, 1925 were not recovered as members of any species group by Parada et al. (2011) as there are not closely related to any other species of the genus. However, given the topology portrayed in Mascheratti et al. (2000), *C. tucounax, C. juris,* and *C. pilarensis* would belong to the *tucumanus* species group. In addition, from that topology, it follows that *C. bergi* and *C. bonettoi* belong to the *mendocinus* group. However, several species still remain with unclear affiliation as they have not been included in any published phylogenetic analysis (Table 2.1). While species groups are relatively well supported (the exception is the *torquatus* species group), relationships among them are poorly supported; the exception being the clade formed by the *mendocinus* and *talarum* groups. This lack of resolution among species groups extends to the base of the radiation of *Ctenomys*.

A somewhat surprising, although not strongly supported, result of the phylogenetic analysis is that the basal dichotomy of the crown group of *Ctenomys* sets *C. sociabilis* as sister to a clade formed by all other species included in that study (Parada et al. 2011). The position of *C. sociabilis* as sister to all other species of the genus was corroborated by the studies of Londoño-Gaviria et al. (2018), who used a distinct sequence of *C. sociabilis* than the one used by Parada et al. (2011), and Sanchez et al. (2018), as well as in the Maximum Likelihood analysis of Leipnitz et al. (2020); there is a problem with the placement of the root of the gene tree portrayed in Fig. 3, where *C. sociabilis* appears as sister to the octodontids and not to the other species of *Ctenomys*). Other studies have recovered another basal dichotomy. The position of *C. sociabilis* as sister to the other tuco-tucos is not corroborated in the Bayesian analysis of Leipnitz et al. (2020) and in the studies of Roratto et al. (2015) and Caraballo and Rossi (2018b); in these topologies, the *frater* species group is sister to the clade formed by all other species of *Ctenomys*, where *C. sociabilis* appears sister to *C. tuconax*. However, these relationships are not significantly supported. In sum, relationships at the base of *Ctenomys* are better portrayed as a polytomy involving several species groups, as well as some lineages formed by single species (e.g., *C. sociabilis*). This basal polytomy has been classically regarded as resulting from an explosive radiation (e.g., Castillo et al. 2005).

Available topologies need to be further tested with a broader taxonomic sampling that includes those species missing from current datasets as well as with the incorporation of other characters, in particular nuclear DNA sequences. In this sense, the study of Castillo et al. (2005) may be regarded as a starting point. However, we note that the two loci employed by Castillo et al. (2005) are relatively short for current sequencing technologies (i.e., more characters could be obtained with the same effort) and that they have not been used in other studies of relationships among caviomorphs (e.g., Upham and Patterson 2015). In addition, a genome-wide sampling approach to tackle tuco-tuco phylogenetics seems feasible now (Lessa et al. 2014). Similarly, the inclusion of morphologic characters is much needed; in this sense, we highlight the analysis recently presented by De Santi et al. (2020), in which patterns of morphological and molecular variation were integrated. In this line, the analysis of morphologic variation, in a phylogenetic context, is much needed to diagnose the recovered species groups.

2.3.4 Subgenera of Ctenomys

Based on trenchant characters, two subgenera have been nominated on the basis of species of *Ctenomys*. Thomas (1916) erected *Haptomys* to contain *C. leucodon*, highlighting its procumbency and unpigmented incisors. Subsequently, Osgood (1946) described *C. conoveri* under the subgenus *Chacomys*, remarking in its overall large size and distinctly grooved incisors. Cabrera (1961), following Osgood (1946), only recognized *Ctenomys* and *Chacomys*. The phylogenetic analyses reviewed above lend no support for classification including three subgenera (i.e., *Chacomys*, *Ctenomys*, and *Haptomys*). Meanwhile, a scheme of two subgenera with *Chacomys* and *Ctenomys* is weakly supported in some studies if *Chacomys* is expanded to encompass the *frater* species group; we note, however, that *Chacomys* would have to be re-diagnosed. In light of available information, we think that it is adequate to maintain a scheme of one genus without subgenera.

2.3.5 Dating

Four studies, using a dense taxonomic sampling, have dated the radiation of *Ctenomys* (Parada et al. 2011; Roratto et al. 2015; Caraballo and Rossi 2018b; Leipnitz et al. 2020). These studies have in common that they are based on cytb gene sequences but differ on several implementation aspects, most notably in calibration points (Table 2.3). Estimates presented by Parada et al. (2011) are older than those of the other studies, and their confidence intervals do not overlap. Those of Roratto et al. (2015), Leipnitz et al. (2020), and Caraballo and Rossi (2018b) are more in line among them, although the estimate for the stem age of *Ctenomys* in the latter study is the oldest of the four studies. The oldest known fossils of *Ctenomys*, gathered at the Uquía Formation in Jujuy, Argentina, have an age of 3.5 Ma (Reguero et al. 2007; Verzi et al. 2010); this age falls within the confidence interval of (2.38–5.14) of the younger molecular estimate that corresponds to that of Leipnitz et al. (2020). As such, this estimate can be discarded as it is contradicted by the fossil record. The remaining three studies imply a ghost lineage for *Ctenomys* of ca. 2.1–5.7 Ma (0.8–9.1 Ma if confidence intervals are considered). Additional fossil evidence, as well as sequencing of new loci, will help resolve these uncertainties.

The study of Cook and Lessa (1998; see also Castillo et al. 2005) focused on diversification rates of *Ctenomys* and with a reduced taxonomic sampling, found that the diversification of *Ctenomys* decreases toward the present. In a broader phylogenetic context, Upham et al. (2019) found that *Ctenomys* is one of the mammal lineages with an accelerated speciation rate.

Table 2.3 Selected results of four dating analyses of *Ctenomys*. Mean estimated ages (Mya) are indicated with 95% confidence intervals in parenthesis

	Parada et al. (2011)	Roratto et al. (2015)	Caraballo and Rossi (2018b)	Leipnitz et al. (2020)
Gene	Cytb	Cytb	Cytb	Cytb
Number of calibration points[#]	1[*]	3[**]	7[***]	2[****]
Stem *Ctenomys*	17.97 (13.5–23.0)	11.17 (10.20–12.13)	21.35 (15.59-26.69)	10.98 (9.08–12.83)
Crown *Ctenomys*	9.22 (6.4–12.6)	5.64 (4.38–6.98)	5.88 (4.32–7.54)	3.71 (2.38–5.14)
All *Ctenomys* minus *sociabilis*	7.74 (5.7–10.2)	–	N/A	N/A
Median crown species groups	1.44–4.97	–	–	0.52–2.18

N/A implies that the given node was not recovered in the phylogenetic analysis (see text for details). A dash (–) indicates information not provided in the paper
[#]for details of the implementation of each calibration point (e.g., type of distribution) as well as the base of each point see the original publications
[*]MRCA for Caviomorpha: 28.5–37
[**]Origin Caviomorpha: 34 (31–37); Ctenomyidae-Octodontidae: 10.65 (9.8–11.5); origin Ctenomyidae: 5 (3.3–7.5)
[***]Crown Octodontoidea: 27.2–41.0; crown Echimyidae 20.5–24.2; crown Abrocomidae: 2.0–4.0 Myr; crown Octodontinae: 6.8–9.0 Myr; the clade form by *Aconaemys* and *Spalacopus*: 5.0–9.0; the clade form by *Octomys*, *Tympanoctomys* (including *Pipanacoctomys*): 2.0–4.0; Crown *Ctenomys*: 3.5–6.0 Myr. A second dating analysis with 4 calibration points was also conducted; results derived from the 4 and 7 calibration point datasets are congruent in terms of overlapping 95% credibility intervals
[****]Ctenomyidae-Octodontidae: 10.65 (9.8–11.5); *Ctenomys*: 5.0 (3.5–6.5)

2.3.6 Trait Evolution

Studies about the evolution of distinct traits of *Ctenomys* are scarce. Several mentions are made in the literature regarding the direction of chromosomal rearrangements, but no study has assessed chromosome evolution at a large scale and with an explicit phylogenetic context.

Verzi and Olivares (2006), using a qualitative and quantitative approach, recognized two morphotypes within *Ctenomys*, for which they used the names of "*C. dorbigny* morphotype" and "*C. fulvus* morphotype." These two groups differ in the size and shape of the postglenoid fossa, auditory meatus, and condyloid and postcondyloid processes of the mandible. Verzi and Olivares (2006) included in the first morphotype the species *C. dorbignyi*, *C. leucodon*, *C. lewisi*, *C. pearsoni*, *C. tucumanus*, and two undescribed forms (*C.* sp. "*bellavista*", *C.* sp. "*perucho*"), while in the second they included *C. fulvus*, *C. australis*, *C. azarae* (here regarded as a synonym of *C. mendocinus*), *C. opimus*, *C. maulinus*, *C. steinbachi*, *C. talarum*, and another undescribed form (*C.* sp. "*trapalco*"). Despite the fact that several species have not been assessed on regard to these features and that none of these groups is monophyletic, it is of interest to note that each of the groups defined by Parada et al. (2011) include only members of one of the morphotypes described by Verzi and Olivares (2006).

Penial morphology was studied by Balbontin et al. (1996), who described three main types of phallus within *Ctenomys*. One morphotype, referred to as "spike bearing species", is present in *C. australis*, *C. azarae* (= *C. mendocinus*), *C. porteousi* (= *C. mendocinus*), *C. rionegrensis*, and *C. talarum*. The second morphotype called "spiny-bulb bearing species," encompasses *C. dorbigny*, *C. pearsoni*, *C. perrensi*, *C. roigi*, and some other populations from Corrientes, Argentina. Finally, the third morphotype is exclusively known from *C.* "*yolandae*" and shows both spikes and spiny bulbs. Broadly viewed, "spike bearing species" correspond to the *mendocinus* and *talarum* species groups, which are sister to each other, while those referred to as "spiny-bulb bearing species" correspond to the *torquatus* species group (sensu Parada et al. 2011).

Species of *Ctenomys* display three sperm morphotypes: simple symmetric, simple asymmetric, and complex asymmetric. Most species are characterized by one of the first two types, whereas the third one is known only from *C.* "*yolandae*" (Feito and Gallardo 1982; Vitullo et al. 1988; Vitullo and Cook 1991). The sperm type of several species remains unknown (see Table 2.1). The asymmetric morphotype was first described by Feito and Barros (1982) and presents a paddle-like head with a postacrosomic process at the base of the head opposite the insertion of the flagellum. This morphotype is not found in other mammals (Vitullo et al. 1988). The symmetric morphotype lacks the postacrosomic process, while the complex asymmetric presents two of these processes. Reconstruction of the evolution of sperm morphs using parsimony agrees with the hypothesis that the symmetric morph is the ancestral character state (Vitullo et al. 1988; Vitullo and Cook 1991); however, the asymmetric sperm may have evolved more than once in the history of *Ctenomys* (Lessa and Cook 1998; D'Elía et al. 1999). As such, the polyphyly of the asymmetric sperm species falsifies the evolutionary scheme suggested by Feito and Gallardo (1982), Vitullo et al. (1988), and Vitullo and Cook (1991), who, without formally postulating the reciprocal monophyly, proposed two major lineages of *Ctenomys* based on these sperm types.

We expect those upcoming denser and more robust phylogenies to provide the framework for the study of the evolution of these and other traits within *Ctenomys*. Chromosomal evolution, which is strikingly large and has played a significant role in the study of the evolutionary biology of *Ctenomys* (e.g., Gallardo 1991; Lopes et al. 2013), is probably the most outstanding of these character sets. The evolution of sociality is a more recent but significant area of great interest (e.g., Lacey 2004, MacManes and Lacey 2012; Tomasco et al. 2019) to include in a phylogenetic framework.

2.3.7 Final Remarks on Phylogenetic Relationships

Much remains to be learned about the phylogenetic relationships of the species of *Ctenomys*, and the same is true about the timing of their diversification. We expect that in the upcoming years this scenario will change. Perhaps paradoxically, one of the major challenges still is to secure samples of all recognized species, including those obtained at or near type localities. Collaborative efforts among research groups seem to be the most promising way to secure a dense taxonomic sampling and associated tissue availability for the genus. Meanwhile, a dense molecular character sampling, even one at the genomic scale, nowadays, appears within reach. Integrating molecular and morphological characters presents additional challenges (but see De Santi et al. 2020). All of these difficulties trace back to the need to assemble adequate specimen series for each species. In this regard, we close this chapter by making a call to our colleagues on the need to keep collecting voucher specimens of *Ctenomys*, an essential step toward the improvement of our understanding of its evolutionary history.

Acknowledgments We express our gratitude to the editors of this book for their invitation to participate. Similarly, we thank all colleagues and students with whom we have exchanged ideas on the systematics of tuco-tucos over several years. Finally, we would like to recognize the effort of those who keep collecting specimens in the field and to those working in the collections that maintain these specimens available for further studies.

References

Agnolin FL, Lucero SO, Chimento NR, Derguy MR, Godoy IN (2020) Catálogo de los ejemplares tipo de la Colección Mastozoológica de la Fundación de Historia Natural "Félix de Azara", Ciudad Autónoma de Buenos Aires, Argentina. Hist Nat (tercera serie) 10:17–52

Aguilar G (1988) Relaciones sistemáticas entre citotipos del género *Ctenomys* (Rodentia, Ctenomyidae), en Chile Central. Tesis de Magister en Ciencias, mención Zoología. Facultad de Ciencias. Universidad Austral de Chile, Chile

Azurdy H (2005) Descripción de una nueva especie de *Ctenomys* (Rodentia, Ctenomyidae) de los valles interandinos de Bolivia. Kempffiana 1:70–74

Balbontin J, Reig S, Moreno S (1996) Evolutionary relationships of *Ctenomys* (Rodentia: Octodontidae) from Argentina, based on penis morphology. Acta Theriol 41:237–253

Bidau CJ (2015) Family Ctenomyidae Lesson, 1842. In: Patton JL, Pardiñas UFJ, D'Elia G (eds) Mammals of South America, volume 2 – rodents. University of Chicago Press, Chicago, pp 818–877

Bidau CJ, Martí DA, Giménez MD (2003) Two exceptional South American models for the study of chromosomal evolution: the tucura *Dichroplus pratensis* and the tuco-tuco of the genus *Ctenomys*. Hist Nat 8:53–72

Bidau, CJ, Giménez MD, Davies YE, and Contreras JR (2005) New Ctenomys karyotypes from lower Chaco, Argentina (Rodentia, Ctenomyidae). Nucleus 48:135–42

Buschiazzo LM, Caraballo DA, Cálcena E, Longarzo ML, Labaroni CA, Ferro JM, Rossi MS, Bolzán AD, Lanzone C (2018) Integrative analysis of chromosome banding, telomere localization and molecular genetics in the highly variable *Ctenomys* of the Corrientes group (Rodentia; Ctenomyidae). Genetica 146:403–414

Cabrera A (1961) Catálogo de los mamíferos de América del Sur. Revista del Museo Argentino de Ciencias Naturales "Bernardo Rivadavia", Instituto Nacional de Investigación Ciencias Naturales. Ciencias Zool 4:308–732

Caraballo DA, Rossi MS (2018a) Integrative lineage delimitation in rodents of the *Ctenomys* Corrientes group. Mammalia 82:35–47. https://doi.org/10.1515/mammalia-2016-0162

Caraballo DA, Rossi MS (2018b) Spatial and temporal divergence of the *torquatus* species group of the subterranean rodent *Ctenomys*. Contrib Zool 87:11–24

Caraballo DA, Abruzzese GA, Rossi MS (2012) Diversity of tuco-tucos (*Ctenomys*, Rodentia) in the Northeastern wetlands from Argentina: mitochondrial phylogeny and chromosomal evolution. Genetica 140:125–136

Castillo AH, Cortinas MN, Lessa EP (2005) Rapid diversification of South American tuco-tucos (*Ctenomys*; Rodentia, Ctenomyidae): contrasting mitochondrial and nuclear intron sequences. J Mammal 86:170–179

Contreras, JR, Roig VG, and Suzarte CM (1977) Ctenomys validus, una nueva especie de "tunduque" de la provincia de Mendoza (Rodentia, Octodontidae). Physis 36:159–62

Contreras JR, Berry LM (1984) Una nueva especie del género *Ctenomys* procedente de la Provincia de Santa Fe (Rodentia, Ctenomyidae). Resúmenes de las VII Jornadas Argentinas de Zoología. Mar del Plata, Argentina, p 75

Contreras JR, Bidau CJ (1999) Líneas generales del panorama evolutivo de los roedores excavadores sudamericanos del género *Ctenomys* (Mammalia, Rodentia, Caviomorpha, Ctenomyidae). Ciencia Siglo XXI 1:1–22

Contreras JR, Contreras A (1984) La situación de *Ctenomys minutus* Nehring, 1887, en la provincia de Entre Ríos, con la descripción de nuevas subespecies (Rodentia: Ctenomyidae). Libro de Resúmenes de las VII Jornadas Argentinas de Zoología, Mar del Plata, p 76

Contreras JR, Maceiras AJ (1970) Relaciones entre tucu-tucus y los procesos del suelo en la región semiárida del sudoeste bonaerense. Agro 12:1–26

Contreras JR, Roig VA (1975) *Ctenomys eremofilus*, una nueva especie de tucu-tuco de la región de Ñancuñán, provincia de Mendoza (Rodentia: Octodontidae). Resúmenes IV Jornada Argentina de Zoología, Corrientes, pp 19–20

Contreras JR, Scolaro JA (1986) Distribución y relaciones taxonómicas entre los cuatro núcleos geográficos disyuntos de Ctenomys dorbignyi en la Provincia de Corrientes, Argentina (Rodentia, Ctenomyidae). Hist Nat 6:21–30

Contreras JR, Manceñido M, Ripa Alsina M (1970) *Ctenomys chasiquensis*. Una nueva especie de tuco-tuco del sudoeste de la provincia de Buenos Aires. en: Resúmenes de Comunicaciones Libres V Congreso Argentino de Ciencias Biológicas. Buenos Aires, p 68

Cook J, Lessa EP (1998) Are rates of diversification in subterranean South American tuco-tucos (genus *Ctenomys*, Rodentia: Octodontidae) unusually high? Evolution 52:1521–1527

Cook JA, Yates TL (1994) Systematic relationships of the tuco-tucos, genus *Ctenomys* (Rodentia: Octodontidae). J Mammal 75:583–599

Cook JA, Anderson S, Yates TL (1990) Notes on Bolivian mammals. 6, The genus *Ctenomys* (Rodentia, Ctenomyidae) in the highlands. Am Mus Novit 2980:1–27

D'Anatro A, D'Elía G (2011) Incongruent patterns of morphological, molecular, and karyotypic variation among populations of *Ctenomys pearsoni* Lessa and Langguth, 1983 (Rodentia, Ctenomyidae). Mamm Biol 76:36–40

D'Elía G (2012) Nomenclatural unawareness, or on why a recently proposed name for Chiloean populations of *Pudu puda* (Mammalia, Cervidae) is unavailable. Rev Chil Hist Nat 85:237–239

D'Elía G, Lessa EP, Cook JA (1999) Molecular phylogeny of tuco-tucos, genus *Ctenomys* (Rodentia: Octodontidae): evaluation of the *mendocinus* species group and the evolution of asymmetric sperm. J Mamm Evol 6:19–38

D'Elía G, Fabre P-H, Lessa EP (2019) Rodent systematics in an age of discovery: recent advances and prospects. J Mammal 100:852–871. https://doi.org/10.1093/jmammal/gyy179

De Santi NA, Verzi DH, Olivares I, Piñero P, Morgan CC, Medina ME, Rivero DE, Tonni EP (2020) A new peculiar species of the subterranean rodent *Ctenomys* (Rodentia, Ctenomyidae) from the Holocene of central Argentina. J S Am Earth Sci 100. https://doi.org/10.1016/j.jsames.2020.102499

Feito R, Barros C (1982) Ultrastructure of the head of *Ctenomys maulinus* spermatozoon with special reference to the nucleus. Gamete Res 5:317–321

Feito R, Gallardo M (1982) Sperm morphology of the Chilean species of *Ctenomys* (Octodontidae). J Mammal 63:658–661

Fernández FA, Fornel R, Freitas TRO (2012) *Ctenomys brasiliensis* Blainville (Rodentia: Ctenomyidae): clarifying the geographic placement of the type species of the genus Ctenomys. Zootaxa 3272:57–68

Fornel R, Cordeiro-Estrela P, Freitas TRO (2018) Skull shape and size variation within and between *mendocinus* and *torquatus* groups in the genus *Ctenomys* (Rodentia: Ctenomyidae) in chromosomal polymorphism context. Genet Mol Biol 1(suppl):263–272. https://doi.org/10.1590/1678-4685-GMB-2017-0074

Freitas TRO (2005) Analysis of skull morphology in 15 species of the genus Ctenomys, including seven karyologically distinct forms of Ctenomys minutus (Rodentia: Ctenomyidae). In: Lacey EA, Myers P (eds) Mammalian diversification: from chromosomes to phylogeography (a celebration of the career of James L. Patton). University of California Publications in Zoology 133, Berkeley, pp 131–154

Freitas TRO, Fernandes FA, Fornel R, Roratto PA (2012) An endemic new species of tuco-tuco, genus *Ctenomys* (Rodentia: Ctenomyidae), with a restricted geographic distribution in southern Brazil. J Mammal 93:1355–1367. https://doi.org/10.1644/12-MAMM-A-007.1

Gallardo MH (1979) Las especies chilenas de *Ctenomys* (Rodentia, Octodontidae). I Estabilidad Cariotípica. Arch Biol Med Exp 12:71–82

Gallardo MH (1991) Karyotypic evolution in *Ctenomys* (Rodentia, Ctenomyidae). J Mammal 72:11–21. https://doi.org/10.2307/1381976

Galliari CA, Pardiñas UFJ, Goin FJ (1996) Lista comentada de los mamíferos argentinos. Mastozool Neotrop 3:39–62

Gardner SL (1991) Phyletic coevolution between subterranean rodents of the genus *Ctenomys* (Rodentia: Hystricognathi) and nematodes of the genus Paraspidodera (Heterakoidea: Aspidoderidae) in the neotropics: temporal and evolutionary implications. Zool J Linnean Soc 102:169–201

Gardner SL, Salazar-Bravo J, Cook JA (2014) New species of *Ctenomys* Blainville 1826 (Rodentia: Ctenomyidae) from the lowlands and central valleys of Bolivia. Spec Publ Mus Texas Tech Univ 62:1–34

Giménez MD, Mirol PM, Bidau CJ, Searle JB (2002) Molecular analysis of populations of *Ctenomys* (Caviomorpha, Rodentia) with high karyotypic variability. Cytogenet Genome Res 96:130–136

Gómez Fernández MJ, Gaggiotti OE, Mirol P (2012) The evolution of a highly speciose group in a changing environment: are we witnessing speciation in the Iberá wetlands? Mol Ecol 21:3266–3282

Honacki JH, Kinman KE, Koeppl JW (1982) Mammal species of the world: a taxonomic and geographic reference. Allen Press and Association of Systematic Collections, Lawrence. ix + 694 pp

ICZN (International Commission on Zoological Nomenclature) (1999) International Code of Zoological Nomenclature, 4th edn. International Trust for Zoological Nomenclature, London, xxix + 306 pp

Kelt DA, Gallardo MH (1994) A new species of tuco-tuco, genus *Ctenomys* (Rodentia: Ctenomyidae) from Patagonian Chile. J Mammal 75:338–348

Kubiak BB, Kretschmer R, Leipnitz LT, Maestri R, Almeida TS, Galiano D, Pereira JC, Oliveira EHC, Ferguson-Smith MA, Freitas TRO (2020) Hybridization between subterranean tuco-tucos (Rodentia, Ctenomyidae) with contrasting phylogenetic positions. Sci Rep 10:1502. https://doi.org/10.1038/s41598-020-58433-5

Lacey EA (2004) Sociality reduces individual direct fitness in a communally breeding rodent, the colonial tuco-tuco (*Ctenomys sociabilis*). Behav Ecol Sociobiol 56:449–457. https://doi.org/10.1007/s00265-004-0805-6

Leipnitz LT, Fornel R, Ribas LEJ, Kubiak BB, Galiano D, Freitas TRO (2020) Lineages of tuco-tucos (Ctenomyidae: Rodentia) from midwest and northern Brazil: late irradiations of subterranean rodents towards the Amazon forest. J Mamm Evol 27:161–176. https://doi.org/10.1007/s10914-018-9450-0

Lessa EP, Cook JA (1998) The molecular phylogenetics of tuco-tucos (genus *Ctenomys*, Rodentia: Octodontidae) suggests an early burst of speciation. Mol Phylogenet Evol 9:88–99

Lessa EP, Cook JA, D'Elía G, Opazo JC (2014) Rodent diversity in South America: transitioning into the genomics era. Front Ecol Evol 2:39. https://doi.org/10.3389/fevo.2014.00039

Londoño-Gaviria M, Teta P, Ríos SD, Patterson BD (2018) Redescription and phylogenetic position of *Ctenomys dorsalis* Thomas 1900, an enigmatic tuco tuco (Rodentia, Ctenomyidae) from the Paraguayan Chaco. Mammalia 83:227–236. https://doi.org/10.1515/mammalia-2018-0049

Lopes C, Ximenes S, Gava A, Freitas TRO (2013) The role of chromosomal rearrangements and geographical barriers in the divergence of lineages in a South American subterranean rodent (Rodentia: Ctenomyidae: *Ctenomys minutus*). Heredity 111:293–305. https://doi.org/10.1038/hdy.2013.49

MacManes MD, Lacey EA (2012) The social brain: transcriptome assembly and characterization of the hippocampus from a social subterranean rodent, the colonial tuco-tuco (*Ctenomys sociabilis*). PLoS ONE 7(9):e45524. https://doi.org/10.1371/journal.pone.0045524

Mapelli FJ, Mora MS, Lancia JP, Gómez Fernández MJ, Mirol MP, Kittlein MJ (2017) Evolution and phylogenetic relationships in subterranean rodents of the *Ctenomys mendocinus* species complex: effects of Late Quaternary landscape changes of Central Argentina. Mamm Biol 87:130–142

Mapelli FJ, Mora MS, Austrich A, Kittlein MJ (2019) *Ctenomys chasiquensis*. En: SAyDS–SAREM (eds) Categorización 2019 de los mamíferos de Argentina según su riesgo de extinción. Lista Roja de los mamíferos de Argentina. Versión digital: http://cma.sarem.org.ar

Martínez PA, Bidau CJ (2016) A re-assessment of Rensch's rule in tuco-tucos (Rodentia: Ctenomyidae: *Ctenomys*) using a phylogenetic approach. Mamm Biol 81:66–72. https://doi.org/10.1016/j.mambio.2014.11.008

Mascheretti S, Mirol P, Gimenez M, Bidau C, Contreras JR, Searle J (2000) Phylogenetics of the speciose and chromosomally variable genus *Ctenomys* (Ctenomyidae, Octodontoidea), based on mitochondrial cytochrome b sequences. Biol J Linn Soc 70:361–376

Massarini AI, Freitas TRO (2005) Morphological and cytogenetics comparison in species of the *mendocinus*-group (genus *Ctenomys*) with emphasis in *C. australis* and *C. flamarioni* (Rodentia- Ctenomyidae). Caryologia 58:21–27

Massarini AI, Barros MA, Ortells MO, Reig OA (1991) Chromosomal polymorphism and small karyotype differentiation in a group of *Ctenomys* species from Central Argentina (Rodentia: Octodontidae). Genetica 83:131–144

Massarini AI, Dyzenchauz FJ, Tiranti SI (1998) Geographic variation of chromosomal polymorphism in nine populations of *Ctenomys azarae*, tuco-tucos of the *Ctenomys mendocinus* group (Rodentia: Octodontidae). Hereditas 128:207–211

Mirol P, Giménez MD, Searle JB, Bidau CJ, Faulkes CG (2010) Population and species boundaries in the South American subterranean rodent *Ctenomys* in a dynamic environment. Biol J Linn Soc 100:368–383

Mora MS, Mapelli FJ, López A, Fernández MJG, Mirol PM, Kittlein MJ (2016) Population genetic structure and historical dispersal patterns in the subterranean rodent *Ctenomys* "*chasiquensis*" from the southeastern Pampas region, Argentina. Mamm Biol 81:314–325

Morgan CC, Verzi DH, Olivares AI, Vieytes EC (2017) Craniodental and forelimb specializations for digging in the South American subterranean rodent *Ctenomys* (Hystricomorpha, Ctenomyidae). Mamm Biol 87:118–124. https://doi.org/10.1016/j.mambio.2017.07.005

Novello AF, Lessa EP (1986) G-band homology in two karyomorphs of the *Ctenomys pearsoni* complex (Rodentia: Octodontidae) of neotropical fossorial rodents. Z Säugetierkund 51:378–380

Ortells MO (1995) Phylogenetic analysis of G-banded karyotypes among the South American subterranean rodents of the genus *Ctenomys* (Caviomorpha: Octodontidae), with special reference to chromosomal evolution and speciation. Biol J Linn Soc 54:43–70

Ortells MO, Contreras JR, Reig OA (1990) New *Ctenomys* karyotypes (Rodentia, Octodontidae) from north-eastern Argentina and from Paraguay confirm the extreme chromosomal multiformity of the genus. Genetica 82:189–201. https://doi.org/10.1007/BF00056362

Osgood WH (1946) A new octodont rodent from the Paraguayan chaco. Fieldiana Zool 31:47–49

Parada A, D'Elía G, Bidau CJ, Lessa EP (2011) Species groups and the evolutionary diversification of tuco-tucos, genus *Ctenomys* (Rodentia: Ctenomyidae). J Mammal 92:671–682

Parada A, Ojeda AA, Tabeni S, D'Elía G (2012) The population of *Ctenomys* from the Ñacuñán Biosphere Reserve (Mendoza, Argentina) belongs to *Ctenomys mendocinus* Philippi, 1869 (Rodentia: Ctenomyidae): molecular and karyotypic evidence. Zootaxa 3402:61–68

Reguero MA, Candela AM, Alonso RN (2007) Biochronology and biostratigraphy of the Uquía Formation (Pliocene- Early Pleistocene, NW Argentina) and its significance in the Great American Biotic Interchange. J S Am Earth Sci 23:1–16

Roratto PA, Fernandes FA, Freitas TRO (2015) Phylogeography of the subterranean rodent *Ctenomys torquatus*: an evaluation of the riverine barrier hypothesis. J Biogeogr 42:694–705

Rosi MI, Cona MI, Roig VG (2002) Estado actual del conocimiento del roedor fosorial *Ctenomys mendocinus* Philippi 1869 (Rodentia: Ctenomyidae). Mastozool Neotrop 9:277–295

Rusconi C (1931) Las especies fósiles del género *Ctenomys* con descripción de nuevas especies. Anales de la Sociedad Científica Argentina 112(129–142):217–236

Sage R, Contreras JR, Roig V, Patton JL (1986) Genetic variation in the South American burrowing rodents of the genus *Ctenomys* (Rodentia: Ctenomyidae). Z Säugetierkund 51:158–172

Sánchez RT, Tomasco HI, Díaz MM, Barquez RM (2018) Contribution to the knowledge of the rare "Famatina tuco-tuco", *Ctenomys famosus* Thomas 1920 (Rodentia: Ctenomyidae). Mammalia 83:11–22. https://doi.org/10.1515/mammalia-2017-0131

Slamovits CH, Cook JA, Lessa EP, Rossi MS (2001) Recurrent amplifications and deletions of satellite DNA accompanied chromosomal diversification in South American tuco-tucos (genus *Ctenomys*, Rodentia: Octodontidae): a phylogenetic approach. Mol Biol Evol 18:1708–1719

Stolz JFB, Gonçalves GL, Leipnitz L, Freitas TRO (2013) DNA-based and geometric morphometric analysis to validate species designation: a case study of the subterranean rodent Ctenomys bicolor. Genet Mol Res 12:5023–5037

Teta P, D'Elía G (2019) The least known with the smallest ranges: analyzing the patterns of occurrence and conservation of South American rodents known only from their type localities. Therya 10:271–278

Teta P, D'Elía G (2020) Uncovering the species diversity of subterranean rodents at the end of the World: three new species of Patagonian tuco-tucos (Rodentia, Hystricomorpha, *Ctenomys*). PeerJ 8:e9259. https://doi.org/10.7717/peerj.9259

Teta P, Abba AM, Cassini GH, Flores DA, Galliari CA, Lucero SO, Ramírez M (2018) Lista revisada de los mamíferos de Argentina. Mastozool Neotrop 25:163–198

Teta P, D'Elía G, Opazo JC (2020) Integrative taxonomy of the southernmost tucu-tucus in the world: differentiation of the nominal forms associated with *Ctenomys magellanicus* Bennet, 1836 (Rodentia, Hystricomorpha, Ctenomyidae). Mamm Biol 100:125–139. https://doi.org/10.1007/s42991-020-00015-z

Thomas O (1916) Two new Argentine rodents, with a new subgenus of *Ctenomys*. Ann Mag Nat Hist 18:303–306

Thomas O (1921) On mammals from the Province of San Juan, western Argentina. Ann Mag Nat Hist Ser 9(8):214–221

Tiranti SI, Dyzenchauz J, Hasson ER, Massarini AI (2005) Evolutionary and systematic relationships among tuco-tucos of the *Ctenomys pundti* complex (Rodentia: Octodontidae): a cytogenetic and morphological approach. Mammalia 69:69–80

Tomasco IH, Lessa EP (2007) Phylogeography of the tuco-tuco *Ctenomys pearsoni*: mtDNA variation and its implication for chromosomal differentiation. In: Kelt DA, Lessa EP, Salazar-Bravo J, Patton JL (eds) The quintessential naturalist: honoring the life and legacy of Oliver P. Pearson. University of California Publications in Zoology 134:v-xii + 1–981, pp 859–882

Tomasco IH, Sánchez L, Lessa EP, Lacey EA (2019) Genetic analyses suggest burrow sharing by Río Negro tuco-tucos (*Ctenomys rionegrensis*). Mastozool Neotrop 26:430–439

Upham NS, Patterson BD (2015) Phylogeny and evolution of caviomorph rodents: a complete timetree for living genera. In: Vassallo AI, Antenucci D (eds) Biology of caviomorph rodents: diversity and evolution. Buenos Aires, Argentina Sociedad Argentina para el Estudio de los Mamíferos (SAREM), pp 63–120

Upham NS, Esselstyn JA, Jetz W (2019) Inferring the mammal tree: species-level sets of phylogenies for questions in ecology, evolution, and conservation. PLoS Biol 17(12):e3000494. https://doi.org/10.1371/journal.pbio.3000494

Verzi DH, Olivares AI (2006) Craniomandibular joint in South American burrowing rodents (Ctenomyidae): adaptations and constraints related to a specialized mandibular position in digging. J Zool 270:488–501

Verzi D, Deschamps C, Tonni EP (2004) Biostratigraphic and palaeoclimatic meaning of the Middle Pleistocene South American rodent *Ctenomys kraglievichi* (Caviomorpha, Octodontidae). Palaeogeogr Palaeoclimatol Palaeoecol 212:315–329. https://doi.org/10.1016/j.palaeo.2004.06.010

Verzi DH, Olivares AI, Morgan CC (2010) The oldest South American tuco-tuco (late Pliocene, northwestern Argentina) and the boundaries of the genus *Ctenomys* (Rodentia, Ctenomyidae). Mamm Biol 75:243–252

Vitullo AD, Cook JA (1991) The role of sperm morphology in the evolution of tuco-tucos, *Ctenomys* (Rodentia, Ctenomyidae): confirmation of results from Bolivian species. Z Säugetierkund 56:359–364

Vitullo AD, Roldan ER, Merani MS (1988) On the morphology of spermatozoa of tuco-tucos, *Ctenomys* (Rodentia: Ctenomyidae): new data and its implications for the evolution of the genus. J Zool 215:675–683

Woods CA (1993) Suborder Hystricognathi. In: Wilson DE, Reeder DM (eds) Mammal species of the world. Smithsonian Institution Press, Washington, DC, pp 771–806

Woods CA, Kilpatrick CW (2005) Infraorder Hystricognathi Brandt, 1855. In: Wilson DE, Reeder DM (eds) Mammal species of the world: a taxonomic and geographic reference, 3rd edn. Johns Hopkins University Press, Baltimore, pp 1538–1600

Chapter 3
Speciation Within the Genus Ctenomys: An Attempt to Find Models

Thales Renato Ochotorena de Freitas

3.1 Introduction

Speciation is a process that results in the emergence of new species, an evolutionary force opposed to extinction that causes species to disappear. As a result, there is a natural balance between these two forces that maintain biodiversity. These processes, speciation and extinction, cause the biodiversity on Earth to be changed or renewed over time.

Although the speciation processes are complex, two major classes can be identified: allopatric and sympatric processes. An allopatric process results from the emergence of extrinsic factors, like a geographical barrier that separates populations of the same species, which, throughout time and due to isolation, turn into two different species (Mayr 1942). On the other hand, sympatric speciation, initially proposed by Darwin (1859) and totally refuted by Mayr (1963), is due to the intrinsic differentiation of a particular species within the same area. Additionally, the sympatric process causes mechanisms of reproductive isolation to develop between sister species leading to differentiation. Noteworthy, a very important condition is that both forms have never undergone an allopatric process (Li et al. 2016).

Chromosomal speciation has been discussed for many years throughout chapters of classic books and articles as, for example, White (1978), King (1993), Rieseberg (2001), and Dobigny et al. (2017), as well as the role of chromosomal rearrangement in this process. Thus, rearrangements such as the Robertsonian translocations have the role of changing the organization of chromosomes within the nucleus, increasing or decreasing the number of chromosomes. Inversions do not change the chromosome number but they change the organization of chromosomes even more because segments within a chromosomal arm are inverted (paracentric inversion).

T. R. O. de Freitas (✉)
Department of Genetics, Federal University of Rio Grande do Sul,
Porto Alegre, Rio Grande do Sul, Brazil
e-mail: thales.freitas@ufrgs.br

© Springer Nature Switzerland AG 2021
T. R. O. de. Freitas et al. (eds.), *Tuco-Tucos*,
https://doi.org/10.1007/978-3-030-61679-3_3

Also, the inverted segment can contain the centromere and thus involve the two chromosomal arms (pericentric inversion).

Hybrid zones in which the hybrids survive beyond the F_1 generation, crossing each other and with their parental types, can form a population with a wide variety of recombinant types (Harrison 1993). Thus, a hybrid zone favors the introgression of alleles, which are incorporated from one taxon to the other (Harrison 1993). This exchange of genes leads to the disappearance of parental differences and, consequently, to the appearance of recombinant species (Barton 2001). Many factors lead to introgressions, such as selective pressure and consequent fitness due to the viability of the hybrid, differentiated dispersion between males and females, and choice of partners. Importantly, these are the factors that determine the size and direction of the introgression (Petit and Excoffier 2009).

The genus *Ctenomys* Blainville, 1826 comprises approximately 65 living species (Bidau 2015; Freitas 2016; D'Elía et al. 2020; Chap. 3 of this volume), sharing some common characteristics, such as small populations distributed in fragments, which are associated with low adult dispersion rates that lead to patterns of low genetic variation within populations and high genetic divergence between populations (Busch et al. 2000; Wlasiuk et al. 2003). Additionally, habitat discontinuities, with numerous barriers to species dispersion, are also responsible for restricting individuals to their native areas. Such conditions can result in increased intraspecific competition within populations and, consequently, in inbreeding (Galiano et al. 2016). These characteristics also favor the setting of new chromosomal rearrangements, which is considered important for the diversification of ctenomids (Reig et al. 1990; Lessa and Cook 1998; Lacey et al. 2000; Freitas 2006; Parada et al. 2011).

The polytypic genus *Ctenomys* was initially divided into groups of species based on biogeographic, morphological, and cytogenetic data. Contreras and Bidau (1999) classified 34 species and divided them into nine groups: Bolivian-Mato Grosso (4 species), Bolivian-Paraguayan (3 species), Patagonian (4 species), Mendocinus (5 species), Oriental (3 species), Chaco (6 species), ancestral (2 species), Corrientes (3 species), Chilean spp. (2 species), and uncertain position (2 species). Parada et al. (2011) used molecular markers by sequencing the mtDNA Cyt-b gene to analyze 39 species and divided them phylogenetically into eight groups: Boliviensis (4 species), Mendocinus (5 species), Torquatus (6 species), Talarum (2 species), Opimus (4 species), Magellanicus (6 species), Tucumanus (4 species), and frater (4 species). Afterward, three phylogenetic trees were determined, including 54 species, (Freitas et al. 2012) and 50 species with 10 species from Bolivia (Gardner et al. 2014). More recently, a third phylogenetic tree included species from the Midwest of Brazil (Leipnitz et al. 2020).

The speciation processes in *Ctenomys* have been always associated with the Allopatric model because the data showed that each species displayed a geographic distribution isolated from the other, the only exception being the sympatric process found between *Ctenomys australis* and *Ctenomys talarum* (Reig et al. 1990). In this chapter, we discuss results that provide evidences about speciation models in *Ctenomys*. First, examples of allopatric speciation in four species of *Ctenomys* are presented: *Ctenomys australis* and *Ctenomys flamarioni*, which inhabit separated regions in South America, and *Ctenomys minutus* and *Ctenomys lami* that share the coastal plain of Rio Grande do Sul, the southernmost state of Brazil. Then, chromosomal speciation in *Ctenomys* is discussed, with an emphasis in *C. lami*, while *C. minutus* is presented as an example of probable sympatric speciation. Furthermore, the role of hybrid zones, both intra and interspecies, in speciation, are discussed. Finally, we report other differentiation processes that have occurred or seem to be starting.

3.2   Allopatric Speciation Model

Two examples related to differentiation events that can be classified as allopatric processes are found in *Ctenomys*. The first involves two species:" *C. flamarioni*, which currently occupies the coast of southern Brazil, and *C. australis*, which inhabits part of the coast of Argentina. The second example involves two other species: *C. lami* and *C. minutus*, both living on the coastal plain of Rio Grande do Sul, in the South of Brazil.

3.2.1   C. flamarioni *and* C. australis

Both species have the same number of chromosomes $2n = 48$, with a varying number of chromosomal arms (FN). The CBG-band pattern is similar between the two species with heterochromatic blocks in the small chromosome arms. The GBG-band pattern is the same in both species, which means that they have homologous chromosomes (Freitas 1994; Massarini and Freitas 2005). These characteristics led Freitas (1994) to consider *C. flamarioni* as belonging to the group of species called Mendocinus, which is formed by *C. mendocinus, C. azarae, C. australis, C. porteousi,* and *C. rionegrensis* (Massarini et al. 1991). At the same time, both species have the same asymmetric sperm shape (Freitas 1995b). However, the skull morphology is different, due to *C. australis* being larger than *C. flamarioni* (Massarini and Freitas 2005; Fornel et al. 2018). Phylogeographic patterns have been described for both species by different authors. Mora et al. (2006) found 25 different haplotypes in the geographical distribution of *C. australis*. Nevertheless, *C. flamarioni* has only nine different haplotypes (Fernández-Stolz 2006). Ecologically, both species live in the first line of dunes, *C. australis*, in Necochea (Argentina), and *C. flamarioni*, on the coastal plain of Rio Grande do Sul (Brazil).

These data suggest that *C. australis* and *C. flamarioni* were initially a single species that occupied a more extensive coastline than the current one, about 18,000 years ago (Massarini and Freitas 2005). Moreover, Mertens et al. (in preparation), through ecological niche modeling, found that in the Last Glacial Maximum (22,000 years ago), both species were distributed in a zone of sympatry that occupied the region that today is Uruguay. The rise of the sea and its advancement over the coastal plain of South America resulted in the Rio de La Plata, which ended up being a barrier that separated the populations of Argentina and southern Brazil. These data lead us to recognize a process of allopatric speciation between *C. flamarioni* and *C. australis*.

3.2.2   C. minutus *and* C. lami

These two species also show an allopatric speciation process. Both have a parallel parapatric distribution in the coastal plain of Rio Grande do Sul (Freitas 1995a). *C. lami* occupies the first line of dunes in the coastal plain, a region that originated in the Plio-Pleistocene, while *C. minutus* occurs in the second line of dunes, which originated in the Pleistocene.

Both *C. lami* and *C. minutus* belong to the Torquatus group, which is formed by ten species. *Ctenomys roigi, Ctenomys perrensis*, and *Ctenomys argentinus* occurs in the Entre Rios region, in Argentina. Separated from these species by the Uruguay river, we found *C. pearsoni*, in Uruguay, *C. torquatus*, in Uruguay and southern Brazil, and *C. ibicuiensis*, only in the south of Rio Grande do Sul, near Uruguay. Finally, inhabiting the coastal plain of Rio Grande do Sul, we found *C. minutus* and *C. lami*. The intriguing question is how *C. minutus* and *C. lami* arrived in the coastal plain, which is separated by the Patos Lagoon from the other regions where all the other species of the Torquatus group inhabit by the Patos Lagoon.

The analysis of the separation process between *C. minutus* and *C. lami* starts with understanding the taxonomic confusion about these species. *C. minutus* was described by Nehring (1887), who recorded the species in a type of vegetation known as "fields." in a location called Taquara do Mundo Novo, in Rio Grande do Sul. Later, Nehring (1900) reported that the animals were found near the beach of Tramandaí, on the banks of the Tramandaí river (29° 53′S; 50° 17′W). Reig et al. (1966) and Reig and Kiblisky (1969), aiming to determine the karyotypes of *Ctenomys* species, described until the 1960s at their respective type-locality, failed to find specimens of *C. minutus* in the place now known as Taquara. However, they found a population in the locality of Santo Antônio da Patrulha, in Rio Grande do Sul, and considered this location as the new type-locality of *C. minutus*. However, reports about cytogenetics (Freitas 1994, 1997) and geographic distribution (Freitas 1995a) of *C. minutus* suggests that the specimens collected by Reig et al. (1966) were, in fact, a separate taxon, the new species described as *C. lami* (Freitas 2001). Both *C. minutus* and *C. lami* have separate, but parallel, geographic distributions, and different chromosomal numbers: *C. minutus* presents $2n = 42$ to 50 (Freitas 1997; Freygang et al. 2004) and *C. lami*, $2n = 54$ to 58 (Freitas 2001). Chromosomally, when comparing the karyotypes $2n = 46a$ of *C. minutus* and $2n = 54$ of *C. lami*, both species are different due to four Robertsonian rearrangements (Freitas 2001). The species are separated by swamp areas and are isolated, which probably led to the beginning of a differentiation, which is still occurring. They are, therefore, two species that are still in a process of differentiation. Nonetheless, at the same time, due to the anthropization of the region, both species are in contact in a hybrid zone, which will be described later in this chapter.

3.3   Sympatric Speciation Model

As mentioned before, sympatric speciation is characterized by the differentiation of a particular species due to some mechanism of reproductive isolation that leads to the appearance of separate species. Herein, we describe findings in *C. minutus* that suggest that this species is undergoing a sympatric speciation process.

3.3.1   C. minutus

Before describing the sympatric speciation that is likely to be occurring in *C. minutus*, it is important to notice the complexity of the facts that occur in this species involving its different karyotypes and their relationship with molecular markers, skull morphology, and the evolution of the coastal plain in southern Brazil.

This species has a linear geographic distribution in the coastal plain of southern Brazil, about 500 km long in a south-north direction (Freitas 1995a). In this direction, *C. minutus* occupies the sandy fields of the coastal plain and then occupies the first line of coastal dunes parallel to the sandy fields until the extreme north of its distribution. In this area, *C. minutus* is currently undergoing an evolutionary process involving variation in the number of chromosomes, high mtDNA variation, and morphological differences in the skull. Throughout its distribution, *C. minutus* has a chromosomal variability of $2n = 42$, 46a and 46b, 48a and 48b, 50a, and 50b (Freitas 1997; Freygang et al. 2004). This variation involves eight pairs of acrocentric chromosomes of the karyotype $2n = 50a$ (#2, #16, #17, #19, #20, #22, #23, and #24). The centric fusions $20 + 17$ and $23 + 19$ form the karyotype $2n = 46a$; the same fusions and the fission of pair #2 in 2p and 2q form the karyotype $2n = 48a$; fusions $20 + 17$, $23 + 19$, *in tandem* fusion $24 + 16$ and the centric fusion of this new segment with #24 forming a chromosome, result in the karyotype $2n = 42$; the centric fusion $20 + 17$, the 2p pericentric inversion, and the in-tandem fusion $24 + 16$ generate the karyotype $2n = 46b$; centric fusions $20 + 17$, $23 + 19$, and pericentric inversion 2p form the karyotype $2n = 48b$; finally, fusion $20 + 17$ and the pericentric inversion 2p form the karyotype $2n = 50b$ (Freitas 1997; Freygang et al. 2004). Other karyotypes were found as $2n = 49a$, and 49b, 47a and 47b, 45 and 43 are products of hybrid zones, which will be described later.

The evolution of chromosome numbers by rearrangements as Robertsonian translocations and *in tandem,* described by Freitas (1997) and Freygang et al. (2004), can be related to the evolution of the coastal plain in the South of Brazil due to the outflow of rivers which formed the channels that subdivided it. Based on the geological data from Weschenfelder et al. (2014), it appears that two different

cut-and-fill events in Patos Lagoon are related with the formation of channels during the Pleistocene and Holocene, which would explain the different karyotypes.
Firstly, the $2n = 50$ karyotype was separated by Araranguá river generating $2n = 50$a
and $2n = 50$b (Fig. 3.1a). The channels of the Camaquã and Jacuí rivers that emerged
during the Pleistocene are likely to have separated the karyotypes $2n = 50$b, $2n = 42$
and $2n = 48$a, respectively. The $2n = 48$a was separated from $2n = 46$a by another

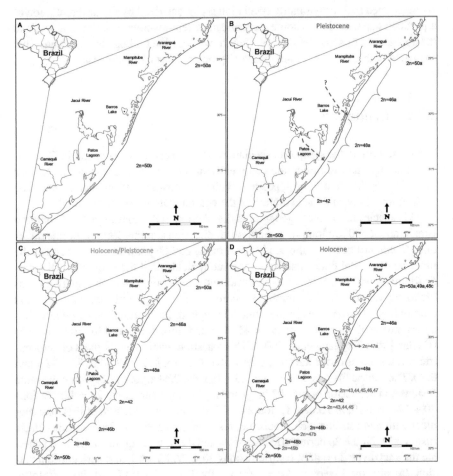

Fig. 3.1 Evolution of the chromosome numbers along the coastal plain of Rio Grande do Sul
related to the geological history of the region with the geographic distribution of diploid numbers.
(**a**) Separation between $2n = 50$ (ancestral) by the Araranguá river in two new karyotypes, $2n = 50$a
and $2n = 50$b. (**b**) During the Pleistocene, new karyotypes appeared: $2n = 46$a, separated by the
Araranguá river from $2n = 50$a, and separated from $2n = 48$a by another undetermined channel.
The Jacuí river separated $2n = 48$a from $2n = 42$, and the Camaquã river separated $2n = 42$ from
$2n = 50$b. (**c**) In the Pleistocene-Holocene, new channels emerged from the Camaquã and Jacuí
rivers and separated two new diploid numbers, $2n = 46$b and $2n = 48$b. (**d**) Nowadays, the channels
have disappeared and hybrid zones between different karyotypes are formed. The Araranguá river
separates $2n = 46$a from $2n = 50$a, $2n = 49$a, and $2n = 48$c

probable channel, which was separated from $2n = 50a$ by the Araranguá river (Fig. 3.1b). Then, in the late Holocene, the same rivers formed new channels, and these separated $2n = 50b$, $2n = 48b$, $2n = 46b$, $2n = 42$, $2n = 48a$, $2n = 46a$, and $2n = 50a$ (Fig. 3.1c). More recently, those river mouths became paleochannels, and contact zones started to be formed between the different diploid numbers (Fig. 3.1d). The Araranguá river is still a barrier that separates $2n = 46a$ from $2n = 50a$, $2n = 49a$, and $2n = 48c$.

Recently, the mtDNA sequences were analyzed with the same individuals from the same populations studied by Freitas (1997) and Freygang et al. (2004), and 52 haplotypes were found, forming six haplogroups, which were named as "Norte," "Litoral," "Lagoa dos Barros," "Mostardas," "Tavares," and "Sul" (Lopes et al. 2013). Firstly, we analyze the relationship between the haplotypes network with the diploid numbers of *C. minutus*. The haplogroup "Norte" has 10 haplotypes that are in animals with karyotypes $2n = 50a$ (7 haplotypes), $2n = 49a$ (2 haplotypes), and $2n = 48c$ (2 haplotypes). All karyotypes share one haplotype, and another haplotype is shared by individuals with $2n = 50a$ and $2n = 49a$. As for the haplogroup "Litoral," this group has 10 haplotypes, all occurring in individuals with $2n = 46a$. The haplogroup "Lagoa dos Barros" has 14 haplotypes, six of which are found in animals with $2n = 46a$, 10 in $2n = 48a$, and three in $2n = 47a$. One haplotype is shared between $2n = 46a$ and $2n = 48a$, and two others are shared between $2n = 47a$ and $2n = 48a$. Another haplotype is shared by $2n = 46a$, $2n = 47a$, and $2n = 48a$. The haplogroup "Mostardas" has 10 haplotypes, seven of them occur in $2n = 48a$, while other two occur in $2n = 42$. One haplotype is shared by $2n = 42$ and $2n = 48a$. The karyotype $2n = 42$ shares one haplotype with the hybrids formed between $2n = 42$ and $2n = 48a$. As for the haplogroup "Tavares," this group has seven haplotypes. Three haplotypes are found in hybrids between $2n = 42$ and $2n = 46b$. One haplotype is found only in individuals with 46b, and another in hybrids with $2n = 47b$. One haplotype is shared by $2n = 46b$, $2n = 47b$, and hybrids between $2n = 42$ and 46b, and another haplotype is shared by the same hybrids and $2n = 42$. Finally, the haplogroup "Sul" has three haplotypes occurring in $2n = 50b$, $2n = 49b$, and $2n = 48b$, respectively. There is one shared haplotype between $2n = 50b$ and $2n = 49b$.

A second analysis reveals the relationships between the haplotypes and the parental karyotypes within their geographic distribution and the coastal plain geological evolution. These relationships are summarized in Fig. 3.2, which presents the diploid numbers of *C. minutus* and the respective haplotypes.

The parental types $2n = 50a$ and $2n = 48c$ have the haplotypes H52 to H47 and H45, respectively. The $2n = 46a$ karyotype has two groups of haplotypes: individuals inhabiting the first line of dunes present haplotypes H44 to H35, while those inhabiting on the fields have haplotypes H34 to H31. Individuals with $2n = 48a$ have haplotypes H30 to H15 and H13. The haplotype H13 is shared with the karyotype $2n = 42$, which has also haplotypes H14, H12, and H11. The karyotype $2n = 46b$ present two haplotypes: H6 and H5, while $2n = 48b$ has only H3, and $2n = 50a$, H2 and H1. Interestingly, with the exception of H13 that is shared between $2n = 42$ and $2n = 48a$, there are no shared haplotypes. At the same time, we can observe that

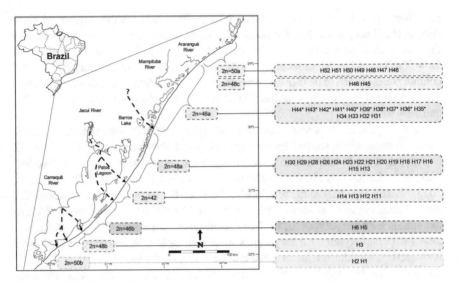

Fig. 3.2 Geographic distribution of *Ctenomys minutus* parental karyotypes with the respective haplotypes. In the Pleistocene/Holocene the coastal plain was divided by river mouths, which are shown as dashed lines separating the inhabiting areas of different karyotypic forms

channels of the Camaquã and Jacuí rivers, the undetermined channel, and the Araranguá river separated parental diploid numbers and haplotypes. This is a consequence of the geological evolution of the coastal plain in the South of Brazil during the Pleistocene and Holocene, where old channels divided the area, resulting in the chromosomal differentiation.

A final analysis allows us to understand the relation between the karyotypes and the haplotypes found along the coastal plain, as a consequence of the geological evolution (Fig. 3.3). At the north of the geographic distribution, the karyotype $2n = 50a$ has haplotypes H52 to H46, while $2n = 49a$ shows H47 and H46, and $2n = 48c$ has H46 and H45. Haplotypes H46 and H47 occur in individuals having any of the three karyotypes. The karyotype $2n = 47a$ shares the haplotype H27 with its parental karyotype $2n = 46a$, and the haplotypes H30 and H28 with the parental $2n = 48a$. Individuals from the hybrid zone between $2n = 48a$ and $2n = 42$ ($2n = 43–47$) have only haplotype H11, which is shared with the parental $2n = 42$. The hybrid zone between $2n = 42$ and $2n = 46b$ ($2n = 43–45$) has the haplotypes H9, H8, and H4. The $2n = 47b$ has the haplotypes H7 and H4, sharing the latter with individuals from the hybrid zone $2n = 42$ and $2n = 46b$. Finally, in the south, the karyotype $2n = 49b$ has the haplotype H2, which is also in $2n = 50b$.

Figure 3.3 shows also the hybridization zones between the *C. minutus* karyotypes, where one can notice the haplotypes being shared between different diploid numbers. It is also observed that the sharing occurs between $2n = 50a$, $2n = 48c$, and $2n = 49a$, between $2n = 46a$, $2n = 48a$, and $2n = 47a$, and between $2n = 50b$ and $2n = 49b$. The haplotype H4 is present only among the hybrids of $2n = 42$ and $2n = 46b$ ($2n = 43$, $2n = 44$, $2n = 45$) as well as in $2n = 47b$.

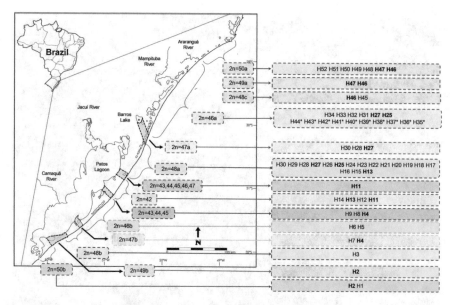

Fig. 3.3 Geographic distribution of the several *Ctenomys minutus* karyotypes and hybrid forms with the respective haplotypes. In the Holocene, the river mouths transformed to paleochannels, and contact zones between the parental types were formed. The hybrid zones are shown as gray areas delimited by dashed lines

C. minutus is the most studied among the species that occur in southern Brazil, and shows the same pattern in three aspects: the karyotypes and the haplogroups share the same North-South distribution, and the same occurs with the shape of the skull in the analysis of geometric morphometry (Fornel et al. 2010). It should be noticed that the different diploid numbers are not considered as intrinsic barriers to each other, because when the river mouths and lagoon exit to the sea disappeared, contact zones started to develop, and sharing started to occur. Again, these facts suggest that chromosomal differences are due to geographical barriers, and so different karyotypes occurred due to the isolation of populations and genetic drift.

Animals with diploid number $2n = 46a$ inhabit both the first line of dunes and the sandy fields of the coastal plain of Rio Grande do Sul (Fig. 3.4) and show an interesting feature regarding the shape of the skull. Freitas (2005), analyzing the skulls of the animals with $2n = 46a$ from parapatric populations, found that the skulls of those animals that occur in the dunes are different from the skulls of the animals that occur in the sandy fields. This difference indicates that *C. minutus* uses different strategies for tunneling, sometimes digging with paws, sometimes digging with teeth, this being a response to soil hardness (Kubiak et al. 2018). So, the strategies are not exclusively evolutionary, being probably a consequence of the harder soil found in sandy fields, which are more difficult to excavate (Kubiak et al. 2018).

These facts suggest two evolutionary alternatives related to the mode of excavation of individuals in populations that occupy different habitats: (i) the presence of directional selection or (ii) strong phenotypic plasticity in the morphology (Kubiak

Fig. 3.4 The environment of the coastal plain of the South of Brazil, with dunes and sandy fields. *Ctenomys minutus* inhabiting the dunes live in large tunnels and show different microsatellite alleles and haplotypes H35 to H44. Those inhabiting sandy fields live in small tunnels and show different microsatellite alleles and haplotypes H25 to H34

et al. 2018). Another interesting feature is the size of the home-range of individuals from two populations of *C. minutus* that inhabit these different habitats. Nineteen adult animals were tracked by radio, and the size of each individual's home-range was estimated using a 2 x 2-meter grid per cell (GCs) and minimum convex polygon (MCP) methods. The results from both methods show that the home-range of *C. minutus* differs between the two habitats, with an average size 1.75-times greater for individuals who inhabit the dunes. This difference in the size of the home-range between habitats is probably associated with differences in resource availability (plant biomass) or soil conditions (Kubiak et al. 2017). All of these data lead to suggest a differentiation in the populations of fields and sandy and coastal dunes. Additionally, Lopes et al. (2013) analyzed 14 microsatellite DNA loci and verified through Bayesian clustering and specimen assignments for the 12 clusters identified by structure that a change occurs from the sandy fields to the dunes. The authors also reported that the haplotypes 35, 36, 37, 38, 39, and 40 were found in individuals living on dunes, while haplotypes 27, 29, 30, 31, 32, 33, and 34 were found on the sandy fields.

The current data about behavior ecology, morphology, mtDNA, and microsatellite data, showing genetic differences between dunes- and sandy fields-inhabiting animals have the same karyotype $2n = 46a$, is a primary indication of a sympatric speciation process in *C. minutus*. The works carried out with the underground rodent Spalax *galili* species shows, with more powerful data (Rad-seq and microRNA), a similar behavior and a strong indication that a sympatric process of differentiation occurs between the two super-species (Hadid et al. 2013; Kexin et al. 2016; Li et al. 2016).

3.4 Chromosomal Speciation Model

Chromosomal speciation is based on the assumption that the fixation of one or more chromosomal rearrangements in a population favors speciation. For speciation to occur after rearrangement, the heterozygote has hybrid sterility with reduced fertility, which can be caused by non-disjunctions, which consequently produces incomplete numbers of chromosomes forming unbalanced gametes (White 1978; King 1993).

For the genus *Ctenomys* with chromosomal multiformity ($2n = 10$–70), there is a classic idea that the chromosomal speciation model (Reig and Kiblisky 1969, Reig et al. 1990) is responsible for the appearance of the various species that constitute the genus. Such an idea is because the *Ctenomys* species meet the conditions expected for this to occur. The occurrence of chromosomal rearrangements in heterozygotes in small isolated demes could generate new karyotypes by genetic drift, which would be tested by selection and gene flow. The genus is known as having the highest chromosomal variation among mammals ($2n = 10$ to 70) and, among rodents, to have the species with the lowest diploid number, for example, *C. steibachi* has $2n = 10$, while *C. argentinus* and *C. pearsoni* show $2n = 70$. In addition, there are species with large chromosomal variation like *C. pearsoni* ($2n = 56, 64,$ and 70); *C. boliviensis* ($2n = 42$–46), *C. rionegrensis* ($2n = 48, 50, 56,$ and 58); *C. minutus* ($2n = 42$–50), *C. lami* ($2n = 54$–58), *C. talarum* ($2n = 44$–48), and *C. perrensis* ($2n = 50, 54, 56,$ and 58) (Novello and Lessa 1986; Cook et al. 1990; Reig et al. 1992; Freitas 1997; Massarini et al. 2002; and Freitas 2007).

In the following, we describe a chromosomal speciation process that is likely to be occurring in *Ctenomys*, in the coastal plain of Rio Grande do Sul.

3.4.1 C. lami

Among *Ctenomys* species, *C. lami* has one of the smallest geographical distributions described. The species inhabits a sandy terrain that was formed at the beginning of the Pleistocene, in the coastal plain of Southern Brazil. The distribution region of this species is only 78 km long and 12 km wide, but Freitas (2007) found a chromosomal variation in which the diploid numbers ($2n = 54, 55a$ and $55b, 56a$ and $56b, 57,$ and 58) associated with the autosomal fundamental numbers (FNs = 74 to 84) form 26 different karyotypes. Such a variation in the diploid numbers is due to Robertsonian-type fusion and centric fission rearrangements involving pairs 1 and 2, while the variation in the FNs was due to pericentric inversions that occur in several pairs in different karyotypes. Considering the variation found in pairs 1 and 2, $2n = 58$ has eight acrocentric chromosomes, which, by the G-band pattern, are the respective short and long arms of pairs 1 and 2 of the form $2n = 54$. Between these two configurations, one finds heterozygotes for both pair 1 and pair 2 ($2n = 55a$ and $2n = 55b$) and homozygotes for four acrocentric chromosomes and a metacentric pair ($2n = 56a$ and $2n = 56b$). There is also $2n = 57$, which is heterozygous for pair 1 and has six acrocentric chromosomes. This chromosomal variation is distributed in four population blocks. In a sequence from southwest to northeast, one finds block A, which shows $2n = 54$, 55a and 56a; block B, $2n = 57$ and 58; block C, $2n = 54$ and 55a, and block D, $2n = 56b$ and 55b (Freitas 2001).

One can hypothesize as to the origin of the chromosomal variation in decreasing the number of chromosomes from $2n = 58$ to $2n = 54$ due to Robertsonian centric fusion rearrangements. An alternative hypothesis is that the basic karyotypes, i.e., $2n = 54$, 58, and 56b, that inhabit the blocks form two contact zones: the first, between blocks A and C, generates the hybrids $2n = 55a$ and 55b; and the second between block B and D generates the hybrid form $2n = 57$. While there is this chromosomal difference between blocks A, B, C, and D, molecular markers of mitochondrial DNA show that there is a genetic flow between blocks A, B, and C, and still between C and D. This sharing of haplotypes are observed between the population blocks: the haplotypes 2, 4, and 7 are surprisingly shared by the diploid numbers $2n = 54$ (block A) and $2n = 58$ (block B); the haplotype 5 is present in blocks A ($2n = 54$), B ($2n = 58$), and C ($2n = 54a$), while the haplotype 15 is shared by blocks C ($2n = 54$) and D ($2n = 56b$) (Fig. 3.5).

The sharing of haplotypes between $2n = 54$ (block A) and $2n = 58$ (block B) shows that there is a crossing system between them, which would result in the double heterozygote $2n = 56$ for pairs 1 and 2, which was never found. The meiosis of $2n = 56$ would result in the formation of two trivalent chromosomes, and for the crossing system to be successful, there must be a normal disjunction between these chromosomes, that is, one metacentric for one pole and two acrocentric chromosomes for the other pole of the cell (Fig. 3.5). Research on meiosis has always focused on explaining how chromosomes are segregated when they form a trivalent due to a Robertsonian translocation. In the 1980s, several articles showed us the pairing of the trivalent through silver staining of the synaptonemal complex (Bogdanov et al. 1986; Hale and Greenbaum 1988). Currently, molecular cytogenetics has shown more evidently how these processes occur in heterozygotes. The alternative pairing of chromosomal rearrangements causes changes in the frequencies of rearrangements and recombination, not being the cause of chromosomal differentiation. This strengthens the idea that, despite the large chromosomal variability, this may not be the direct cause of speciation, but a consequence of it. If this were the case, *C. lami* would not hybridize with *C. minutus*, and therefore, there would be no hybrid zone between $2n = 46a$ and $2n = 56$ and the formation of F1, F2, and F3 hybrids (Gava and Freitas 2003).

Lopes and Freitas (2012) did not find genetic structure associated with the four different karyotype blocks separately, or even with the chromosomal rearrangements found in this species. Their findings suggest that the karyotype blocks

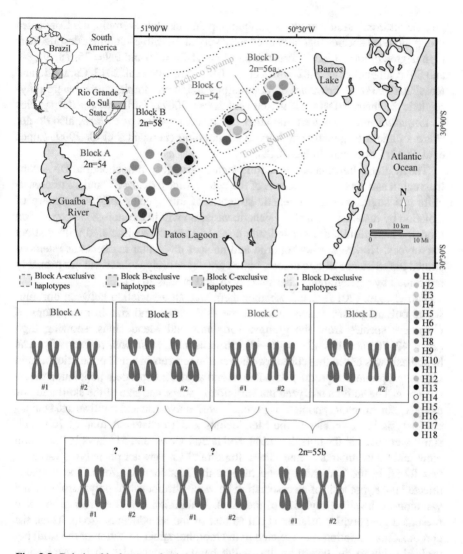

Fig. 3.5 Relationship between haplotypes and diploid numbers in *Ctenomys lami* population blocks. Four haplotypes are shared by different diploid numbers. The mating of individuals with different diploid numbers generates double heterozygotes in relation to pairs 1 and 2

described by Freitas (2007) proved to be inconsistent for mtDNA and microsatellite data, and that chromosomal rearrangements probably do not prevent reproduction among individuals with distinct karyotypes. Thus, they do not act as reproductive barriers. The fixation of new chromosomal rearrangements seems to be frequent in species of the genus *Ctenomys* and is generally not followed by sterility, reduction in fitness, or negative heterosis of heterozygous carriers (Tomasco and Lessa 2007). There are some examples of chromosomal polymorphic ctenomids that do not show genetic differentiation among karyotypic populations. For example, the Corrientes group includes species that are genetically very similar to each other, despite their high degree of karyotype differentiation ($2n = 41$-$2n = 70$ and FN = 76, 78, 80, 84, and 86) (Giménez et al. 2002; Mirol et al. 2010, Caraballo et al. 2012, Caraballo and Rossi 2017). For *C. pearsoni*, all populations and karyomorphs studied were poly-phyletic in their mtDNA (Tomasco and Lessa 2007). Furthermore, the species *C. torquatus* and *C. minutus*, the latter being sister species of *C. lami*, also do not share a pattern of genetic and karyotype variation (Fernandes et al. 2009; Lopes et al. 2013; Lopes and Freitas 2012).

The available information on chromosomal variability, mitochondrial DNA control region and subunit I sequences of the cytochrome c oxidase I, and 14 microsatellite loci suggests that the genetic structure of this species follows a pattern of isolation by distance and a clinal genetic variation within the step-stone population model. These results did not indicate a genetic structure associated with distinct karyotypes. However, mitochondrial and nuclear molecular markers have demonstrated the existence of two demes, which are not entirely isolated but are probably reinforced by a geographical barrier (Lopes and Freitas 2012).

Fornel et al. (2018) tested whether there was an association between chromosomal polymorphism and variation in the skull shape and size in two groups of *Ctenomys* species from the groups: Torquatus and Mendocinus, showing high ($2n = 42$–70) and low ($2n = 46$–48) chromosomal variations, respectively. The hypothesis was based that chromosomal rearrangements in small populations, as in *Ctenomys*, produce reproductive isolation and allows independent population diversification. The authors analyzed the variation in shape and size of the skull and jaw using a geometric morphometric approach, with univariate and multivariate statistical analysis in 12 species of the Mendocinus and Torquatus. A total of 763 adult skulls were used in the dorsal, ventral, and lateral views, and 515 jaws in the lateral view, and 93 landmarks in four views. Importantly, no greater phenotypic variation was found in the Torquatus group than in the Mendocinus group. These results rejected the hypothesis of an association between chromosomal polymorphism and variation in the shape and size of the skull. In addition, the Torquatus group also presented phenotypic variation equal to that of the Mendocinus group. Thus, the chromosomal variation is not related to the morphological evolution of the skull but probably due to the heterogeneity of the habitat associated with biomechanical restrictions and other factors such as geography, phylogeny, and demography (Fornel et al. 2018).

3.5 Hybrid Zones

Hybrid zones are places where two different forms co-exist and originate hybrids. The forms may differ according to molecular, chromosomal, morphological, physiological, or behavioral markers. Using one or several markers at the same time, it is possible to assess the level of hybridization that is occurring within or between species. Hybrid zones have two important roles in the evolutionary process: they increase genetic diversity or trigger an introgressive process, with genes being exchanged between populations. Nevertheless, a hybrid zone can homogenize populations (Barton 2001).

The occurrence of hybrid zones in *Ctenomys* is still not entirely known. There are 65 species (Teta and D'Elía 2020) described for the genus, and we can hypothesize that there are hybrid zones both between and within species that have not been found yet.

So far, few hybrid zones were described between species. A hybrid zone between *C. minutus* and *C. lami* was described by Gava and Freitas (2003), and others between *C. flamarioni* and *C. minutus* reported by Kubiak et al. (2020). What draws the attention of these two cases is their difference in phylogenetic terms. In the first case, two closely related sister species are involved. In the second one, both species are very distant phylogenetically (Parada et al. 2011) and belong to two groups: *C. flamarioni* is part of the Mendocinus group, while *C. minutus* belongs to the Torquatus group.

3.6   Intraspecific Hybrid Zones

Regarding hybrid zones within species, *C. minutus* is the one with the largest number of contacts already described for the genus. These contacts have caused the great chromosomal variability previously described. In the geographic distribution of *C. minutus,* from north to south, we find $2n = 50a$, 48a, and 49c, but the Araranguá river isolates these three diploid numbers. Then, it is possible to observe $2n = 46a$ and $2n = 48a$ distributed parapatrically in the coastal plain, forming a hybrid zone where the hybrid $2n = 47a$ occurs. This hybrid shows a stable chromosomal polymorphism in a dynamic and a narrow hybrid zone between chromosomal races (Gava and Freitas 2002; Gava and Freitas 2004).

Gava and Freitas (2002) studied 132 specimens of *C. minutus* from 22 sites in that contact zone and adjacent areas, where seven polymorphic sites were identified. A single chromosomal rearrangement is involved in the observed variation, which served as a Mendelian diallelic character in a Hardy-Weinberg analysis. Additionally, the clustering of polymorphic populations revealed that the mating system is random among the genotypes. Moreover, the observed frequency of the metacentric chromosome follows a clinal variation. Importantly, the hybrid zone is located along an ecotone, which could imply that exogenous selection is important for its maintenance.

With the same populations, the same hybrid zone was evaluated with six microsatellite loci in 107 specimens (Gava and Freitas 2003). Of the 56 alleles discovered in the study, 39.2% are exclusive to alternative cytotypes or contact populations. The clinal variation is not obvious because there are no fixed microsatellite differences between divergent chromosomal populations, but the variation in the Hai 2 locus is gradual in the zone. Local populations are highly differentiated and structured (Gava and Freitas 2003).

Castilho et al. (2012) found variation in microsatellite loci and in chromosomal polymorphisms of a parental population with $2n = 42$ and $2n = 48a$ and hybridized populations of *Ctenomys minutus*. The cytogenetic analysis and genetic variations of six microsatellite loci were included in 101 specimens with $2n/AN = 42/74$ and $48a/76$ and animals of contact zone. The cytogenetic analysis found 26 different karyotypes in 50 individuals from the hybrid population in an area of 7 km^2. Of the 26 karyotypes, only 14% had a parental configuration, and none had the expected $2n$ and AN combination for an F1 hybrid. The remaining karyotypes were alternative hybrid forms, with 2n ranging from 42 to 46, and AN from 68 to 80. These results suggest that chromosomal rearrangements are of little importance in establishing reproductive barriers for this species (Castilho et al. 2012). All of these data indicate that there is no effective barrier between the diploid numbers nor gene flow between them (Gava and Freitas 2003; Castilho et al. 2012).

Introgressions exist between different chromosomal numbers in the hybrid zone of $2n = 46a$ with $2n = 48a$, where the microsatellite alleles occur in both parental diploid numbers. The same is true for the haplotypes H26, H27, and H28. A similar process was found in the hybrid zone between $2n = 42$ and $2n = 48a$, where Castilho et al. (2012) observed the exchange of several microsatellite alleles between these diploid numbers. Introgression also occurs with the H13 haplotype in the same hybrid zone.

3.7 Interspecific Hybrid Zones

Contact zones between species in the South of Brazil were described in *Ctenomys*, one between *C. minutus* and *C. lami* and the most recently found between *C. flamarioni* and *C. minutus*.

The first interspecific contact zone was reported by Gava and Freitas (2003) and Ximenez (2009) between divergent populations of *Ctenomys minutus* ($2n = 48a$) distributed parapatrically with *C. lami* ($2n = 56b$). In the western region of Barros Lake, 21 different karyotypes were found between the parents $2n = 48a$ and $2n = 56b$. They were recorded in a sample of 26 specimens: $2n = 48a$, NA = 76; $2n = 49$, NA = 74, 76; $2n = 50$, NA = 74, 76; $2n = 51$, NA = 76, 77, 78, and 80; $2n = 52$, NA = 79; $2n = 53$, NA = 74, 80; $2n = 54$, NA = 78, 80; $2n = 55$, NA = 76; $2n = 56$, NA = 78, 80. It is worth noting the wide chromosomal variability throughout an area of about 2 km^2. Moreover, at the same time, only one individual was found with $2n = 52$, NA = 79, which corresponds to the F1 hybrid; so, the other 20 karyotypes are from F_2 onward. The variation is a result of Robertsonian mechanisms of chromosomal evolution and *in tandem* fusions. Importantly, polymorphisms have been considered as the result of a secondary contact between populations after divergence in allopatry. As described earlier in this chapter, the geomorphological evolution of the coastal plain provides clues to the existence of past geographical barriers that modulated the populations of *Ctenomys*, during the Holocene.

The second interspecific contact zone in southern Brazil is between *C. minutus* and *C. flamarioni*, and this is the second case of sympatry found for the genus. The first one was described between *C. australis* and *C. talarum* in the coastal dunes in southern Argentina, (Malizia and Busch 1991). The sympatry between these two species is very complex due to the interaction between them. Kubiak et al. (2015) investigated the vegetation structure, plant biomass, and soil hardness selected by these two species when distributed in sympatry and allopatry. These species show segregation in their selection of microhabitats, differing in relation to soil hardness, biomass, and vegetation cover. *C. flamarioni* and *C. minutus* show habitat segregation in the area where they occur in sympatry. Nevertheless, *C. flamarioni* shows a distinction in habitat selection when it occurs in allopatry and sympatry, while *C. minutus* selected the same habitat characteristics in both conditions. A possible explanation for the observed pattern is that these species have acquired different adaptations over time, which allows them to explore different resources and, thus, avoid competitive interactions (Kubiak et al. 2015).

Species with similar ecological requirements coexisting in the same geographic region are likely to competitively exclude each other or, alternatively, can coexist. For this analysis, two methodological approaches were used (ecological niche modeling [ENM] and geometric morphometry) to test two hypotheses: (i) if, given their behavioral, morphological, and ecological similarities, one species competitively excludes the other or (ii) if the character changes allow their coexistence in two places where species are known to occur in sympatry. The results of the ENM-based approach did not provide evidence of competitive exclusion. However, geometric morphometric analyses showed the change in the skull size of *C. minutus*. Interestingly, this result suggests that *C. minutus* can exclude *C. flamarioni* from areas with softer soils and greater food availability (Kubiak et al. 2017).

While there is interaction between these two species, regardless of their separate spaces, different morphologies, completely different karyotypes, divergence in evolutionary history, and form of sperm, they are still capable of generating natural hybrids. Also, this ability to generate natural hybrids demonstrated that females of both species are capable of generating hybrids with males of the two species. In addition, the specific chromosome probes isolated from *Ctenomys flamarioni* showed chromosomes of each species in the hybrid karyotype, and the homeology relationship between the chromosomal arms of both species. Moreover, microsatellites also showed the hybridization process at the population level, where through population assignment, individuals of both parental species were shown, as well as hybrids. Furthermore, it is worth mentioning that mtDNA analyses allowed to detect that hybrids have mtDNA from one species or from the other, showing that the crosses are clearly bidirectional (Kubiak et al. 2020).

3.8 Other Processes

A very recent differentiation was found in the Midwest region of Brazil with some species lineages. Species from Argentina, Uruguay, Bolivia, and southern Brazil are well studied at the phylogenetic level. However, the taxonomic status of the species that inhabit the Mid-western and Northern Brazil remains scarcely known. For the three species of *Ctenomys* that occur in these two regions (*C. nattereri*, Wagner 1848; *C. bicolor*, De Miranda-Ribeiro 1914; and *C. rondoni*, De Miranda-Ribeiro 1914), it can only be found one article for each, and a formal description of these taxa is nonexistent (De Miranda-Ribeiro 1914). Additionally, the geographical distribution of these species is also unknown in the Brazilian states of Acre, Rondônia, and Mato Grosso.

Ctenomys bicolor, an endemic species from the northwest of Brazil, is present in southern Amazonia, specifically at the edge of the Amazon ecoregion. The type was collected in 1912 by zoologist Alipio Miranda Ribeiro, during the Roosevelt-Rondon scientific expeditions (Avila-Pires 1968) in a remote area of the Amazon region, and the specimen described is deposited in the National Museum of Rio de Janeiro, under the collection I.D. MNRJ-2052. Afterward, the description of this individual was published 2 years later (De Miranda-Ribeiro 1914), with no indication of the type locality; but, Bidau and Avila-Pires (2009) redefined it as in the state of Rondônia, in the locality of Pimenta Bueno. Stolz et al. (2013) confirmed the location based on a sample of 10 animals. That report determined $2n = 40$, NF = 68, their species status through phylogenetic analysis with 820 bp of Cyt-b, and the skull was analyzed with geometric morphometry.

Noteworthy, these findings showed a new behavior for *Ctenomys*. These populations were found living inside the forest and feeding from tree roots, contradicting the existing knowledge that practically all species would live in open formations and feeding on grasses.

It can be hypothesized that the invasion of the Amazonian group in the forest area seems to have occurred at a time (before 850 thousand years ago, in the quaternary) when large extensions of savannah existed where today the forest is located (Haffer and Prance 2002). Such an ancestor, common to the three species, in a more recent period of high temperatures and humidity, was constrained by the advancement of forest expansion.

In this changing environment, individuals that could take advantage of the forest environment to survive were selected and differentiated themselves into a new morphotype, adapted to life and food on the soil of the forests. Leipnitz et al. (2020) studied the samples from the mid-west and north of Brazil, with the intent to analyze these populations. The authors built a phylogeny based on maximum-likelihood and Bayesian inference methods with haplotypes of the *Ctenomys* cytochrome b gene and with haplotypes representative of the genus *Ctenomys*. Also, they evaluated skull morphology using geometric morphometric analysis to assess whether the skull morphology findings agree with the observed phylogenetic patterns. The results showed the occurrence of two species, namely, *Ctenomys bicolor*, in the state of Rondônia, and *Ctenomys nattereri*, in Mato Grosso and Bolivia. Additionally, the results revealed two lineages of *Ctenomys* that are distinct from *Ctenomys bicolor* and *C. nattereri*, which they called *Ctenomys* sp. *B Xingu* and *Ctenomys* sp. *B Central*. Importantly, both species and lineages share a more recent ancestor common with *C. boliviensis* and are part of the species group known as *boliviensis*.

Named as Corrientes group, three species, *C. roigi*, *C. perrensis*, and *C. dorbignyi*, are tuco-tucos that inhabit the Iberá region. This area has sandy soils under the influence of the second largest wetland in South America, located in the northeast of Argentina, in the province of Corrientes. The phylogeny of these species was studied under cytogenetic point of view by Ortells and Barrantes (1994). These species are characterized as a "species complex" with one of the largest chromosomal variability within the genus. To analyze the evolutionary relationships in the Corrientes group and its chromosomal variability, the partial sequences of three mitochondrial markers were obtained by Caraballo et al. (2012, Caraballo and Rossi 2017). These authors found that the Corrientes group is monophyletic, and they divided the complex into three main clades that grouped the related karyomorphs. Also, in Argentina there is another group of species, belonging to the Mendocinus group, that shows interesting evolutive conditions (see Chap. 5).

3.9 Final Comments

Considering the existing 65 species of *Ctenomys* (Teta and D'Elía 2020), we found that little is known about the evolutionary patterns and processes that are still shaping this genus. Analyzing the literature, we notice that only about 15 species were studied from an integrative point of view, that is, chromosomal, morphological, and molecularly. In this way, we only have the evolutionary panorama of these 15 species, i.e., the evolutionary response to the region where these species live. Notably, this process is observed in the Pampa region, coastal plain of Southern Brazil, the coast of Uruguay, the region of Entre Rios, in Argentina, and Midwest region of Brazil. In these regions, evolution is occurring in real time but at different evolutionary phases. Species such as *C. perrensis*, *C. roigi*, and *C. dorbignyi* live in the Iberá region in Argentina, in a sandy area that is influenced by the water regime of the second largest wetland in South America, and they are undergoing a recent process of differentiation, although it is not possible (yet) to see a certain evolutionary pattern. Furthermore, the species of Mato Grosso, Brazil, *C. bicolor*, *C. rondoni*, and *C. nattereri* are undergoing a similar process, with low chromosomal or morphological variability, also in relation to molecular markers. Moreover, these species do not have a well-defined pattern, because the process is very recent; however, it is possible to know that all the variation originated from a colonization of the region by one or more species came from the Bolivian region. In the coastal plain of southern Brazil, we have three species, *C. flamarioni*, *C. minutus*, and *C. lami*, that are in an "older" process of differentiation and, thus, we can recognize patterns of chromosomal variation, morphology, and molecular markers. However, hybrid zones between and within species indicate that they are still in a long process of evolution. Similarly, the same can be observed in *C. pearsoni,* on the Uruguayan coast. On the other hand, *C. torquatus* exhibits a much more defined pattern, as if it had already found homeostasis in relation to morphology, chromosomes, and molecular markers.

Acknowledgments I am deeply grateful to my undergraduate, MSc, and PhD students whose works provided much of the results that I used to write this chapter and the collaborators I had throughout the years studying *Ctenomys*. I also thank Carla Freitas for suggestions and reviewing the text and Raquel Freitas for help with figures and final text revision. Thanks also to Gislene Gonçalves for valuable discussions. This study was financed in part by Conselho Nacional de Desenvolvimento Científico e Tecnológico (CNPq), Fundação do Amparo à Pesquisa do Estado do Rio Grande do Sul (FAPERGS), and Coordenação do Aperfeiçoamento de Pessoal de Nível Superior (CAPES) – Finance Code 001. I also acknowledge the support from the Graduate Programs PPGBM, PPGBAN, and PPGEcologia-UFRGS.

Literature Cited

Avila-Pires FD (1968) Tipos de mamíferos recentes no Museu Nacional, Rio de Janeiro. Arch Mus Nac 53:161–191

Barton NH (2001) The role of hybridization in evolution. Mol Ecol 10:551–568

Bidau CI (2015) Ctenomyidae. *Ctenomys*. In: Patton J, Pardiñas FU, D'Elía G (eds) Mammals of South America, Rodents, vol 2. University of Chicago Press, Chicago, pp 818–877

Bidau CJ, Avila-Pires FD (2009) On the type locality of *Ctenomys bicolor* Miranda-Ribeiro 1914 (Rodentia: Ctenomyidae). Mastozool Neotrop 16:445–447

Bogdanov YF, Kolomiets OL, Lyapunova EA, Yanina IY, Mazurova TF (1986) Synaptonemal complex and chromosome chains in the rodent Ellobius talpinus heterozygous for 10 robertsonian translocation. Chromosoma 94:94–102

Busch C, Antinuchi CD, del Valle JC, Kittlein MJ, Malizia AI, Vassallo AI, Zenuto RR (2000) Population ecology of subter- ranean rodents. In: Lacey EA, Patton JL, Cameron GN (eds) Life underground: the biology of subterranean rodents. University of Chicago Press, Chicago, pp 183–226

Caraballo DA, Rossi MS (2017) Integrative lineage delimitation in rodents of the *Ctenomys* Corrientes group. Mammalia 82:35–47

Caraballo DA, Abruzzese GA, Rossi MS (2012) Diversity of tuco-tucos (Ctenomys, Rodentia) in the Northeastern wetlands from Argentina: mitochondrial phylogeny and chromosomal evolution. Genetica 140:125–136

Castilho CS, Gava A, Freitas TRO (2012) A hybrid zone of the genus *Ctenomys*: A case study in southern Brazil. Genet Mol Biol 35:990–997

Contreras JR, Bidau CJ (1999) Líneas generales del panorama evolutivo de los roedores excavadores sudamericanos del género *Ctenomys* (Mammalia, Rodentia, Caviomor- pha: Ctenomyidae). Ciencia Siglo, XXI 1:1–22

Cook JA, Anderson S, Yates TL (1990) Notes on Bolivian mammals: 6. The genus *Ctenomys* (Rodentia, Ctenomyidae) in the highlands. Am Mus Novit 2980:1–27

D'Elía G, Teta P, Lessa EP (2020) A short overview of the systematics of *Ctenomys*: species limits and phylogenetic relationships. In: Freitas TRO, Gonçalves GL, Maestri R (eds) Tuco-tucos – an evolutionary approach to the diversity of a Neotropical rodent. Springer, Cham

Darwin C (1859) On the origins of species by means of natural selection. John Murray, London

De Miranda-Ribeiro A (1914) Zoologia. Commisão de Linhas Telegráficas Estratégicas de MattoGrosso ao Amazonas. Annexo 5, Historia Natural; publ no 17, Mammíferos. 49 pp + Append, 3 pp + 25 pls Miranda-Ribeiro, 1914

Dobigny G, Britton-Davidian J, Robinson TJ (2017) Chromosomal polymorphism in mammals: an evolutionary perspective. Biol Rev 92:1–21

Fernandes FA, Fornel R, Cordeiro-Estrela P, Freitas TRO (2009) Intra- and interespecific skull variation in two sister species of the subter- ranean rodent genus *Ctenomys* (Rodentia, Ctenomyidae): coupling geometric morphometrics and chromosomal polymorphism. Zool J Linnean Soc 155:220–237

Fernández-Stolz G (2006) Estudos evolutivos, filogeográficos e de conservação em uma espécie endêmica do ecossistema de dunas costeiras do sul do Brasil, *Ctenomys flamarioni* (Rodentia – Ctenomyidae), através de marcadores moleculares microssatélites e DNA mitocondrial. Ph.D. thesis, Universidade Federal do Rio Grande do Sul. Porto Alegre, Brasil

Fornel R, Cordeiro-Estrela P, Freitas TRO (2010) Skull shape and size variation in *Ctenomys minutus* (Rodentia: Ctenomyidae) in geographical, chromosomal polymorphism, and environmental contexts. Biol J Linn Soc 101:705–720

Fornel R, Cordeiro-Estrela P, Freitas TRO (2018) Skull shape and size variation within and between *mendocinus* and *torquatus* groups in the genus *Ctenomys* (Rodentia: Ctenomyidae) in chromosomal polymorphism context. Genet Mol Biol 41:263–272

Freitas TRO (1994) Geographic variation of heterochromatin in *Ctenomys flamarioni* (Rodentia: Octodontidae) and its cytogenetic relationship with other species of the genus. Cytogenet Cell Genet 67:193–198

Freitas TRO (1995a) Geographic distribution and conservation of four species of the genus *Ctenomys* in southern Brazil. Stud Neotropical Fauna Environ 30:53–59

Freitas TRO (1995b) Geographic distribution if sperm forms in the genus *Ctenomys* (Rodentia: Octodontidae). Revista Brasileira de Genética 18:43–46

Freitas TRO (1997) Chromosome polymorphism in *Ctenomys minutus* (Rodentia–Octodontidae). Rev Bras Genética 20:1–7

Freitas TRO (2001) Tuco-tucos (Rodentia, Octodontidae) in Southern Brazil: *Ctenomys lami* spec. nov. Separated from *C. minutus* Nehring 1887. Stud Neotrop Fauna Environ 36:1–8

Freitas TRO (2005) Analysis of skull morphology in 15 species of the genus *Ctenomys*, including seven karyologically distinct forms of *Ctenomys minutus* (Rodentia:Ctenomyidae). In: Lacey E, Myers P (eds) Mammalian diversification: from chromosomes to phylogeography. University of California Publications in Zoology, Berkeley, pp 131–154

Freitas TRO (2006) Cytogenetics status of four *Ctenomys* species in the south of Brazil. Genetica 126:227–235

Freitas TRO (2007) *Ctenomys lami*: the highest chromosome variability in *Ctenomys* (Rodentia, Ctenomyidae) due to a centric fusion/fission and pericentric inversion system. Acta Theriol 52:171–180

Freitas TRO (2016) Family Ctenomyidae. In: Wilson DE, Lacher TE Jr, Mittermeier RA (eds) Handbook of the Mammals of the World: Lagomorphs and Rodents I. Lynx Editions, Barcelona, pp 499–534

Freitas TRO, Fernandes FA, Fornel R, Roratto PA (2012) An endemic new species of tuco-tuco, genus *Ctenomys* (Rodentia: Ctenomyidae), with a restricted geographic distribution in southern Brazil. J Mammal 93(5):1355–1367

Freygang CC, Marinho JR, Freitas TRO (2004) New karyotypes and some considerations about the chromosomal diversication of. *Ctenomys minutus* (Rodentia: Ctenomyidae) on the coastal plain of the Brazilian state of Rio Grande do Sul. Genetica 121:125–132

Galiano D, Kubiak BB, Menezes LS, Overbeck GE, de Freitas TRO (2016) Wet soils affect habitat selection of a solitary subterranean rodent (Ctenomys minutus) in a Neotropical region. J Mammal 97:1095–1101

Gardner SL, Salazar-Bravo J, Cook JA (2014) New species of *Ctenomys* Blainville 1826 (Rodentia: Ctenomyidae) from the lowlands and central valleys of Bolivia. Special Publications of the Museum of the Texas Tech University 62:1–34

Gava A, Freitas TRO (2002) Characterization of a hybrid zone between chromosomally divergent populations of *Ctenomys minutus* (Rodentia: Ctenomyidae). J Mammal 83:843–851

Gava A, Freitas TRO (2003) Inter and intra-specific hybridization in tuco-tucos (*Ctenomys*) from Brazilian Coastal Plains (Rodentia: Ctenomyidae). Genetica 119:11–17

Gava A, Freitas TRO (2004) Microsatellite analysis of a hybrid zone between chromosomally divergent populations of *Ctenomys minutus* from southern Brazil (Rodentia: Ctenomyidae). J Mammal 85:1201–1206

Giménez MD, Mirol PM, Bidau CJ, Searle JB (2002) Molecular analysis of populations of *Ctenomys* (Caviomorpha, Rodentia) with high karyotypic variability. Cytogenet Genome Res 96:130–136

Hadid Y, Tzur S, Pavlicek T, Sumbera R, Skliba J, Loevy M, Fragman-Sapir O, Beiles A, Arieli R, Raz S, Nevo E (2013) Possible incipient sympatric ecological speciation in blind mole rats (*Spalax*). Proc Natl Acad Sci U S A 110(7):2587–2592

Haffer JE, Prance GT (2002) Impulsos climáticos da evolução na Amazônia durante o Cenozóico: sobre a teoria dos Refúgios da diferenciação biótica. Estudos avançados 16:175–206

Hale DW, Greenbaum IF (1988) Synapsis of a chromosomal pair heterozygous of a pericentric-inversion and the presence of a heterochromatic short arm. Cytogent Cells Genet 48:55–57

Harrison RG (1993) Hybrids and hybrid zone: historical perspective. In: Harrison RG (ed) Hybrid zones and the evolutionary process. Oxford University Press, New York, pp 3–12

Kexin L, Wang LY, Knisbacher BA, Xu Q, Levanon EY, Wang HH, Frenkel-Morgenstern M, Tagore S, Fang XD, Bazak L, Buchumenski I, Zhao Y, Lovy M, Li XF, Han LJ, Frenkel Z, Beiles A, Cao YB, Wang ZL, Nevo E (2016) Transcriptome, genetic editing, and micro RNA divergence substantiate sympatric speciation of blind mole ret., *Spalax*. Proc Natl Acad Sci U S A 113:7584–7589

King M (1993) Species evolution: the role of chromosome change. Cambridge University Press, Cambridge

Kubiak BB, Galiano D, Freitas TRO (2015) Sharing the space: Distribution, habitat segregation and delimitation of a new sympatric area of subterranean rodents. PLoS One 10:e0123220

Kubiak BB, Gutiérrez EE, Galiano D, Maestri R, Freitas TRO (2017) Can niche modeling and geometric morphometrics document competitive exclusion in a pair of subterranean rodents (Genus *Ctenomys*) with Tiny Parapatric Distributions? Sci Rep 7:16283

Kubiak BB, Maestri R, Almeida TS, Borges LR, Galiano D, Fornel R, de Freitas TRO (2018) Evolution in action: soil hardness influences morphology in a subterranean rodent (Rodentia: Ctenomyidae). Biol J Linn Soc 20:1–11

Kubiak BB, Kretschmer R, Leipnitz LT, Maestri R, Almeida TS, Borges LR, Galiano D, Pereira JC, Oliveira EHC, Ferguson-Smith MA, Freitas TRO (2020) Hybridization between subterranean tuco-tucos (Rodentia, Ctenomyidae) with contrasting phylogenetic positions. Sci Rep 10:1502

Lacey EA, Patton JL, Cameron GN (2000) Life underground: the biology of subterranean rodents. University of Chicago Press, Chicago/London, p 449

Leipnitz LT, Fornel R, Ribas LEJ, Kubiak BB, Galiano D, Freitas TRO (2020) Lineages of tuco-tucos (Ctenomyidae: Rodentia) from midwest and northern Brazil: late irradiations of subterranean rodents towards the Amazon Forest. J Mamm Evol 27:161–176

Lessa EP, Cook JA (1998) The molecular phylogenetics of tuco-tucos (genus *Ctenomys*, Rodentia: Octodontidae) suggests an early burst of speciation. Mol Phylogenet Evol 9:88–99

Li KX, Wang LY, Knisbacher BA, Xuv, Levanon EY, Wang HH, Frenkel-Morgenstern M, Tagore S, Fang XD, Bazak L, Buchumenski I, Zhao Y, Lovy M, Li XF, Han LJ, Frenkel Z, Beiles A, Bin Cao Y, Wang ZL, Nevo E (2016) Transcriptome, genetic editing, and microRNA divergence substantiate sympatric speciation of blind mole rat, Spalax. Proceedings of the National Academy of Sciences of the United States of America, 113:7584–7589

Lopes CM, Freitas TRO (2012) Human impact in naturally patched small populations: genetic structure and conservation of the burrowing rodent. J Hered 103:672–681

Lopes CM, Ximenes SSF, Gava A, Freitas TRO (2013) The role of chromosomal rearrangements and geographical barriers in the divergence of lineages in a South American subterranean rodent (Rodentia: Ctenomyidae: *Ctenomys minutus*). Heredity 111:293–305

Malizia AI, Busch C (1991) Reproductive parameters and growth in the fossorial rodent *Ctenomys talarum* (Rodentia: Octodontidae). Mammalia 55:293–305

Massarini AI, Freitas TRO (2005) Morphological and cytogenetics comparison in species of the mendocinus-group (genus *Ctenomys*) with emphasis in *C. australis* and *C. flamarioni* (Rodentia: Ctenomyidae). Caryologia 58:21–27

Massarini AI, Barros MA, Ortells MO, Reig OA (1991) Chromossomal polymorphism and small karyotypic diferentiation in a group of *Ctenomys* species from central Argentina (Rodentia: Octodontidae). Genetica 83:131–144

Massarini AI, Mizrahi D, Tiranti S, Toloza A, Luna F, Schleich EC (2002) Extensive chromosomal variation in Ctenomys talarum talarum from the Atlantic coast of Buenos Aires province, Argentina (Rodentia-Octodontidae). Mastozool Neotrop 9:199–207

Mayr E (1942) Systematic and the origin of species. Columbia University Press, New York

Mayr E (1963) Animal species and evolution. Belknap Press of Harvard Univ Press, Cambridge, MA

Mirol P, Giménez MD, Searle JB, Bidau CJ, Faulkes CG (2010) Population and species boundaries in the South American subterranean rodent *Ctenomys* in a dynamic environment. Biol J Linn Soc Lond 100:368–383

Mora MS, Lessa EP, Kittlein MJ, Vassallo AI (2006) Phylogeography of the subterranean rodent *Ctenomys australis* in sand-dune habitats: evidence of population expansion. J Mammal 87:1192–1203

Nehring A (1900) Uber dis Scha¨ del von *Ctenomys minutus* Nhrg., Ct. torquatus Licht und Ct. pundti Nhrg. Stzumgsb Ges NaturfFr 9:201–210

Novello AF, Lessa EP (1986) G-Band homology in 2 karyomorphs of the Ctenomys pearsoni complex (Rodentia; Octodontidae) of neotropical fossorial rodents. Z Säugetierkd 51:378–380

Ortells MO, Barrantes GE (1994) A study of genetic distances and variability in several species of the genus *Ctenomys* (Rodentia, Octodontidae) with special reference to probable causal role of chromosome in speciation. Biol J Linn Soc Lond 53:189–208

Parada A, D'Elía G, Bidau CJ, Lessa EP (2011) Species groups and the evolutionary diversification of tuco-tucos, genus *Ctenomys* (Rodentia: Ctenomyidae). J Mammal 92:671–682

Petit RJ, Excoffier L (2009) Gene flow and species delimitation. Trends Ecol Evol 24:386–393

Reig OA, Kiblisky P (1969) Chromosome multiformity in the genus *Ctenomys* (Rodentia, Octodontidae). Chromosoma 28:211–244

Reig OA, Contreras JR, Piantanida MJ (1966) Contribución a la elucidación de la sistematica de las entidades del genero *Ctenomys* (Rodentia-Octodontidae) Cont. Cient Fac Exactas y Nat Univ de Buenos Ayres (Zool) 6:297–352

Reig OA, Busch C, Ortells MO, Contreras JR (1990) An overview of evolution, systematics, population biology, cytogenetics, molecular biology and speciation in *Ctenomys*. In: Nevo E, Reig OA (eds) Evolution of subterranean mammals at the organismal and molecular levels. Wiley-Liss, New York, pp 71–96

Reig OA, Massarini AI, Ortells MO, Barros MA, Tiranti SI, Dyzenchauz FJ (1992) New karyotypes and C-banding patterns of the subterranean rodents of the genus *Ctenomys* (Caviomorpha, Octodontidae) from Ar- gentina. Mammalia 56:603–623

Rieseberg LH (2001) Chromosomal rearrangements and speciation. Trends Ecol Evol 16:351–358

Stolz JFB, Gonçalves GL, Leipnitz LT, Freitas TRO (2013) DNA-based and geometric morphometric analysis to validate species designa- tion: a case study of the subterranean rodent *Ctenomys bicolor*. Genet Mol Res 12:5023–5037

Teta P, D'Elía G (2020) Uncovering the species diversity of subterranean rodents atthe end of the world: three new species of Patagonian tuco-tucos (Rodentia, Hystricomorpha, *Ctenomys*). PeerJ 8:e9259

Tomasco IH, Lessa EP (2007) Phylogeography of the tuco–tuco Ctenomys pearsoni: mtDNA variation and its implication for chromosomal differentiation. In: Kelt DA, Lessa EP, Salazar-Bravo JA, Patton JL (eds) The quintessential naturalist: honoring the life and legacy of Oliver P. Pearson. University of California Publications in Zoology, Berkeley, pp 859–882

Wagner JA (1848) Beiträge zur Kentnniss der Arten von *Ctenomys*. *Arch*. Naturgesch 14:72–78

Weschenfelder J, Baitelli R, Corrêa ICS, Bortolin EC, Santos CB (2014) Quaternary incised valleleys in southern Brazil coastline. J S Am Earth Sci 55:83–93

White MJD (1978) Modes of speciation. W.H. Freeman and Co., New York
Wlasiuk G, Garza JC, Lessa EP (2003) Genetic and geographic differentiation in the Rio Negro tuco-tuco (*Ctenomys rionegrensis*): inferring the roles of migration and drift from multiple genetic markers. Evolution 57:913–926
Ximenez SSF (2009) Análises citogenéticas em uma zona híbrida interespecífica entre *Ctenomys minutus* e *Ctenomys lami* (Rodentia:Ctenomyidae) na planície costeira do Sul do Brasil. Trabalho de Conclusão, Ciências Biológicas-UFRGS

Part II
Geographic Patterns

Chapter 4
Geographical and Macroecological Patterns of Tuco-Tucos

Renan Maestri and Bruce D. Patterson

4.1 Introduction

Tuco-tuco is the vernacular name given to subterranean rodents of the genus *Ctenomys*, which are widely distributed over the southern half of South America. Species are distributed from Peru and the Brazilian state of Rondônia in the north (11°S) to Tierra del Fuego in the south, and from the Atlantic to the west coast (Bidau 2015; de Freitas 2016). Within those geographic limits, 65–70 species are currently recognized (mammaldiversity.org; D'Elía et al. 2020). Even more impressively, all of this diversity arose over the last 5 Ma, indicating explosive speciation compared to other genera of Caviomorpha (Upham and Patterson 2015). Rapid speciation and geographic colonization gave rise to a large number of allopatric species (Bidau 2015; de Freitas 2016), with only a few known cases of syntopy (Galiano and Kubiak 2020). Contiguous allopatry with little overlap between species has been attributed to the subterranean habits and behavior (Nevo 1979), but a formal comparison of the exclusivity in the geographic ranges of tuco-tucos is lacking. In addition, overall patterns in species and range size distribution – a legacy of their speciation and colonization processes – remain to be explored.

Patterns in the distribution of species richness, geographic range size, and body size have been common themes in the mammalian literature (Ruggiero and Kitzberger 2004; Smith and Lyons 2011; Lyons et al. 2019). Macroecological analyses of those characteristics helped to identify general patterns and uncover some of

R. Maestri (✉)
Department of Ecology, Federal University of Rio Grande do Sul, Porto Alegre,
Rio Grande do Sul, Brazil
e-mail: renan.maestri@ufrgs.br

B. D. Patterson
Negaunee Integrative Research Center, Field Museum of Natural History, Chicago, IL, USA
e-mail: bpatterson@fieldmuseum.org

© Springer Nature Switzerland AG 2021
T. R. O. de. Freitas et al. (eds.), *Tuco-Tucos*,
https://doi.org/10.1007/978-3-030-61679-3_4

the underlying processes behind mammalian evolution (Brown and Maurer 1989; Stevens 1989; Brown et al. 1993; Smith et al. 2008). For example, geographical patterns such as Rapport's rule – the tendency of geographic range size to be smaller at lower latitudes than at higher latitudes – and Bergmann's rule – the tendency of body size to be bigger at higher than at lower latitudes – were first described using mammals (Bergmann 1847; Rapoport 1982; Stevens 1989; Blackburn et al. 1999). To date, only a few macroecological investigations of species' range and/or body size distribution patterns have been focused on tuco-tucos. Among them, Medina et al. (2007) found that tuco-tuco body size follows the converse of Bergmann's rule, Bidau et al. (2011) found increases in relative tail length with increases in latitude (the converse of Allen's Rule), and Caraballo et al. (2020) explored the overlap of species range maps with protected areas.

This chapter will explore spatial patterns in the distribution of (i) species diversity, (ii) geographic range size, and (iii) body size of tuco-tucos in the hope of offering some light for future investigations on *Ctenomys* biogeography. The chapter is organized in three main sections devoted to exhibit the general spatial patterns on each theme (i, ii, and iii above) and to compare the geographic range size of tuco-tucos with other caviomorphs.

4.2 Patterns of Species Distribution

Species of tuco-tucos are unevenly distributed in the southern half of South America. Most species occur in the northern portions of Argentina (Bidau 2015; de Freitas 2016). The pattern of species richness for *Ctenomys* species is shown in Fig. 4.1. The range maps of each species (Bidau 2015) were overlaid in a gridded map a $1° \times 1°$ scale to generate the map.

The hotspot of species richness is located between 23° and 29° degrees S latitude, where the range maps of 7–10 species overlap. Outside of this hotspot, between 1 and 6 range maps overlap over the remaining tuco-tuco distribution. Most regions have only one or two species. It is important to note that an overlap in range maps should not be interpreted as if the species live in syntopy within those geographic limits. Tuco-tucos have predominantly allopatric distributions with only three or four cases of syntopy currently documented (see Galiano and Kubiak 2020), although more studies investigating the spatial relationships of species living in close proximity are urgently needed. However, range maps are useful to visualize broad richness patterns that may steer inferences of large-scale spatial and temporal processes shaping the biogeographical history of tuco-tucos.

A weak linear relationship was found in the correlation of species richness with latitude (r = 0.13), longitude (r = −0.04), and elevation (r = 0.17), suggesting an absence of clear richness gradients along these three axes (Fig. 4.2). The most range overlaps for tuco-tucos coincide with intermediate latitude, longitude, and elevation values.

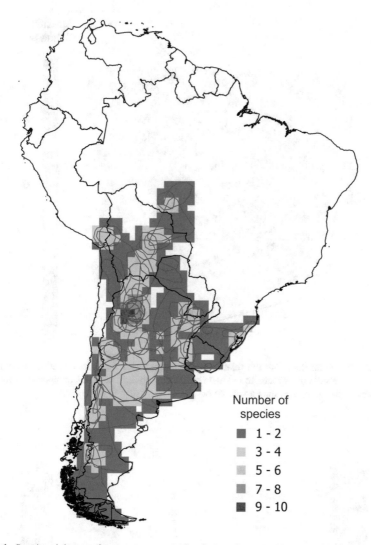

Fig. 4.1 Species richness of tuco-tucos across South America on a 1°× 1° scale. Range maps of each species are shown as grey polygons

4.3   Patterns of Geographic Range Size Distribution

Most species of tuco-tuco have small geographic range sizes (Fig. 4.3), a frequency distribution that resembles the all-mammalian pattern (Brown et al. 1996; Gaston 1998). However, compared to other families of caviomorphs (Patton et al. 2015; Maestri & Patterson 2016), the tuco-tucos (sole living members of the family Ctenomyidae) have one of the smallest geographic range sizes (Fig. 4.4), with the median of the geographic range size virtually identical to members of its sister

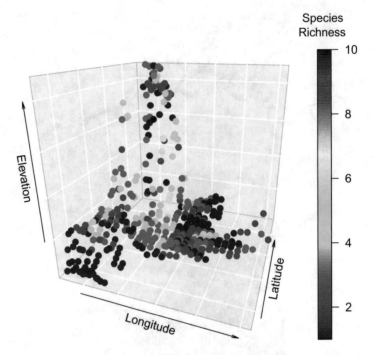

Fig. 4.2 Scatterplot depicting the relationship among species richness, elevation, latitude, and longitude. Each dot corresponds to a 1°× 1° cell. Geographical variables composed of the three main axes while richness is shown with colors

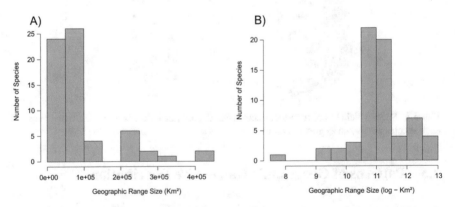

Fig. 4.3 Histograms of (**a**) geographic range size (km²) and (**b**) log of geographic range size (km²)

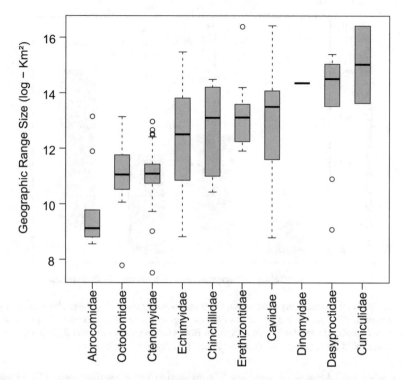

Fig. 4.4 Boxplot representing variation in geographic range size of species from different cavio-morph families. Boxes represent the third and first quartiles, plus the median (bold line), and whiskers represent upper and lower limits; dots indicate outliers

family Octodontidae, and trailing only the Abrocomidae – a poorly known family of rodents with few extant species (Patton and Emmons 2015). Moreover, as can be seen from Fig. 4.4, species of tuco-tucos vary little in geographic range size when compared to species in other families, which may suggest constraints on the size of the range around some modal value.

Examining patterns of range overlap, a surprising result was found in the comparisons of species-rich genera of caviomorphs (Fig. 4.5). *Ctenomys* species are commonly perceived as being strongly allopatric, with species essentially replacing one another across the landscape with little overlap. We calculated geographic range exclusivity as the proportion of a species range that does not overlap with any other species of its own genus. Remarkably, variation in geographic range exclusivity of 65 species of *Ctenomys* is statistically indistinguishable from that shown by *Dasyprocta* (Dasyproctidae, n = 10), *Proechimys* (Echimyidae, n = 22), and *Coendou* (Erethizontidae, n = 14) (Fig. 4.5). Only *Coendou* and *Dasyprocta* show significant differences in range exclusivity (F = 3.02; P = 0.03), probably driven by the strong degree of contiguous allopatry among species of agoutis.

The geographic range size (log) of tuco-tucos has no obvious relationship with the mean latitudinal point of each species ($R^2 = 0.01$; $P > 0.05$) (Fig. 4.6Aa),

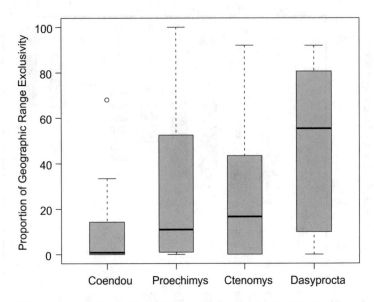

Fig. 4.5 Boxplot representing variation in the proportion of geographic range size exclusivity – the proportion of a species' range that does not overlap with any other species of its genus. Four species-rich genus of caviomorph are compared. Boxes represent the third and first quartiles, plus the median (bold line), and upper and lower limits; dots represent outliers

suggesting the absence of Rapoport's rule at this interspecific scale. Geographic range size (log) was also linearly uncorrelated with mean elevation ($R^2 = 0.005$; $P > 0.05$) and species' log body size ($R^2 = 0.02$; $P > 0.05$) (Fig. 4.6b, c). The relationship between geographic range size and body size takes the approximate form of a triangle, where species with large body sizes and small geographic range sizes are lacking, similar to the general mammalian pattern (Smith et al. 2008; Lyons et al. 2019). Such a triangular envelope was hypothesized to result from the higher extinction rates of large-sized species; their low densities over a restricted area of geographic occurrence result in vulnerably small population sizes (Brown and Maurer 1989; Diniz-Filho et al. 2005; Tomiya 2013).

Nevertheless, at an assemblage scale, a positive relationship between average geographic range size (average of range size for all species occurring on each $1 \times 1°$ cell) and average cell latitude (S) is apparent ($R^2 = 0.19$; $P < 0.001$) (Fig. 4.7a). This positive relationship is in accordance with the predicted by Rapoport's rule (Stevens 1989), suggesting the prevalence of small range-sized species at lower latitudes, even though the relationship between richness and latitude is weak. No trend was found for the relationship between average geographic range size and mean cell elevation ($R^2 = 0.001$; $P > 0.05$), and a weakly negative relationship was found between average geographic range size and average body size across cells ($R^2 = 0.05$; $P < 0.01$) (Fig. 4.7).

Fig. 4.6 Scatterplot depicting the relationship of log geographic range size (km²) with (**a**) latitude, (**b**) elevation, and (**c**) body size – log of skull centroid size. Each dot represents a species

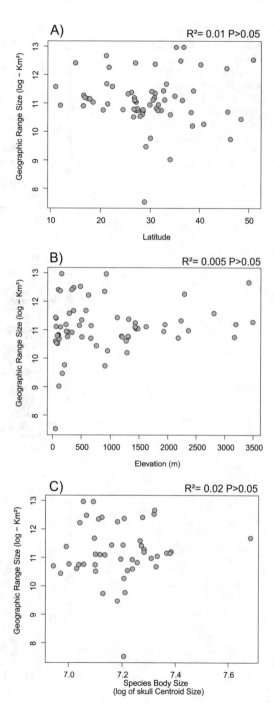

Fig. 4.7 Scatterplot depicting the relationship of average geographic range size by cell (km²) with (**a**) average latitude, (**b**) average elevation, and (**c**) average body size by cell – log of skull centroid size. Each dot represents a 1°× 1° cell

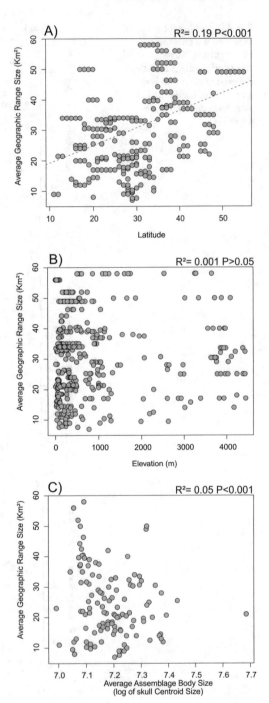

4.4   Patterns of Body Size Distribution

Species' body size distribution is right skewed (Fig. 4.8), which is a common pattern in many groups of organisms (Brown et al. 1993). In this chapter, the skull centroid size was used as a proxy for body size. Skull centroid size was measured through geometric morphometric techniques applied on 1359 specimens of tuco-tucos (see Fornel et al. 2020 for a complete methodological description). Centroid size measures usually correlate well with body length and weight across rodents, moreover, these results are in accordance with those of Medina et al. (2007) that used tuco-tucos body weight and length.

A negative relationship was observed between species body size and their average S latitude ($R^2 = 0.17$; $P < 0.002$) (Fig. 4.9a). This pattern of larger sizes closer to the equator (the antithesis of Bergmann's Rule) was previously reported by Medina et al. (2007) using body length and weight as a dependent variable. The subterranean habits of tuco-tucos, which decouple them from air temperature, and increased resource availability (NPP) at lower latitudes (Blackburn et al. 1999; Blackburn and Hawkins 2004; Alhajeri et al. 2020) may jointly explain the relationship between body size and latitude in tuco-tucos (Medina et al. 2007). Moreover, it has been suggested that soil properties are correlated to humidity which in turn is correlated with latitude (Medina et al. 2007). Soil excavation is an important part of

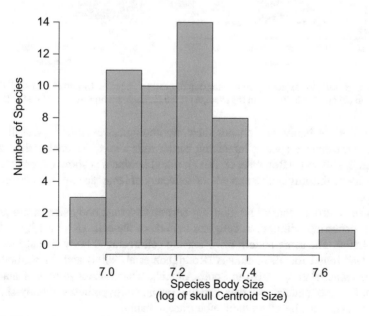

Fig. 4.8 Histogram representing variation of species' body size – log of skull centroid size as in Fornel et al. 2020

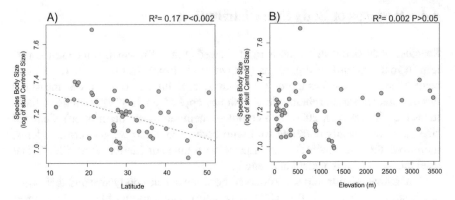

Fig. 4.9 Scatterplot depicting the relationship of species' body size (log of skull centroid size) with (**a**) latitude and (**b**) elevation. Each dot represents a species

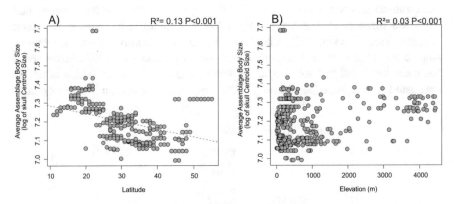

Fig. 4.10 Scatterplot depicting the relationship of average body size by cell (log of skull centroid size) with (**a**) average latitude and (**b**) average elevation. Each dot represents a $1° \times 1°$ cell

tuco-tuco's life habits and shapes their morphological evolution (Vassallo et al. 2020). For example, species inhabiting harder soils tend to be smaller than species inhabiting soft soils (Borges et al. 2017); if soil hardness is shown to be correlated with latitude this might also explain the tendency of larger species to occur at lower latitudes.

There is also a negative relationship between average body size (average of all species occurring within each cell) and latitude of the cells ($R^2 = 0.13$; $P < 0.001$) (Fig. 4.10a). The same pattern of increasing cell-averaged body size at lower latitudes was found for all mammals (Rodríguez et al. 2008) and for sigmodontine rodents (Maestri et al. 2016) in South America. That various groups of mammals exhibit the same pattern may decrease the cogency of hypotheses of body-size variation in tuco-tucos based on their subterranean habits.

At both the interspecific and the assemblage levels, there was little relationship between body size and elevation ($R^2 = 0.02$, $P > 0.05$; and $R^2 = 0.03$, $P < 0.001$) (Figs. 4.9b and 4.10b).

4.5   Concluding Remarks

We have shown general patterns in the distribution of species diversity, range size, and body size of *Ctenomys*. The center of richness for tuco-tucos lies between 23° and 29° degrees S latitude in northern Argentina, a region that is intermediate in elevation, longitude, and latitude compared to the entire geographical distribution of tuco-tucos. The center of richness may be also the center of initial diversification for tuco-tucos because it harbors one of the oldest fossils of *Ctenomys*, *C. uquiensis* (Verzi et al. 2020). This hypothesis should be tested with phylogenetic and paleontological data, and if supported would provide a striking historical explanation for the tuco-tucos' richness gradient. Patterns in the distribution of species' richness and range size confirmed that tuco-tucos have mostly allopatric distributions and species of *Ctenomys* have small range sizes when compared to other caviomorphs. However, when tuco-tucos are compared with other species-rich genera of caviomorphs in terms of the exclusivity of their geographic ranges, they are unexceptional. The conventional wisdom that suggested tuco-tucos have more exclusive ranges than other mammals is fallacious. Moreover, environmental correlates of the latitude may influence the geographic range size of *Ctenomys* species, judging from the assemblage-level occurrence of Rapoport's rule. Body size patterns confirmed that tuco-tuco species tend to have larger body size in lower latitudes, and this tendency for *Ctenomys* to defy Bergmann's Rule was evident at the assemblage level as well. Future investigations that include phylogenetic information to frame macroecological and biogeographical associations may provide a historical perspective to existing patterns, and help to understand the processes generating such patterns.

Acknowledgments We thank Claudio Bidau, in memory, for generating the geographic range maps used here. RM was supported by UFRGS, CAPES, and CNPq (150391/2017-0), BP was supported by FMNH.

Literature Cited

Alhajeri BH, Porto LMV, Maestri R (2020) Habitat productivity is a poor predictor of body size in rodents. Curr Zool 66:135–143

Bergmann C (1847) Über die Verhältnisse der Wärmeokönomie der Thiere zu ihrer Grösse. Göttinger Studien 3:595–708

Bidau CJ (2015) Family Ctenomyidae. In: Patton JL, Pardiñas UFJ, D'Elía G (eds) Mammals of South America, Vol. 2: Rodents. The University of Chicago Press, Chicago, pp 818–876

Bidau CJ, Martí DA, Medina AI (2011) A test of Allen's rule in subterranean mammals: the genus *Ctenomys* (Caviomorpha, Ctenomyidae). Mammalia 75:311–320

Blackburn T, Hawkins B a (2004) Bergmann's rule and the mammal fauna of northern North America. Ecography 27:715–724

Blackburn TM, Gaston KJ, Loder N (1999) Geographic gradients in body size: a clarification of Bergmann's rule. Divers Distrib 5:165–174

Borges LR, Maestri R, Kubiak BB, Galiano D, Fornel R, Freitas TRO (2017) The role of soil features in shaping the bite force and related skull and mandible morphology in the subterranean rodents of genus *Ctenomys* (Hystricognathi: Ctenomyidae). J Zool 301:108–117

Brown JH, Maurer BA (1989) Macroecology: the division of food and space among species. Science 243:1145–1150

Brown JH, Marquet PA, Taper ML (1993) Evolution of body size: consequences of an energetic definition of fitness. Am Nat 142:573–584

Brown JH, Stevens GC, Kaufman DM (1996) The geographic range: size, shape, boundaries, and internal structure. Annu Rev. Ecol Syst 27:597–623

Caraballo DA, López SL, Carmarán AA, Rossi MS (2020) Conservation status, protected area coverage of *Ctenomys* (Rodentia, Ctenomyidae) species and molecular identification of a population in a national park. Mamm Biol 100:33–47

D'Elía G, Teta P, Lessa EP (2020) A short overview of the systematics of *Ctenomys*: species limits and phylogenetic relationships. In: Freitas TRO, Gonçalves GL, Maestri R (eds) Tuco-tucos – An evolutionary approach to the diversity of a Neotropical rodent. Springer, Cham

de Freitas TRO (2016) Family Ctenomyidae (Tuco-tucos). In: Wilson D, Lacer T, Mittermeier RA (eds) Handbook of the mammals of the world lagomorphs and rodents I, vol 6. Lynx Edicions Publications, Barcelona, pp 498–534

Diniz-Filho JAF, Carvalho P, Bini LM, Tôrres NM (2005) Macroecology, geographic range size-body size relationship and minimum viable population analysis for new world carnivora. Acta Oecol 27:25–30

Fornel R, Maestri R, Cordeiro-Estrela P, de Freitas TRO (2020) Morphological evolution in tuco-tucos: skull shape and size diversification in the genus *Ctenomys* (Rodentia: Ctenomyidae). In: Freitas TRO, Gonçalves GL, Maestri R (eds) Tuco-tucos – an evolutionary approach to the diversity of a Neotropical rodent. Springer, Cham

Galiano D, Kubiak BB (2020) Environmental and ecological features of the genus *Ctenomys*. In: Freitas TRO, Gonçalves GL, Maestri R (eds) Tuco-tucos – an evolutionary approach to the diversity of a Neotropical rodent. Springer, Cham

Gaston, K. J. 1998. Species-range size distributions: products of speciation, extinction and transformation, Phil Trans R Soc B Biol Sci 353:219–230

Lyons SK, Smith FA, Ernest SKM (2019) Macroecological patterns of mammals across taxonomic, spatial, and temporal scales. J Mammal 100:1087–1104

Maestri R et al (2016) Geographical variation of body size in sigmodontine rodents depends on both environment and phylogenetic composition of communities. J Biogeogr 43:1192–1202

Medina AI, Martí DA, Bidau CJ (2007) Subterranean rodents of the genus *Ctenomys* (Caviomorpha, Ctenomyidae) follow the converse to Bergmann's rule. J Biogeogr 34:1439–1454

Nevo E (1979) Adaptive convergence and divergence of subterranean mammals. Annu Rev. Ecol Syst 10:269–308

Patton, J. L. et al. 2015. Mammals of South America, vol. 2: rodents. – Univ. of Chicago Press

Patton JL, Emmons LH (2015) Family Abrocomidae. In Mammals of South America, Vol. 2: Rodents, pp 805–814

Rapoport EH (1982) Areography: geographic strategies of species. Pergamon P, Oxford

Renan Maestri, Bruce D. Patterson (2016) Patterns of Species Richness and Turnover for the South American Rodent Fauna. PLOS ONE 11 (3):e0151895

Rodríguez MÁ, Olalla-Tárraga MÁ, Hawkins BA (2008) Bergmann's rule and the geography of mammal body size in the Western hemisphere. Glob Ecol Biogeogr 17:274–283

Ruggiero A, Kitzberger T (2004) Environmental correlates of mammal species richness in South America: effects of spatial structure, taxonomy and geographic range. Ecography 27:401–417

Smith FA, Lyons SK (2011) How big should a mammal be? A macroecological look at mammalian body size over space and time. Phil Trans R Soc B Biol Sci 366:2364–2378

Smith FA, Lyons SK, Ernest SKM, Brown JH (2008) Macroecology: more than the division of food and space among species on continents. Prog Phys Geogr 32:115–138

Stevens GC (1989) The latitudinal gradient in geographical range: how so many species coexist in the tropics. Am Nat 133:240–256

Tomiya S (2013) Body size and extinction risk in terrestrial mammals above the species level. Am Nat 182

Upham NS, Patterson BD (2015) Evolution of caviomorph rodents: a complete phylogeny and timetree for living genera. Biol Caviomorph Rodents Diversity Evol 1:63–120

Vassallo AI, Echeverría AI, Becerra F, Buezas GN, Díaz A, Longo V, Cohen M (2020) Biomechanics and strategies of digging. In: Freitas TRO, Gonçalves GL, Maestri R (eds) Tuco-tucos – An evolutionary approach to the diversity of a Neotropical rodent. Springer, Cham

Verzi, D. H., N. A. D. Santi, A. I. Olivares, C. C. Morgan, and A. Álvarez. 2020. The history of *Ctenomys* in the fossil record: a young radiation of an ancient family. In Tuco-tucos – An evolutionary approach to the diversity of a Neotropical rodent (T. R. O. Freitas, G. L. Gonçalves & R. Maestri, eds.). Springer Cham

Chapter 5
Phylogeography and Landscape Genetics in the Subterranean Rodents of the Genus *Ctenomys*

Fernando Javier Mapelli, Ailin Austrich, Marcelo Javier Kittlein, and Matías Sebastián Mora

5.1 An Overview of Phylogeography and Landscape Genetics

During the last two decades, since the publication of *"Phylogeography: The History and Formation of Species"* by John C. Avise (2000), studies of population genetics at broader time scales gave rise to so-called phylogeography. This discipline establishes that most species in nature exhibit some degree of genetic structure linked to geography (Avise 2000; Richards et al. 2007). A species' phylogeographic structure is the expected result of the interaction between demographic and genealogical processes, in line with the dynamics of the Earth's processes (geological or climatic; Hare 2001; Lessa et al. 2003). Phylogeographic studies have made possible to test biogeographic hypotheses, describe demographic and evolutionary processes that led to the formation of discrete evolutionary units, and also fostered inferences about the processes that determined the origin, distribution, and maintenance of biodiversity (Avise 2000); essential information for conservation and taxonomy.

Alongside, recent and accelerating fragmentation and degradation of natural environments advanced the development of methodological tools in landscape genetics. This discipline directly associate landscape characteristics and anthropic disturbances with parameters that quantify population genetic diversity. In particular, landscape genetics is a relatively new discipline that focuses on how landscape characteristics (e.g., topography, suitable habitats vs. unsuitable habitats, barriers

F. J. Mapelli (✉)
División Mastozoología, Museo Argentino de Ciencias Naturales "Bernardino Rivadavia", Buenos Aires, Ciudad Autónoma de Buenos Aires, Argentina

A. Austrich · M. J. Kittlein · M. S. Mora
Departamento de Biología, Facultad de Ciencias Exactas y Naturales, Universidad Nacional de Mar del Plata, Instituto de Investigaciones Marinas y Costeras (IIMyC), CONICET – UNMdP, Mar del Plata, Buenos Aires, Argentina
e-mail: kittlein@mdp.edu.ar; msmora@mdp.edu.ar

© Springer Nature Switzerland AG 2021
T. R. O. de. Freitas et al. (eds.), *Tuco-Tucos*,
https://doi.org/10.1007/978-3-030-61679-3_5

such as rivers and mountains, corridors, etc.), and environmental variables (e.g., temperature, humidity, and precipitations) mold gene flow. In accordance with its objectives, this discipline has involved smaller spatial and temporal scales and the use of highly variable molecular markers (e.g., multi-locus approaches with micro-satellites or SNPs). In this way, phylogeography and landscape genetics have served to identify evolutionarily independent units in natural populations providing back-ground for decisions about the conservation and viability of species.

5.2 Why Study Underground Rodents as Models from Population Genetic Approaches?

Tuco-tucos inhabit all regions of Southern South America, including Bolivia, Brazil, Peru, Uruguay, Paraguay, Chile, and Argentina; with 68 species within the genus (that is 45% of all species of subterranean rodents; Lacey 2000; Bidau 2015; Freitas 2016; Teta and D'Elia 2020; Teta et al. 2020; see also D'Elia et al. Chap. 3 in this volume). These animals are found in a multiplicity of habitats, from the Andes Mountains to the coastal dunes of the Atlantic and from the humid steppes of the Pampas to the dry deserts of the Chaco (Mapelli et al. 2017; Torgashevaa et al. 2017; Kubiak et al. 2020; Teta and D'Elia 2020; Teta et al. 2020). The genus *Ctenomys* is probable one of the most extreme cases of high chromosomal diversity (with diploid chromosome numbers ranging from 10 to 70; Reig and Kiblisky 1969; Freitas and Lessa 1984; Reig et al. 1990; Freitas 1995; Ortells 1995; Fernandes et al. 2009) and has experienced a rapid and relatively recent radiation during the Middle/Late Pleistocene, which resulted in the outburst of the many contemporary species (Reig et al. 1990; Lessa and Cook 1998; Cook and Lessa 1998; Slamovits et al. 2001; Reguero et al. 2007; Verzi et al. 2010, 2014; Parada et al. 2011).

Tuco-tucos presents a set of life history traits that position them as excellent models for the use of population genetic approaches (Reig et al. 1990; Busch et al. 2000; Wlasiuk et al. 2003; Lopes and Freitas 2012; Lopes et al. 2013; Esperandio et al. 2019). First, tuco-tucos have a high specificity in habitat use; the high cost of burrowing activities determines that they only occupy environments with sandy, fri-able, and permeable soils (Busch et al. 2000; Luna and Antinuchi 2007). These soil types are frequently patchily distributed, fragmented, and inserted in a landscape matrix with multiple hostile environments for the occupation for tuco-tucos (water-bodies, flooded areas with marsh vegetation, rocks, dense forests or grasslands, human impacted areas, etc.), which determines that populations are also commonly fragmented.

Tuco-tucos are mostly confined to their burrows, have a strong phylopatric behavior, and move little aboveground as compared to surface-dwelling rodents (Zenuto and Busch 1995, 1998; Busch et al. 2000; Lacey et al. 2000; Lessa 2000; Cutrera et al. 2005). Moreover, tuco-tucos are highly territorial and it has been gen-erally assumed that once they establish their territories, after natal dispersal, they make few additional dispersal movements (Zenuto and Busch 1998; Busch et al. 2000; Cutrera et al. 2005). Most individuals taken by aerial predators (e.g., owls) consist of subadult individuals leaving, presumably, their natal burrows (Vassallo et al. 1994; Kittlein et al. 2001).

Low population densities, habitat specialization, and low dispersal abilities are aspects that greatly impact on the scales at which genetic differences in tuco-tucos are exhibited (Mora and Mapelli 2010; Mora et al. 2010). Low population densities could arise, in part, because tuco-tucos tend to be solitary, with each adult preserving its own exclusive-use territory (Busch et al. 2000; Zenuto and Busch 1995, 1998; Cutrera et al. 2005). This mode of life results in small population units, sparsely distributed, with low genetic variation and high inter-population divergence brought about by genetic drift and low levels of gene flow (Busch et al. 2000; Lacey 2000, Cutrera et al. 2005; Fernández-Stolz et al. 2007; Mora et al. 2006, 2007, 2010; Cutrera and Mora 2017; Mapelli et al. 2017). A particular combination of behavioral and eco-physiological characteristics determines that, even at small spatial scale (at very few kilometers), tuco-tucos show high levels of genetic structure with strong phylogeographic patterns (Cutrera et al. 2006, 2010; Mora et al. 2006, 2007, 2013; Mapelli et al. 2012a).

High habitat specialization of tuco-tucos has conditioned that population demographic trajectories were strongly affected by changes in the extent of the environments that they occupy (Mora et al. 2016). In this regard, historical variations in the extension and connectivity of their habitats have strongly impacted the evolution and diversification of populations (Mora et al. 2006; Mirol et al. 2010; Roratto et al. 2014; Caraballo and Rossi 2017; Mapelli et al. 2012a, 2017). Thus, patterns of genetic variation at the geographical level have allowed to infer the evolutionary dynamics of the populations and, by extension, the dynamics of the environments they occupy.

5.3 Patterns of Isolation by Distance

Limited dispersal in *Ctenomys* species has important consequences on the spatial distribution of genetic variation (Lessa 2000; Fernández-Stolz et al. 2007; Lopes et al. 2013). If dispersal distances are small, a pattern of spatial autocorrelation emerges in the distribution of genetic variation: individuals that are close to each other are likely to be more related and therefore genetically more similar than individuals that are spaced farther apart (Mora et al. 2010). Something similar occurs at larger spatial scales, where the limited dispersal results in relatively higher values of gene flow between neighboring populations, which finally translate into a positive relationship between genetic and geographic distances. Consequently, a balance between the loss of genetic variation due to genetic drift and its gain because of the immigration of individuals is established. This equilibrium between these evolutionary forces determines that the spatial distribution of genetic variation in the species conforms to an isolation by distance (IBD) pattern, where the genetic dissimilarity among populations is proportional to the distance among them. The genetic population structure of many species (particularly those that have lower dispersion rates and are widely distributed) is characterized by an IBD pattern, and it has been frequently observed in nature (Vekemans and Hardy 2004).

Considering the life history traits of tuco-tucos (e.g., low dispersal capabilities, small population sizes), it is expected that the spatial distribution of the genetic variation will fit to an IBD pattern. However, there are several studies on population genetics in *Ctenomys* showing the absence of an IBD pattern or illustrating conflicting information between different regions of the genome. In this regard, mutation rate on different molecular markers must be considered in the analyses.

Phylogeographic studies in *Ctenomys* were primarily based on mitochondrial DNA (mtDNA) sequence analyses, whereas population genetic approaches at most recent ecological time scales have analyzed the nuclear DNA variation at microsatellite loci. Different DNA fragments of the genome have different rates of mutation, with mtDNA rates being from 3–5 times lower than those reported for microsatellite loci. These differences in mutation rates among different molecular markers have evident implications in how the genetic variation is recovered in natural populations. Particularly, it is expected that molecular markers with slower mutation rates require more time to reach a new equilibrium between genetic drift and migration after a demographic change event. Thus, for an IBD pattern to arise, it is necessary that the balance between gene flow and genetic drift be sustained for enough time. Because of differences in their effective population sizes and mutational rates, it is expected that historical demographic events affect mitochondrial genetic variation more strongly than variation at microsatellite loci. Several phylogeographic studies in *Ctenomys* show that mtDNA still reflects the historical demographic processes (population expansions or contractions) that occurred during the Pleistocene and Holocene (see the next section in this chapter); consequently, the spatial distribution of genetic variation does not fit to an IBD pattern. For instance, some species of tuco-tucos such as *C. rionegrensis*, *C. australis*, *C. magellanicus*, and *C. torquatus* have experienced recent range expansions (Wlasiuk et al. 2003; Mora et al. 2006; Gonçalves and Freitas 2009; Fasanella et al. 2013; Roratto et al. 2014) and genetic estimates of gene flow using mitochondrial DNA are higher therefore, than current levels, concealing the IBD pattern.

Following Slatkin (1993), D'Elía et al. (1998) suggested that the current distribution of *C. rionegrensis* in Uruguay is the result of a recent range expansion, suggesting that genetic estimates of gene flow should be higher than current levels. Wlasiuk et al. (2003) proposed a scenario in which *C. rionegrensis* expanded in the recent past from a more limited geographic range and has subsequently differentiated in near isolation, with genetic drift most likely playing a major role in overall genetic differentiation (see also D'Anatro et al. 2011). The recent range expansion in this species would hide the IBD pattern using mitochondrial and nuclear molecular markers. Mora et al. (2006) in *C. australis*, Fasanella et al. (2013) in *C. magellanicus*, and Roratto et al. (2014) in *C. torquatus* also showed that genetic differentiation was not consistent with a simple IBD pattern when mtDNA was considered, possibly evidencing a lack of equilibrium between gene flow and local genetic drift. Like *C. rionegrensis*, these species experienced a recent demographic expansion as shown by their pattern of mtDNA variation.

On the other hand, several species of *Ctenomys* have shown patterns implying historical IBD, emphasizing the idea that populations have remained long enough in demographic equilibrium between drift and migration (see Mora et al. 2007, 2013, 2016; Tomasco and Lessa 2007; Lopes and Freitas 2012; Lopez et al. 2013). Two examples using mtDNA are reported in Mora et al. (2016) and Lopes and Freitas (2012), in *C. chasiquensis* and *C. lami*, respectively. These authors considered the entire distributional ranges of these species and suggested moderate to high population genetic structure at a regional level accompanied by IBD patterns. Equilibrium between genetic drift and gene flow at larger spatial scales has also been reported in other species of ctenomyids that are almost linearly distributed (*C. talarum*, in Mora et al. 2007, 2013; *C. pearsoni*, in Tomasco and Lessa 2007; *C. minutus*, in Lopes 2011 and Lopes et al. 2013). In these examples, the narrow distribution of the species limits dispersal and gene flow to a dominant spatial direction as compared to what is possible in a widespread and two-dimensional distributional range. As was suggested by Mora et al. (2006), one-dimensional spatial systems usually have a reduced variance in geographical genetic attributes, as compared to two-dimensional arrangements, increasing the chance to detect such IBD pattern (see also Slatkin and Barton 1989 for a more comprehensive discussion).

There is a strong relationship between the mutational rate of molecular markers, the spatial arrangements of populations (i.e., one-dimensionally vs two-dimensionally distributed species) and the recovery rate of genetic variation due to random mutational processes in natural populations following demographic changes (e.g., bottlenecks and/or population expansions). This means that species that have undergone recent demographic expansions (accompanied or not by expansion in their geographical ranges) are more likely to depict IBD patterns in their most variable DNA portions of their genomes (e.g., microsatellite loci in *C. australis* and *C. porteousi*; see Austrich et al. 2020 and Mapelli et al. 2012b, respectively).

Using microsatellite loci, several studies in population genetics of tuco-tucos have tried to assess the relative importance of geographic distances among populations and those landscape elements that potentially could mold gene flow patterns. For some species, distance appears to be the main factor structuring genetic variation among different populations (Mapelli et al. 2012b; Austrich et al. 2020). In other cases, landscape features that impose potential barriers to gene flow or are associated to suitable habitat conditions for the occurrence of a species are more important than distance among populations (Wlasiuk et al. 2003; Mora et al. 2010, 2017). Thus, close populations can diverge more quickly if there is a strong barrier to the movement of individuals, whereas populations located in areas of more continuous habitat can exchange more migrants, promoting the increase of genetic similarity among them. On the other hand, the divergence rate among populations can also be variable, hindering the establishment of an IBD pattern. Basically, differentiation among populations is a consequence of genetic drift and this process is directly dependent upon population size. Populations of smaller sizes are subject to stronger effects of genetic drift, which leads to a faster fixation of allelic variants. It is thus expected that these populations have reduced genetic variability in comparison to larger and more stable populations. For instance, the lack of an IBD pattern in *C. chasiquensis* denote that differentiation was not related to geographical distances, but to the landscape or habitat characteristics where the populations were sampled (see Mora et al. 2017). In this species, habitat fragmentation (both anthropic and natural) has caused the isolation of some populations in the eastern area of its distribution, with the consequent accentuation of genetic drift. This pattern could explain, at least in recent times, the lack of equilibrium between gene flow and genetic drift across the entire distribution of the species. A similar pattern was observed in *C. rionegrensis*, where genetic drift due to population isolation has been strong at recent times, minimizing the effects of geographical distances. In these examples, the lack of an IBD pattern seems to indicate that population divergence is not related to geographical distances but rather to landscape characteristics.

5.4 Phylogeographic Patterns

As most species of tuco-tucos are endemics, their complete distributional ranges were covered in several studies, disclosing the factors that have brought about the observed phylogeographic patterns. In general, these studies show a close association between population size variations and historical changes in habitat availability; suggesting that diversification patterns both among and within species, were closely related to historical habitat connectivity (Mora et al. 2006, 2007, 2013, 2016; Tomasco and Lessa 2007; Mirol et al. 2010; Gómez Fernández et al. 2012; Mapelli et al. 2012a, 2017; Lopes et al. 2013; Roratto et al. 2014; Cutrera and Mora 2017).

The glacial-interglacial cycles of the Late Quaternary produced extensive disturbances not only in different biogeographic areas throughout South America, but also in many regions of the world (Rabassa et al. 2000, 2011). In South America, these changes largely conditioned the availability and connectivity of suitable habitats for tuco-tucos, profoundly affecting their phylogeographic patterns (Chan et al. 2005; Mapelli et al. 2012a, 2017; Tammone et al. 2018a, 2018b). In many species of *Ctenomys*, demographic expansions and contractions thus seem to have mainly responded to landscape changes associated with climatic perturbations occurred in the Late Pleistocene and Holocene (Tammone et al. 2018a, 2018b). These changes have led to the formation of sandy accumulations, in the form of coastal and continental dunes, which have allowed the permanence or range expansions of the populations of many species of tuco-tucos (Mora et al. 2006, 2007). In this regard, not only the species in Andean regions have altered their distributional ranges as a result of the glacial-interglacial cycles (Gutiérrez-Tapia and Palma 2016), but species distributed in many areas distant from the Andes, such as the South American Pampas and coastal areas, have also been largely affected (Tonni et al. 1999; Mora et al. 2006, 2007, 2016; Mapelli et al. 2012a).

Expansion of sandy areas affected not only the demography and evolution of tuco-tucos populations, but also impacted on the diversification patterns between species (Mapelli et al. 2017). Caraballo and Rossi (2018) used phylogeographic approaches to understand the diversification of the "*torquatus*" group, a set of closely phylogenetically related species of tuco-tucos distributed along Uruguay, eastern Argentina and southern Brazil (*C. torquatus, C. lami, C. minutus, C. ibicuiensis, C. pearsoni*) and the "Corrientes" group (tuco-tucos distributed around and between the Iberá marshes in Corrientes province, Argentina, whose taxonomy is not yet well established). These authors suggest that the geographic expansion of these lineages and the colonization of the Argentine areas occurred during the most arid stages of the Pleistocene when sandy soils expanded throughout the region (between 500,000-1,000,000 ybp) and the effectiveness of the Uruguay River as a barrier would have diminished considerably.

Mapelli et al. (2017) studied the "*mendocinus*" species complex (or "*mendocinus*" group; D'Elía et al. 1998, 1999; Wlasiuk et al. 2003; Massarini and Freitas 2005; Stolz 2006; Fernández-Stolz 2006; Fernández-Stolz et al. 2007; Parada et al. 2011; Mora et al. 2016) in central Argentina and suggest that the dispersion of the group was associated with the driest stages of the Late Quaternary. The rapid cladogenesis experienced by the "*mendocinus*" group was associated with the formation of large sandy areas during the Late Pleistocene. Paleoclimate estimates indicate that this period was the driest and coldest time for the last glacial-interglacial cycle (Iriondo and Kröhling 1995). These climatic conditions generated an increase in the extension of arid environments and may have permitted a more continuous distribution of tuco-tucos of *C. mendocinus* species complex throughout the Pampas in Central Argentina (Mapelli et al. 2017). On the other hand, the interglacial periods were characterized by increased temperature and humidity (Quattrocchio et al. 2008). Suitable environments for these tuco-tucos retracted during humid periods and expanded during arid pulses (Mapelli et al. 2017). Thus, during periods of habitat retraction and lower overall population sizes, the combined effects of natural selection and genetic drift would have favored the divergence among lineages. In this form, the divergence among different lineages of the "*mendocinus*" species group may have taken place during humid periods, when climate conditions resulted in environmental discontinuities that isolated populations (Mapelli et al. 2017).

At the Holocene-Pleistocene boundary (about 12,000–9000 ybp) an important increase in temperature and humidity occurred in the area of the current distributional range of "*mendocinus*" group (Tonni et al. 1999). These climatic changes, in conjunction with a rise of sea level and the flooding of the riverbeds favored the spread of inland grasslands, increasing plant cover (Quattrocchio et al. 2008) and decreasing the extent of habitat available for these tuco-tucos. The easternmost range of this area have been studied in detail from phylogeographic perspectives and the results obtained highlight the strong imprint of the environmental change during the Pleistocene-Holocene boundary on the population demography of these species.

Three of the species belonging to the "*mendocinus*" group currently occupy remnant and fragmented habitats. *Ctenomys rionegrensis* is restricted to very fragmentary sandy areas associated with sedimentary deposits formed by the Rio Negro, Uruguay (D'Elía et al. 1998; Wlasiuk et al. 2003; Kittlein and Gaggiotti 2008; D'Anatro et al. 2011). *C. porteousi* is restricted to sandy soils associated with deflation basins in the Lagunas Encadenadas del Oeste, Argentina (Mapelli and Kittlein

2009; Mapelli et al. 2012a, 2012b) and *C. chasiquensis* is restricted to a series of sandy paleo valleys in southern Buenos Aires and La Pampa provinces, Argentina (Mora et al. 2016, 2017). Their geographical patterns of mitochondrial genetic variation show local populations evolving in relative isolation, with a strong effect of genetic drift at local level and low gene flow between populations (Wlasiuk et al. 2003; Mapelli et al. 2012a; Mora et al. 2016). However, it is interesting to highlight that the differences between the phylogeographic patterns of these species are very subtle and are related to the history of the environments they currently occupy. *C. rionegrensis* shows very low genetic variation by locality, with very few haplotypes shared among populations. This suggests a strong effect of genetic drift locally, and low values of gene flow between localities; as expected for animals with low vagility evolving in relatively isolated populations (Wlasiuk et al. 2003). *C. porteousi* also shows strong effects of genetic drift and low levels of gene flow between populations (Mapelli et al. 2012a); but the effects do not seem to be as marked as in *C. rionegrensis*. *C. porteousi*, however, showed a greater number of shared haplotypes among populations, suggesting a less marked effect of genetic drift. Population differentiation seems not to have been generated by a combination of unique haplotypes as in *C. rionegrensis*, but by distinctive assortments of shared haplotypes. This pattern suggests a greater connectivity between populations in the past, with strong subsequent effects of genetic drift at local level.

The phylogeographic pattern of *C. chasiquensis* was similar to that observed in *C. porteousi*, but with higher past connectivity between populations and greater genetic variation by locality (Mora et al. 2016). In addition, populations located towards the East of the range of *C. chasiquensis* (a more fragmented area with higher intrusions of grasslands) presented evidence of higher past isolation as compared with populations in the western range of the species (Mora et al. 2016, 2017). Historical demographic reconstructions for *C. porteousi* (Mapelli et al. 2012a) and *C. chasiquensis* (Mora et al. 2016) show that their population dynamics were strongly impacted by environmental changes at the Holocene-Pleistocene boundary; in both species there was a significant reduction in population sizes associated with these historical events. Accordingly, the loss and fragmentation of their natural habitats would have had a profound impact on the spatial distribution of mitochondrial genetic variation, reducing connectivity between populations and increasing the effects of genetic drift.

On the other hand, *C. australis* (sand-dune tuco-tuco), another species included in the "*mendocinus*" group that occupies the sand dunes along the Atlantic coast of Buenos Aires province, Argentina, presents a completely different phylogeographic pattern to that observed in its sister species (Mora et al. 2006, 2010; Cutrera and Mora 2017). This endemic and endangered subterranean rodent presents an ancestral haplotype distributed in high frequency throughout the entire range of its distribution, which indicates a marked process of population growth associated with a notable geographic population expansion (see Fig. 5.1). The coastal sand dunes occupied by this species are a relatively recent habitat originated during the sea-level fluctuations produced by oceanic regressions and transgressions in the Middle Holocene (6500–3000 ybp; Isla 1998; Isla et al. 2001). *C. australis* is an extreme

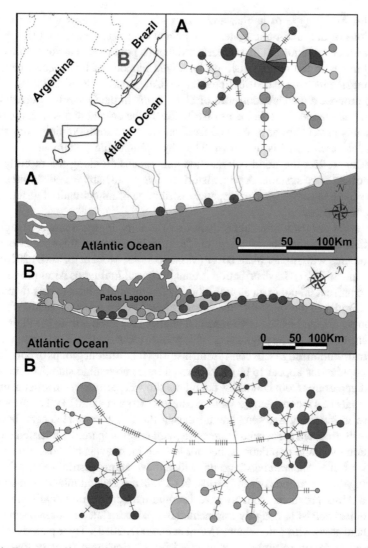

Fig. 5.1 Examples of haplotype networks depicting the contrasting patterns in the spatial distribution of genetic variation of mitochondrial DNA in two species of tuco-tuco from Argentina and Brazil: (**a**) *Ctenomys australis* (modified from Mora et al. 2006), (**b**) *C. minutus* (modified from Lopes et al. 2013). Circles correspond to different mtDNA haplotypes, and their size is proportional to their frequency in the samples. Colors of the circles designate their spatial location (locality or group of localities) on the maps (**a** and **b**), and lines connecting the circles represent the most likely genealogical relationships between haplotypes (the crossed marks represent the number of mutational steps among haplotypes). Light green polygons on the map represent suitable habitat for the species. Different localities in *C. minutus* have different groups of haplotypes, which diverge by several mutational steps. This pattern is indicative that populations have occupied the area in demographic equilibrium long enough to establish an isolation by distance (IBD) pattern. In contrast, *C. australis* shows an ancestral haplotype with much higher frequency than the other haplotypes in the network and occurring in all localities. The other haplotypes diverge from the ancestral haplotype in one or only a few mutational steps and occur only in some localities. This pattern is indicative of a recent demographic expansion, possibly accompanied by a geographical range expansion

specialist to this type of habitat and shows strong associations with the unstable conditions of the sand-dune system (Malizia et al. 1991; Zenuto and Busch 1995, 1998; Mora et al. 2006). It can be assumed that its populations are at least as old as coastal dunes in the region since no fossilized remains of this species have been found outside its current distributional range (Contreras and Reig 1965). Mora et al. (2006) propose a recent occurrence of *C. australis* in the Quaternary coastal dune system; but whether it extended its distribution to the coastal dunes with the formation of this novel habitat or derived from an ancestral form coming from another geographic area remains unknown. The phylogeographic pattern in *C. australis* would thus reflect the rapid colonization and expansion along the recently originated coastal dune systems. Again, historical changes in habitat availability seem to have molded, in some extent, the spatial distribution of mitochondrial genetic variation in this species.

Ctenomys talarum (the Talas tuco-tuco) occurs in the Pampean region of Argentina, and unlike the species of the "*mendocinus*" group, this species occupies environments with higher plant cover (Vassallo 1998; Mora et al. 2007, 2013; Luna and Antinuchi 2007). Its distribution is mainly coastal in Buenos Aires province, but it is also present inland in areas with a highly fragmented habitat matrix (Mora et al. 2013). Similar to the species of the "*mendocinus*" group, their populations seem to be strongly affected by the environmental changes occurred during the Pleistocene-Holocene boundary. The phylogeographic pattern recovered for *C. talarum* throughout its distribution range tells contrasting histories for different groups of populations. Some populations appear to be characterized by demographic stability and no significant departures from neutrality, but others show departures from strict neutrality, with signals of a recent demographic expansion (Mora et al. 2013). The progressive increase in humidity and temperature during the Pleistocene-Holocene boundary and, the consequent associated increase in plant cover, apparently initiated the fragmentation of inland populations, but not to such high levels as those affecting the species of the "*mendocinus*" group which have more restrictive requirements regarding the occurrence of sandy soils. In *C. talarum*, several inland populations in Buenos Aires province seem to have become extinct recently (Quintana 2004), while others persist in a highly fragmented matrix, associated to stream sand banks and inland areas with paleo-dunes (Mora et al. 2007, 2013). Their patterns of mitochondrial variation resemble that reported for *C. rionegrensis*: very low genetic variability by location and almost no haplotypes shared between locations, suggesting that they have remained in relative isolation (Mora et al. 2013). Genetic diversity and differentiation of coastal populations tell a different story. At present, these populations occupy sand-dune environments that originated as a result of sea-level fluctuations during the Holocene. The patterns of genetic variation in these areas indicate population expansions, evidently associated with the colonization of these novel environments of recent formation (Mora et al. 2007, 2013).

Interestingly, estimates of the main demographic changes observed in several *Ctenomys* species from the central Argentina seem to be temporarily coincident. It seems that while some species adapted to environments with low plant cover underwent population retractions (e.g., *C. porteousi* and *C. chasiquensis*; Mapelli et al.

2012a; Mora et al. 2016), other species adapted to environments with higher plant cover (e.g., *C. talarum*) remained demographically stable in some regions, and underwent population expansions in others (Mora et al. 2007, 2013). This shows some differences in the way environmental change affected the demography of each species.

The Patagonian tuco-tucos *Ctenomys sociabilis* and *Ctenomys haigi* offer other interesting example of differential demographical responses to concomitant changes in the environment. *C. sociabilis* is highly endemic to the upper valley of the Limay River (Chan et al. 2005; Chan and Hadly 2011; Tammone et al. 2018a, 2018b); while *C. haigi* is widely distributed in the eastern Limay Valley and adjacent areas of Río Negro province, Argentina (Lacey et al. 1997, 1998). Chan et al. (2005) and Tamnone et al. (2018a, b) quantified the mitochondrial genetic variation in current and fossil samples and observed important evidence of a marked reduction in the genetic variability of *C. sociabilis*, but not in *C. haigi*, during the last 12,000 years. The reduction in genetic variability observed in *C. sociabilis* is of such magnitude that its populations are currently characterized by a single haplotype throughout its entire distributional range (see Fig. 5.2). Differential demographic responses observed between these species could have been influenced by species-specific attributes such as the documented differences in habitat specialization. *C. sociabilis*

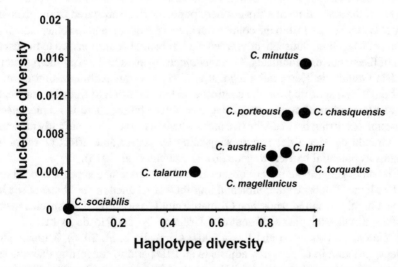

Fig. 5.2 Relationship between haplotype and nucleotide diversity for mitochondrial DNA in some representative species of *Ctenomys*. Haplotype diversity is a measure of how many haplotypes are in the gene pool of a population or species, while nucleotide diversity is a measure of how different these alleles are from each other. In this figure, species that underwent recent population expansions present low values of nucleotide diversity and high values of haplotype diversity. Only those studies that include the entire or almost complete distribution of a species are given: *C. australis* (Mora et al. 2006), *C. chasiquensis* (Mora et al. 2016), *C. lami* (Lopes and Freitas 2012), *C. magellanicus* (Fasanella et al. 2013), *C. minutus* (Lopes et al. 2013), *C. porteousi* (Mapelli et al. 2012a), *C. sociabilis* (Chan et al. 2005), *C. talarum* (Mora et al. 2013) and *C. torquatus* (Roratto et al. 2014)

inhabits a more restricted subset of habitats compared to the ecologically more generalized *C. haigi* (Lacey et al. 1997, 1998; Lacey and Wieczorek 2004). During the Middle-Holocene, the area occupied by these species was spatially and temporally dynamic and this variability might have contributed to the decline of *C. sociabilis* but not of *C. haigi* at the study sites (Chan et al. 2005; Tammone et al. 2018a, 2018b). Consequently, the greater habitat specialization of *C. sociabilis* would have made this species more sensitive than *C. haigi* to different environmental changes (Tammone et al. 2018a, 2018b).

The Late Quaternary environmental changes have also had significant effects on the historical demography of *Ctenomys magellanicus* (Fasanella et al. 2013), the southernmost species of this genus. The species is fragmentarily distributed both in the steppe and in the forest ecotone of the Isla Grande de Tierra del Fuego in the southern extreme of Argentina and Chile, although there is fossil evidence that the species previously occupied the final extreme of continental Patagonia in Santa Cruz Province, Argentina (Pardiñas 2013). The phylogeographic pattern recovered for the species suggests a recent history of demographic expansion with a recent colonization of the northern area of its distribution. Possibly one or a few haplotypes of the southern distribution have colonized the northern area during the Pleistocene when rivers had low-flow regimes (Fasanella et al. 2013). These authors have proposed this species might have lived at a Pleistocene refuge during the adverse environmental conditions in Tierra del Fuego (see also Rabassa et al. 2000, 2011). At the same time that the northern populations disappeared from the island, they also were extinct from the continental areas of Argentina in the other side of the Straits of Magellan. Subsequently, when environmental conditions led to the expansion of the steppe environments, *C. magellanicus* recolonized the northern sector of the Isla Grande de Tierra del Fuego. It is likely that this recolonization occurred after the Island split-up from the continent, so that the Strait of Magellan prevented the recolonization of continental Patagonia. Interestingly, there is no present-day migration occurring between the two areas where tuco-tucos are currently present in Isla Grande de Tierra del Fuego, suggesting an appreciable effect of rivers and streams as potential barriers to gene flow (Fasanella et al. 2013).

The phylogeographic pattern observed in *C. pearsoni* also appears to be associated with the Holocene environmental imprint that resulted in variations of sea levels in a highly dynamic landscape (Tomasco and Lessa 2007). This species mainly inhabits the coastal plains of southern Uruguay by the Río de la Plata and the Atlantic Ocean (Tomasco and Lessa 2007; Caraballo et al. 2016). Mitochondrial genetic variation in *C. pearsoni* adjusts to an IBD pattern, suggesting that this species conserved a stable regime of differentiation by genetic drift under limited gene flow for a long term. Additionally, *C. pearsoni* shows no signs of possible departures from neutrality; in fact, this regional stability in its historical demography might have been retained in spite of Holocene marine transgressions and other environmental and climatic perturbations that affected the distribution of the species (Tomasco and Lessa 2007). Locally, however, populations may have shifted their distributions in response to sea-level changes, especially on the coast in the northwest area of Uruguay, where it occupies systems of coastal dunes of very recent formation (Tomasco and Lessa 2007). Possibly, more relaxed restrictions associated to the type of habitat that this species occupy (not certainly restricted to the first strip of coastal dunes as *C. australis* and *C. flamarioni*; see Mora et al. 2006 and Fernández-Stolz et al. 2007, respectively) have prevented local extinctions or drastic population reductions on the Uruguayan coast.

Several phylogeographic studies have also been carried out in species of *Ctenomys* from southern Brazil at different spatial scales (El Jundi and Freitas 2004; Stolz 2006; Fernández-Stolz et al. 2007; Gonçalves and Freitas 2009; Gonçalves et al. 2012; Lopes and Freitas 2012; Lopes et al. 2013). For instance, *C. torquatus* (a species with one of the widest geographical ranges, occurring in the lowland grasslands of southern Brazil and northern Uruguay; Freitas 1995; Fernandes et al. 2009) shows a phylogeographic pattern explained by the habitat availability for the species and the historical patterns of colonization of this area. Mitochondrial genetic variation shows a process of population expansion and a recent colonization of the peripheral areas of its global distribution (Fig. 5.2). This population expansion has been associated with greater environmental availability during the Late Pleistocene (Lopes and Freitas 2012; Roratto et al. 2014). Conditions of greater aridity prevailing during this period would have allowed a greater expansion of the grassland environments with a concomitant reduction of gallery forests along the rivers; which would have allowed the population expansion of *C. torquatus* and the colonization of novel areas (Roratto et al. 2014). The formation of coastal dune systems during the Holocene would also have allowed the expansion of this species into these areas. Demographic reconstructions of populations occupying these environments support this hypothesis and indicate that colonization and expansion in these areas would date back to the last 5000 years, when the most recent sand-dune system was formed (Tomazelli et al. 2000; Roratto et al. 2014).

C. lami and *C. minutus*, two very related species, occur in areas close to *C. torquatus*, but their phylogeographic patterns are somewhat contrasting: these two species are characterized by IBD patterns and their populations do not appear to have been so affected by the climatic oscillations of the Late Quaternary (El Jundi and Freitas 2004). *C. lami* is highly endemic with a very restricted distribution; this species only occupy an area of 936 km^2 (Lopes and Freitas 2012). This zone consists of sandy fields formed by fluctuations of the sea level in the beginning of the Pleistocene, approximately 400,000 years ago (Tomazelli et al. 2000). A clinal pattern of genetic variation could be observed in *C. lami*, since the haplotypes are commonly shared between geographically close localities. The absence of any evidence of population expansion and the fact that populations adjust to an IBD pattern suggests demographic stability and low effects of Late Quaternary environmental changes on their populations.

A similar pattern can be observed in *C. minutus*; this species has a coastal distribution in the Southeastern of Brazil, occupying sand-dune systems formed during the marine transgressions and regressions in the Pleistocene and Holocene (Lopes 2011; Lopes et al. 2013). Despite the marked environmental dynamics of the region, *C. minutus* shows demographic stability and one of the strongest phylogeographic breaks within the genus: in general, most haplotypes are separated by several mutational steps, and were limited to single area (see Fig. 5.1). This pattern of spatial distribution of genetic variation suggests that populations of this species have remained stable for a considerable period and their demographic trajectories have been little affected by environmental changes (Lopes et al. 2013). These latter authors used Bayesian approaches and showed that the effective population sizes were mostly constant, with only some minor sudden retractions observed from the last 13,000 ybp (see also Lopes 2011).

Phylogeographic approaches have also been applied for the study of the species of the "Corrientes" group distributed in the northeast of Argentina (Mirol et al. 2010; Caraballo et al. 2012; Gómez Fernández et al. 2012). The "Corrientes" group represent a typical case of uncertainty on species boundaries; including three named species (*C. roigi*, *C. perrensi*, and *C. dorbignyi*), and several other populations whose taxonomical status has not been determined yet (Giménez et al. 2002; Mirol et al. 2010; Caraballo and Rossi 2017). The diversification of the "Corrientes" group also seems to be due to the historical availability of the habitat; although in this case habitat availability is not seem to be associated with the glacial and interglacial cycles but to variations in the hydrological regimen in the region. Tuco-tucos from this group occupy sandy soils located in elevated sites inside a large wetland, influenced by the effects of the Paraná River and the Iberá Wetland (Giménez et al. 2002; Mirol et al. 2010). The flooded systems typically observed in this extensive wetland create a highly dynamic and unstable environment for the presence of tuco-tucos. This is because water-level fluctuates depending on rainfall and the soil composition as well as underground rivers flow. Thus, the wetlands changes in time and space, altering the availability of suitable habitat for tuco-tucos, connecting or isolating populations at different times and in different areas (Mirol et al. 2010; Gómez Fernández et al. 2016). Although very few haplotypes in this group of species are shared between locations, they appear closely related, differing in few mutational steps. At the same time there is no clear geographical structure and there is evidence of an important historical gene flow between some sites (Caraballo et al. 2012; Gómez Fernández et al. 2012). The highly dynamic hydrological regime seems to be responsible for temporal connection and disconnection among populations, affecting the patterns of genetic differentiation at regional level.

In summary, some studies in *Ctenomys* have shown concordant phylogeographic patterns across species. There are very few cases (e.g., *C. minutus*, *C. lami*, *C. pearsoni*; Lopes and Freitas 2012; Lopes et al. 2013; Tomasco and Lessa 2007) in which species of tuco-tucos have remained stable sufficiently long to permit a demographic stability. These species adjusts to an IBD pattern, suggesting that the populations have remained stable long enough time to have achieved a balance between genetic drift and migration. On the contrary, many species of *Ctenomys* have suffered recent demographic processes (population growth or decrease) that have profoundly impacted the distribution of genetic variation. Most species (*C. rionegrensis*, *C. australis*, *C. porteousi*, *C. chasiquensis*, *C. talarum*, *C. torquatus*) show strong demographic signals associated with historical variations in the extend of the natural habitats. The glacial-interglacial cycles of the Late Quaternary have strongly impacted the distribution and demography of these tuco-tucos, particularly the environmental change occurred in the Pleistocene-Holocene boundary, both for their magnitude and temporal proximity. These climatic and environmental processes have left deep traces not only in the distribution of mitochondrial genetic variation, but also in nuclear DNA (see Cutrera and Mora 2017).

5.5 Spatial Scales of Population Genetic Structure

During the last three decades, researchers have used diverse genetics tools to define population boundaries. In this regard, the use of highly variable molecular markers in studies of assigning individuals to populations has revolutionized population genetics. SNPs and microsatellite loci have been particularly efficient in elucidating population boundaries at fine spatial scales and to define significant evolutionary units in hundreds of species (Lacey 2000; Helyar et al. 2011).

The use of microsatellite loci (Lacey et al. 1999; Lacey 2001; Roratto et al. 2011) in addition to statistical Bayesian approaches have allowed to infer putative genetic populations and assign individuals to them in several *Ctenomys* species. These approaches have permitted to understand, to some extent, the factors and processes shaping the population structure and spatial distribution of genetic diversity in this genus (Wlasiuk et al. 2003; Gava and Freitas 2004; El Jundi and Freitas 2004; Fernández-Stolz et al. 2007; Gonçalves and Freitas 2009; Mirol et al. 2010; Mora et al. 2010; Gómez Fernández et al. 2012; Lopes and Freitas 2012; Mapelli et al. 2012b; Fasanella et al. 2013; Lopes et al. 2013; Mora et al. 2017; Esperandio et al. 2019). In general, these studies describe the tuco-tucos as species very prone to be spatially structured. High habitat specificity, a phylopatric behavior and low dispersal capabilities of tuco-tucos determine that gene flow occurs most probably within the same local population. Consequently, gene flow between nearby localities is low and strongly conditioned by the connectivity among populations, which leads to high interpopulation divergence.

But not only the distances between localities determine the population connectivity in *Ctenomys*. Several studies (Mora et al. 2010; Lopes and Freitas 2012; Mapelli et al. 2012b; Lopes et al. 2013; Mora et al. 2017; Esperandio et al. 2019) have shown that habitat discontinuities strongly condition gene flow patterns between populations, so numerous landscape elements can become important barriers to population connectivity (in addition to geographical distances between populations). Species of tuco-tucos inhabit more fragmented habitats have shown the strongest patterns of population structure than species distributed in landscapes with continuous and more conserved habitats (Mapelli et al. 2012b, 2020; Mora et al. 2017). Similarly, within the same species, populations located in areas with low habitat connectivity show greater divergence relative to areas where the habitat of tuco-tucos is less fragmented (see Mora et al. 2017; Mapelli et al. 2020). This can also be illustrated in several studies, where populations occupy fragmented habitats (both by natural and anthropic action) and where sampling locations only 10–15 km away from each other constitute independent genetic units (Wlasiuk et al. 2003 in

C. rionegrensis; Gonçalves and Freitas 2009 in *C. torquatus*; Mapelli et al. 2012b in *C. porteousi*). The case of *C. rionegrensis* is paradigmatic since populations located at these distances seem to evolve in almost complete isolation with extremely reduced gene flow (Wlasiuk et al. 2003; D'Anatro et al. 2011). In species occurring in a more continuous habitat the genetic population units are defined at larger spatial scales. This can be observed in *C. lami* (Lopes and Freitas 2012), in *C. minutus* (Castilho et al. 2012; Lopes et al. 2013), and in *C. chasiquensis* (Mora et al. 2017). All these studies show how patterns of population genetic structure are strongly conditioned by the landscape configuration.

Another aspect to consider is how genetic variation recovers after an event of demographic change. Because population bottlenecks often accompany colonization events that occur when a species expands its range, peripheral populations frequently have lower genetic diversity than central populations. Greater effective population sizes in central populations than in peripheral populations suggest also an important role of genetic drift after colonization (Swaegers et al. 2013). In this regard, these patterns are recurrent in tuco-tucos: greater genetic divergence among the populations located in peripheral areas relative to those populations from the core area, and lower genetic diversity in the peripheral populations.

Several authors have shown how populations of tuco-tucos found in the peripheral areas are characterized by low genetic diversity, and in some cases tend to establish independent genetic units (Gómez Fernández et al. 2012; Lopes and Freitas 2012; Mapelli et al. 2012b; Lopes et al. 2013; Mora et al. 2017; Austrich et al. 2020). As was suggested previously, this pattern could be explained by the greater effect of genetic drift in peripheral populations (recurrent founder events lead to greater fluctuations in population sizes; Mapelli et al. 2012b). Some extreme cases of this scenario can be observed in *C. rionegrensis* (Wlasiuk et al. 2003), in *C. australis* (Austrich et al. 2020), and in the species of the "Corrientes" group (Gómez Fernández et al. 2012, 2016), where the most isolated populations completely lack genetic variability in several microsatellite loci. Another example is *C. magellanicus,* a species that expanded its range into the northernmost areas of Isla Grande de Tierra del Fuego after the Pleistocene glaciation, and populations in these relatively recently colonized areas have considerably lower genetic diversity than populations within the areas of former glacial refuge (Fasanella et al. 2013).

In general, tuco-tucos are structured following a classical metapopulation model (Hanski and Ovaskainen 2000; Hanski and Gaggiotti 2004), with local populations exchanging individuals depending on the habitat connectivity among sites. In this way, landscape characteristics and distances among populations have a significant effect in limiting the movement of individuals and structuring the genetic variation of tuco-tucos (Mora et al. 2017), perhaps to a greater extent than social factors (e.g., encounter of couples during reproduction, maintenance of home-range, etc.). In such a context, several species have shown that some minor landscape features such as changes in soil texture and porosity, watercourses, roads, afforestations, and some other barriers appear to play a significant role in limiting the dispersion and gene flow (Wlasiuk et al. 2003; Mapelli et al. 2012b; Mora et al. 2010, 2017; Esperandio et al. 2019). In these studies, the spatial distribution of genetic clusters

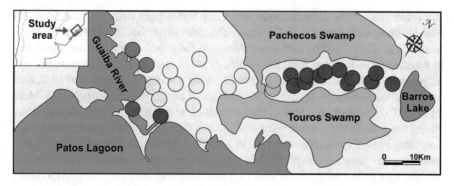

Fig. 5.3 Population genetic structure in *Ctenomys lami* along its complete distributional range (modified from Lopes and Freitas 2012). Circles on the map correspond to sampling sites and their colors denote membership to the one of the main genetic clusters identified using microsatellite markers. The figure shows how the spatial distribution of genetic clusters was shaped by major habitat discontinuities. Connections between Pachecos and Taurus Swamps divide the main genetic clusters; there are other four genetic clusters confirmed by only one or two neighboring localities. These small populations are in peripheral areas of the species' distributional range and have less connectivity, thus subject to greater effects of genetic drift. Consequently, they diverged faster than populations in the central areas of the species distribution

appears to be limited by the availably of the suitable habitat for the species. Here, the main landscape discontinuities generally coincide with the boundaries between populations (see Fig. 5.3; Lopes and Freitas 2012, Lopes et al. 2013). Although sample locations distanced by 20–40 km can constitute a single genetic population in areas of greater habitat connectivity, in those areas with greater fragmentation the population genetic structure usually is given at smaller spatial scales (see Mora et al. 2010).

5.6 Landscape Genetics: Which Factors Structure the Genetic Variation?

Landscape genetics is a novel discipline focuses on how geographical and environmental features structure the genetic variation on populations (Manel et al. 2003, 2005). Although still incipient, it has allowed to analyze jointly population genetic and geospatial data derived from satellite images and several types of thematic maps (altitude, vegetation, degree of anthropic impact, etc.) in order to understand how the landscape configuration affect the connectivity among populations (Storfer et al. 2007; Waits et al. 2016; Bradburd et al. 2018).

Even though there are few approaches that have used tools derived from landscape genetics in tuco-tucos, these studies show some common factors as responsible in structuring populations at the genetic level. As mentioned above, these studies revealed that populations located in areas with greater habitat fragmentation are more genetically differentiated, indicating that habitat discontinuities strongly restrict gene flow patterns in *Ctenomys* (Mapelli et al. 2012b; Mora et al. 2017; Austrich et al. 2020). The landscape effects on the population structure at fine spatial scale can be inferred from the studies of Mora et al. (2010) in *C. australis* and Esperandio et al. (2019) in *C. minutus*. Studying the dispersion patterns in *C. australis* and considering a sampling design at small geographic scales (< 4 km), Mora et al. (2010) showed how habitat discontinuities have had more relevance in structuring the genetic variation at microsatellite loci than geographical distances among sampling sites. Esperandio et al. (2019) evaluated the degree of genetic differentiation between pairs of sampling sites distanced to approximately 1 km, one pair divided by a road, and the other with no potential barrier between them. In this study is demonstrated, although incipient, an effect of the road restricting the gene flow between sampling units.

Kittlein and Gaggiotti (2008) using a landscape genetic approach and combining geospatial and genetic data (information of microsatellite loci extracted from Wlasiuk et al. 2003) assessed the potential causes that have promoted the genetic differentiation among populations in *C. rionegrensis*. These authors showed how sampling locations at lower altitudes exhibited signs of having experienced strong genetic drift. Thus, elevation of sampling locations strongly conditions the genetic differentiation, leading to a faster divergence of the populations located at lower sites. Despite populations of *C. rionegrensis* are distributed in rather flat areas, lower areas could be relatively more affected by flood from nearby rivers than populations at relatively higher elevations. Under this scenario, floods would reduce habitat availability and, consequently, population sizes in areas of low altitude. Thus, populations located in areas of low altitude will be subject to greater effects of genetic drift and, therefore, would diverge more quickly (Kittlein and Gaggiotti 2008; see also D'Anatro et al. 2011).

Later, Mapelli et al. (2012b, in *C. porteousi*) and Mora et al. (2017, in *C. chasiquensis*) derived landscape variables using satellite images, elevation, and soil maps, and analyzed the effect of these features on the population genetic structure. Among several environmental variables analyzed in *C. porteousi*, the main landscape factor that has affected the population differentiation was the geographic location that the local populations occupy relative to the most suitable habitats for this species. Populations located in peripheral areas, where the suitable habitat of tuco-tucos is distributed more fragmentary, showed greater genetic differentiation, denoting strong effects of gene drift in the recent past (Mapelli et al. 2012b). The results obtained in *C. chasiquensis* show something similar to *C. porteousi* (Mora et al. 2017). This species is distributed in a series of paleo-valleys that were filled by wind sediments (Mora et al. 2016); so their habitat is then basically linear, thinning toward the East, as the paleo-valleys enter the Pampas Grassland. As a result, populations located further East showed a greater degree of genetic differentiation. Thus, again, it is observed that the degree to which local populations are structured is a consequence of the position that they occupy in relation to the distribution of the most suitable habitat for the species. Consequently, it is expected that populations occupying areas of lower environmental availability have suffered stronger effects of gene drift and have diverged more quickly. Interestingly, the degree of vegetation cover was a strong determinant of genetic differentiation among populations of *C. chasiquensis*. These tuco-tucos are notoriously affected by vegetation cover, showing very little tolerance to soils with a high percentage of vegetation. Consequently, Mora et al. (2017) showed that genetic differentiation in *C. chasiquensis* was more pronounced in those areas where the vegetation cover was higher. This result highlights the importance of habitat quality, and not just its quantity, on the pattern of population structure in *Ctenomys*. Thus, it is expected that areas of lower habitat quality will sustain population of smaller size, increasing the rate of genetic drift and favoring a faster differentiation.

At a larger spatial scale, Gómez Fernández et al. (2016) showed in the "Corrientes species complex" of *Ctenomys* how the distribution of genetic variation among lineages of this group was primarily shaped by the habitat distribution. Populations of these tuco-tucos are distributed on sandy soils where its permanence depends of the historical variations of the course of Parana River and the subsequent environmental evolution of the region. Moreover to the distance among localities, the main determinant of population differentiation was also related to the occurrence of the most suitable habitats of these species. Thus, sites with greater habitat availability support populations of larger sizes which retain greater genetic variability (see also Mirol et al. 2010).

Although the species studied by Mapelli et al. (2012b), Gómez Fernández et al. (2016) and Mora et al. (2017) inhabit an anthropically fragmented landscape and, their studies included landscape metrics that describe this fragmentation, this process does not seem to have much effect on the pattern of population-genetic structure in *Ctenomys*. All these studies covered the entire distributional range of the species under study, and therefore they only captured the most important factors that have somehow affected their population genetic structure. At these spatial scales (> 15 km), the most suitable habitat for the occupation of the species seems to be the main factor limiting the population genetic structure. At lower spatial scales, however, the effects of anthropic activities begin to become more evident. Mapelli et al. (2020) studied how different landscape elements have affected the population genetic structure in an area of approximately 150 km² occupied by populations belonging to the "Corrientes" group. Using the observed genetic data, these authors applied a novel tool and assessed the cost that different landscape elements offer to gene flow. The habitat occupied by these species sits within a landscape matrix dominated by water bodies and flood-prone habitats. Consequently, the altitude was the main landscape factor that conditioned the genetic differentiation among localities, with low and flooded areas presenting very high friction values to the movement of the tuco-tucos. It should be noted, however, the second cause of differentiation between sampling locations was related to the anthropic impacts in the area. Only the forested areas (mainly with exotic tree species such as *Pinus* and *Eucalyptus* destined to commercial forestry) older than 20 years have strongly impacted the population genetic structure of tuco-tucos, limiting the gene flow between among localities. This is the first study that documents the negative effects of anthropic habitat fragmentation on the genetic connectivity among populations of tuco-tucos. Due to the negative effects of the anthropic activities on the tuco-tuco populations reported in the last decades, understanding how the landscape and environmental features have molded the genetic variability in natural populations will be very significant for the development of proper management and conservation plans.

5.7 What About Conservation Genetics?

The inferences of dispersal patterns and population genetic structure provides essential knowledge for the conservation of threatened species (Hanski and Gaggiotti 2004; Manel et al. 2005). Consequently, a crucial objective of conservation genetics is not only to recognize the overall genetic structure, diversity, and connectivity between populations, but also to understand the factors that shape such patterns (Apodaca et al. 2012). In this form, assessing the magnitude of these threatening processes is critical in conservation management and has become a priority in studies of conservation biology (Ciofi et al. 1999).

As was suggested by Waples and Gaggiotti (2006), the management of endangered species requires the identification of units that behave independently in terms of population dynamics. One of the means to study the population dynamics of species and delimit spatially discrete units in highly fragmented landscapes is to quantify the population connectivity and gene flow based on inferences about migration rates (Hanski and Gaggiotti 2004). The movement of individuals (and their linked genes) will mostly depend on the characteristics of the environment they inhabit, since a fragmented landscape offers a greater degree of heterogeneity that can restrict the dispersion (Hanski and Gaggiotti 2004; Crooks and Sanjayan 2006). Thus, the loss or reduction of genetic variability, as a result of habitat fragmentation, leads to a decrease in the connectivity and a reduction in effective population sizes (Hanski and Gaggiotti 2004), increasing the effect of genetic drift and levels of inbreeding and decreasing the evolutionary potential of the species (Crispo et al. 2011).

Although some species of *Ctenomys* have wide distributions over distinct ecoregions, many have narrow distributional ranges and occupy severely fragmented environments, either by natural or anthropogenic causes (Mapelli and Kittlein 2009; Freitas 2016). One interesting example is the recent described species *C. ibicuensis* from sandy soils on the western slopes of the state of Rio Grande do Sul in southern Brazil (Freitas et al. 2012). Populations of this species have a narrow geographic distribution in a small area (approximately 500 km²) that has been suffering from anthropogenic pressure from soybean, pine, and eucalyptus plantations, as well as desertification. All of these situations have led the researchers to consider *C. ibicuiensis* as endangered. Other similar cases are *C. australis* and *C. porteousi*, which currently have minor ranges than 500 km² (Mora et al. 2006; Mapelli and Kittlein 2009). *C. australis* occupy a restricted geographic range immerse in a highly fragmented sand dune landscape in the Southeast of Buenos Aires province, Argentina (Mora et al. 2006; Cutrera and Mora 2017). The habitat of this species is being continuously fragmented due to the development of coastal towns and cities oriented to the exploitation of beach tourism, which generates strong pressure on their populations (Mora et al. 2010). *C. porteousi* is an endangered rodent with a very narrow distributional range in central Argentina. Their suitable habitat was estimated in only 509 km^2, which represents less than 10% of its geographical range (Mapelli and Kittlein 2009, Mapelli et al. 2012b). The habitat of this species is naturally fragmented, but in recent years the degree of fragmentation has notably increased because of the extraordinary expansion of soybean cultivation in the region, a very similar situation currently observed in many species of *Ctenomys*.

The continuous exploitation of natural habitats for agriculture, livestock, forestry, and urbanization has a strong negative impact on the natural habitats of many species of tuco-tucos. Anthropic fragmentation of their natural habitats placed most *Ctenomys* species in a highly vulnerable situation. As the recent recategorization of mammals species from Argentina shows, more than 50% of the total species of the genus Ctenomys are classified in some risk category (Red list of the mammals of Argentina; http://cma.sarem.org.ar/).

As was mentioned previously, several population genetic studies showed that peripheral and isolated populations have low genetic diversity and low effective population sizes (Lopes and Freitas 2012; Mapelli et al. 2012b; Gómez Fernández et al. 2012; Lopes et al. 2013; Mora et al. 2017; Austrich et al. 2020). Furthermore, Mora et al. (2010), Esperandio et al. (2019), and Mapelli et al. (2020) showed that small discontinuities in the suitable habitat can establish a strong barrier to the movement of tuco-tucos, and can increase isolation among populations. Thus, these isolated populations would be more vulnerable to the effects of demographic and environmental stochasticity and, consequently, more likely to experience local extinction processes. A significant conclusion of these studies is that habitat fragmentation strongly conditions gene flow patterns and population structure in *Ctenomys*, even at small spatial scales.

On the other hand, phylogeographic studies have showed how the history and evolution of the landscapes have intensely conditioned the mode in which genetic variation has been geographically partitioned (Mora et al. 2006, 2013, 2016; Mapelli et al. 2012b; Roratto et al. 2014). Thus, these studies illustrate how the population demography of tuco-tucos was strongly impacted by environmental changes that have modified the extension and fragmentation of their natural habitats.

In this context, the availability of suitable habitat for tuco-tucos seems to be the main variable to consider in order to design future conservation plans. Even for widely distributed species, management plans must be strictly aimed to protect those environments having suitable habitat characteristics (e.g., sandy and well-drained soils) for the long-term permanence of tuco-tucos. Due to the high habitat specificity and low vagility that characterize these species, the main efforts should be aimed at securing the connectivity and suitability of the environments they occupy. Several species of *Ctenomys* (e.g., *C. australis*, *C. flamarioni*, and *C. minutus*) occur exclusively in coastal environments, associated with sand-dune systems originated during the Late Quaternary. Other species mainly occurring on coastal dune environments (*C. talarum* and *C. pearsoni*) from Argentina and Uruguay are also found in some fragmentary inland populations (Mora et al. 2013; Tomasco and Lessa 2007). The occupation of these coastal environments determines linear distributions, so that even punctual anthropic impacts can easily fragment the tuco-tuco populations.

Literature Cited

Apodaca JJ, Rissler LJ, Godwin JC (2012) Population structure and gene flow in a heavily disturbed habitat: implications for the management of the imperilled Red Hills salamander (*Phaeognathus hubrichti*). Conserv Genet 13:913–923

Austrich A, Mora MS, Mapelli FJ, Fameli A, Kittlein MJ (2020) Influences of landscape characteristics and historical barriers on the population genetic structure in the endangered sand-dune subterranean rodent *Ctenomys australis*. Genetica. https://doi.org/10.1007/s10709-020-00096-1

Avise JC (2000) Phylogeography: the history and formation of species. Harvard University Press, Cambridge, MA

Bidau CI (2015) Ctenomyidae. Ctenomys. In: J Patton, FU Pardiñas, G D'Elía (eds) Mammals of South America. Vol 2. Rodents. University of Chicago Press, Chicago, pp. 818–877

Bradburd GS, Coop GM, Ralph PL (2018) Inferring continuous and discrete population genetic structure across space. Genetics 210:33–52

Busch C, Antinuchi CD, del Valle JC, Kittlein MJ, Malizia AI, Vassallo AI, Zenuto RR (2000) Population ecology of subterranean rodents. In: Lacey EA, Patton JL, Cameron GN (eds) Life underground: the biology of subterranean rodents. University of Chicago Press, Chicago, pp 183–226

Caraballo DA, Rossi MS (2017) Integrative lineage delimitation in rodents of the *Ctenomys* Corrientes group. Mammalia 82:35–47

Caraballo DA, Rossi MS (2018) Spatial and temporal divergence of the torquatus species group of the subterranean rodent *Ctenomys*. Contrib Zool 87:11–24

Caraballo DA, Abruzzese GA, Rossi MS (2012) Diversity of tuco-tucos (Ctenomys, Rodentia) in the Northeastern wetlands from Argentina: mitochondrial phylogeny and chromosomal evolution. Genetica 140:125–136

Caraballo DA, Tomasco IH, Campo DH, Rossi MS (2016) Phylogenetic relationships between tuco-tucos (*Ctenomys*, Rodentia) of the Corrientes group and the *C. pearsoni* complex. Mastozool Neotrop 23:39–49

Castilho CS, Gava A, de Freitas TRO (2012) A hybrid zone of the genus *Ctenomys*: a case study in southern Brazil. Genet Mol Biol 35:990–997

Chan YL, Hadly EA (2011) Genetic variation over 10,000 years in *Ctenomys*: comparative phylochronology provides a temporal perspective on rarity, environmental change and demography. Mol Ecol 20:4592–4605

Chan YL, Lacey EA, Pearson OP, Hadly EA (2005) Ancient DNA reveals Holocene loss of genetic diversity in a south American rodent. Biol Lett 1:423–426

Ciofi C, Beaumont MA, Swingland IR, Bruford MW (1999) Genetic divergence and units for conservation in the Komodo dragon *Varanus komodoensis*. Proc R Soc B Biol Sci 266:2269–2274

Contreras JR, Reig OA (1965) Datos sobre la distribución del género *Ctenomys* (Rodentia: Octodontidae) en la zona costera de la Provincia de Buenos Aires entre Necochea y Bahia Blanca. Physis 25:169–186

Cook JA, Lessa EP (1998) Are rates of diversification in subterranean south American tuco-tucos (genus *Ctenomys*, Rodentia: Octodontidae) unusually high? Evolution 52:1521–1527

Crispo E, Moore JS, Lee-Yaw JA, Gray SM, Haller BC (2011) Broken barriers: human-induced changes to gene flow and introgression in animals. BioEssays 33:508–518

Crooks KR, Sanjayan M (2006) Connectivity conservation. Cambridge University Press, Cambridge

Cutrera AP, Mora MS (2017) Selection on MHC in a context of historical demographic change in 2 closely distributed species of Tuco-tucos (Ctenomys australis and C. talarum). J Hered 108:628–639

Cutrera AP, Lacey EA, Busch C (2005) Genetic structure in a solitary rodent (Ctenomys talarum): implications for kinship and dispersal. Mol Ecol 14:2511–2523

Cutrera AP, Lacey EA, Bush C (2006) Intraspecific variation in effective population size in tuco-tucos (Ctenomys talarum): the role of demography. J Mammal 87:108–116

Cutrera AP, Mora MS, Antenucci CD, Vassallo AI (2010) Intra and interspecific variation in home-range size in sympatric Tuco-Tucos, Ctenomys australis and C talarum. J Mammal 91(6):1425–1434

D'Anatro, A., G. Wlasiuk, and E. P. Lessa. 2011. Historia climática del Cuaternario tardío y estructura poblacional del tucu-tucu de Río Negro Ctenomys rionegrensis Langguth y Abella. Pp. 155-172 in El Holoceno en la zona costera del Uruguay (Departamento de Publicaciones, Unidad de Comunicación, Universidad de la República). Universidad de la República,Montevideo, Uruguay

D'Elía G, Lessa EP, Cook JA (1998) Geographic structure, gene flow and maintenance of melanism in Ctenomys rionegrensis (Rodentia: Octodontidae). Zeitschrift für Säugertierkunde 63:285–296

D'Elía G, Lessa E, Cook J (1999) Molecular phylogeny of Tuco-Tucos, genus Ctenomys (Rodentia: Octodontidae): evaluation of the mendocinus species group and the evolution of asymmetric sperm. J Mamm Evol 6:19–38

El Jundi TARJ, Freitas TRO (2004) Genetic and demographic structure in a population of Ctenomys lami (Rodentia-Ctenomyidae). Hereditas 140(1):18–23

Esperandio IB, Ascensão F, Kindel A, Kindel A, Tchaicka L, Freitas TRO (2019) Do roads act as a barrier to gene flow of subterranean small mammals? A case study with Ctenomys minutus. Conserv Genet 20:385–393

Fasanella M, Bruno C, Cardoso YP, Lizzarralde MS (2013) Historical demography and spatial genetic structure of the subterranean rodent Ctenomys magellanicus in Tierra del Fuego (Argentina). Zool J Linnean Soc 169:697–710

Fernandes FA, Gonçalves GL, Ximenes SSF, Freitas TRO (2009) Karyotypic and molecular polymorphisms in the Ctenomys torquatus (Rodentia: Ctenomyidae): taxonomic considerations. Genetica 136:449–459

Fernández-Stolz, G. 2006. Estudos evolutivos, filogeográficos e de conservação em uma espécie endêmica do ecossistema de dunas costeiras do sul do Brasil, Ctenomys flamarioni (Rodentia – Ctenomyidae), através de marcadores moleculares microssatélites e DNA mitocondrial. M.S. thesis, Universidade Federal do Rio Grande do Sul. Porto Alegre, Brasil

Freitas TRO (1995) Geographic distribution and conservation of four species of the genus Ctenomys in southern Brazil. Stud Neotropical Fauna Environ 30:53–59

Freitas, T. R. O. 2016. Family Ctenomyidae. In: DE Wilson, TE Jr Lacher, RA Mittermeier (eds) Handbook of the mammals of the world: lagomorphs and rodents I. Lynx editions, Barcelona, pp 499–534

Freitas TRO, Lessa EP (1984) Cytogenetics and morphology of Ctenomys torquatus (Rodentia, Octodontidae). J Mammal 65:637–642

Freitas TRO, Fernandes FA, Fornel R, Roratto PA (2012) An endemic new species of tuco-tuco, genus Ctenomys (Rodentia: Ctenomyidae), with a restricted geographic distribution in southern Brazil. J Mammal 5(19):1355–1367

Gava A, Freitas TRO (2004) Microsatellite analysis of a hybrid zone between chromosomally divergent populations of Ctenomys minutus from southeastern Brazil (Rodentia, Ctenomyidae). J Mammal 85:1201–1206

Giménez MD, Mirol P, Bidau CJ, Searle JB (2002) Molecular analysis of populations of *Ctenomys* (Caviomorpha, Rodentia) with high karyotypic variability. Cytogenet Genome Res 96:130–136

Gómez Fernández MJ, Gaggiotti OE, Mirol P (2012) The evolution of a highly speciose group in a changing environment: are we witnessing speciation in the Iberá wetlands? Mol Ecol 21:3266–3282

Gómez Fernández MJ, Boston ES, Gaggiotti OE, Kittlein M-J, Mirol PM (2016) Influence of environmental heterogeneity on the distribution and persistence of a subterranean rodent in a highly unstable landscape. Genetica 144(6):711–722

Gonçalves GL, Freitas TRO (2009) Intraspecific variation and genetic differentiation of the collared tuco-tuco (*Ctenomys torquatus*) in southern Brazil. J Mammal 90(4):1020–1031

Gonçalves GL, Hoekstra HE, Freitas TRO (2012) Striking coat colour variation in tuco-tucos (Rodentia: Ctenomyidae): a role for the melanocortin-1 receptor? Biol J Linn Soc 105:665–680

Gutiérrez-Tapia P, Palma RE (2016) Integrating phylogeography and species distribution models: cryptic distributional responses to past climate change in an endemic rodent from the Central Chile hotspot. Divers Distrib 22:638–650

Hanski, I, and O. E. Gaggiotti. 2004. Ecology, genetics, and evolution of metapopulations. Elsevier Academic Press. San Diego

Hanski I, Ovaskainen O (2000) The metapopulation capacity of a fragmented landscape. Nature 404:755–758

Hare MP (2001) Prospects for nuclear gene phylogeography. Trends Ecol Evol 16(12):700–706

Helyar SJ, Hemmer-Hansen J, Bekkevold D, Taylor MI, Ogden R, Limborg MT, Cariani A, Maes GE, Diopere E, Carvalho GR, Nielsen EE (2011) Application of SNPs for population genetics of nonmodel organisms: new opportunities and challenges. Mol Ecol Resour 11:123–136

Iriondo M, Kröhling (1995) El sistema eólico pampeano. Com. Mus. Cs. Naturales Florentino Ameghino 5:1–79

Isla FI (1998) Holocene coastal evolution of Buenos Aires. Quaternary of South America and Antarctic peninsula, A. a. Balkema 11:297–321

Isla FI, Cortizo LC, Turno OH (2001) Dinámica y Evolución de las Barreras Medanosas, Provincia de Buenos Aires, Argentina. Revista Brasileira de Geomorfol 2:73–83

Kittlein MJ, Gaggiotti OE (2008) Interactions between environmental factors can hide isolation by distance patterns: a case study of *Ctenomys rionegrensis* in Uruguay. Proc R Soc B Biol Sci 275:26–33

Kittlein MJ, Vassallo AI, Busch C (2001) Differential predation upon sex and age classes of tuco-tucos (*Ctenomys talarum*, Rodentia: Octodontidae) by owls. Mamm Biol 66:281–289

Kubiak BB, Kretschmer R, Leipnitz LT et al (2020) Hybridization between subterranean tuco-tucos (Rodentia, Ctenomyidae) with contrasting phylogenetic positions. Sci Rep 10:1502

Lacey, E. A. 2000. Spatial and social systems of subterranean rodents. Pp. 257–293 in life underground: the biology of subterranean rodents (E. A. Lacey, J. L. Patton, and G. N. Cameron, eds.). University of Chicago Press. Chicago and London

Lacey EA (2001) Microsatellite variation in solitary and social tuco-tucos: molecular properties and population dynamics. Heredity 86(5):628–637

Lacey EA, Wieczorekb JR (2004) Kinship in colonial tuco-tucos: evidence from group composition and population structure. Behav Ecol 6(15):988–996

Lacey EA, Braude SH, Wieczorek JR (1997) Burrow sharing by colonial tuco-tucos (*Ctenomys sociabilis*). J Mammal 78:556–562

Lacey EA, Braude SH, Wieczorek JR (1998) Solitary burrow use by adult Patagonian tuco-tucos (*Ctenomys haigi*). J Mammal 79:986–991

Lacey, E. A., J. E Maldonado, J. P. Clabaugh, and M. D. Matocq. 1999. Interspecific variation in microsatellites isolated from tuco-tucos (Rodentia: Ctenomyidae). Mol Ecol 8(10):1754–1756

Lacey EA, Patton JL, Cameron GN (2000) Life underground: the biology of subterranean rodents. University of Chicago Press, Chicago, Illinois

Lessa EP (2000) The evolution of subterranean rodents: a synthesis. In: Life underground: the biology of subterranean rodents (E. A. Lacey, J. L. Patton, and G. N. Cameron (ed) University of Chicago Press. Illinois, Chicago, pp 389–420

Lessa EP, Cook JA (1998) The molecular phylogenetics of tuco-tucos (genus *Ctenomys*, Rodentia: Octodontidae) suggests an early burst of speciation. Mol Phylogenet Evol 9:88–99

Lessa EP, Cook JA, Patton JL (2003) Genetic footprints of demographic expansion in North America, but not Amazonia, during the late quaternary. PNAS 100:10331–10334

Lopes, C. M. 2011. História evolutiva de *Ctenomys minutus* e *Ctenomys lami* na planície costeira do Sul do Brasil. M. S. thesis, Universidade Federal do Rio Grande do Sul. Porto Alegre, Brasil

Lopes C, Freitas TRO (2012) Human impact in naturally patched small populations: genetic structure and conservation of the burrowing rodent, Tuco-Tuco (*Ctenomys lami*). J Hered 103(5):672–681

Lopes C, Ximenes S, Gava A et al (2013) The role of chromosomal rearrangements and geographical barriers in the divergence of lineages in a south American subterranean rodent (Rodentia: Ctenomyidae: *Ctenomys minutus*). Heredity 111:293–305

Luna F, Antinuchi CD (2007) Energy and distribution in subterranean rodents: sympatry between two species of the genus *Ctenomys*. Compar Biochem Phys Part A 147:948–954

Malizia AI, Vassallo AI, Busch C (1991) Population and habitat characteristics of two sympatric species of *Ctenomys* (Rodentia: Octodontidae). Acta Theriol 36:87–94

Manel S, Schwartz MK, Luikart G, Taberlet P (2003) Landscape genetics: combining landscape ecology and population genetics. Trends Ecol Evol 18(4):189–197

Manel S, Gaggiotti OE, Waples RS (2005) Assignment methods: matching biological questions with appropriate techniques. Trends Ecol Evol 20(3):136–142

Mapelli FJ, Kittlein MJ (2009) Influence of patch and landscape characteristics on the distribution of the subterranean rodent *Ctenomys porteousi*. Landsc Ecol 24(6):726–733

Mapelli F, Mora M, Mirol P, Kittlein M (2012a) Effects of quaternary climatic changes on the phylogeography and historical demography of the subterranean rodent *Ctenomys porteousi*. J Zool 286:48–57

Mapelli FJ, Mora MS, Mirol PM, Kittlein MJ (2012b) Population structure and landscape genetics in the endangered subterranean rodent *Ctenomys porteousi*. Conserv Genet 13:165–181

Mapelli FJ, Mora MS, Lancia JP, Gómez Férnandez MJ, Mirol PM, Kittlein MJ (2017) Evolution and phylogenetic relationships in subterranean rodent of *Ctenomys mendocinus* species complex: effects of late quaternary landscape changes of Central Argentina. Mamm Biol 87:130–142

Mapelli FJ, Boston ESM, Fameli A, Gómez Fernández MJ, Kittlein MJ, Mirol P (2020) Fragmenting fragments: landscape genetics of a subterranean rodent (Mammalia, Ctenomyidae) living in a human impacted wetland. Landsc Ecol 35:1089–1106

Massarini A, Freitas T (2005) Morphological and cytogenetics comparison in species of the mendocinus group (genus *Ctenomys*) with emphasis in *C. australis* and *C. flamarioni* (Rodentia: Ctenomyidae). Caryologia 58:21–27

Mirol P, Giménez MD, Searle JB, Bidau C, Faulkes CG (2010) Population and species boundaries in the south American subterranean rodent *Ctenomys* in a dynamic environment. Biol J Linn Soc 100:368–383

Mora, M. S., and F. J. Mapelli. 2010. Conservación en médanos: Fragmentación del hábitat y dinámica poblacional del tuco–tuco de las dunas. Pp. 161–181 in Manual de manejo de barreras medanosas de la Provincia de Buenos Aires (F. I. Isla, and C. A., Lasta eds.). Universidad de Mar del Plata, Mar del Plata, Buenos Aires

Mora MS, Lessa EP, Kittlein MJ, Vassallo AI (2006) Phylogeography of the subterranean rodent *Ctenomys australis* in sand-dune habitats: evidence of population expansion. J Mammal 87:1192–1203

Mora MS, Lessa EP, Cutrera AP, Kittlein MJ, Vassallo AI (2007) Phylogeographic structure in the subterranean tuco-tuco *Ctenomys talarum* (Rodentia: Ctenomyidae): contrasting the demographic consequences of regional and habitat-specific histories. Mol Ecol 16:3453–3465

Mora M, Mapelli F, Gaggiotti O, Kittlein M, Lessa E (2010) Dispersal and population structure at different spatial scales in the subterranean rodent *Ctenomys australis*. BMC Genet 11:9

Mora MS, Cutrera AP, Lessa EP, Vassallo AI, D'Anatro A, Mapelli FJ (2013) Phylogeography and population genetic structure of the Talas tuco-tuco (*Ctenomys talarum*): integrating demographic and habitat histories. J Mammal 94(2):459–476

Mora MS, Mapelli FJ, López A, Gómez Fernández MJ, Mirol PM, Kittlein MJ (2016) Population genetic structure and historical dispersal patterns in the subterranean rodent *Ctenomys "chasiquensis"* from the southeastern pampas region, Argentina. Mamm Biol 81(3):314–325

Mora MS, Mapelli FJ, López A, Gómez Fernández MG, Mirol PM, Kittlein MJ (2017) Landscape genetics in the subterranean rodent *Ctenomys "chasiquensis"* associated with highly disturbed habitats from the southeastern pampas region, Argentina. Genetica 145:575–591

Ortells MO (1995) Phylogenetic analysis of G-banded karyotypes among the south American subterranean rodents of the genus *Ctenomys* (Caviomorpha: Octodontidae), with special reference to chromosomal evolution and speciation. Biol J Linn Soc 54:43–70

Parada A, D'Elía G, Bidau CJ, Lessa EP (2011) Species groups and the evolutionary diversification of tuco-tucos, genus *Ctenomys* (Rodentia: Ctenomyidae). J Mammal 92:671–682

Pardiñas UFJ (2013) Localidades típicas de micromamíferos en Patagonia: el viaje de J. Hatcher en las nacientes del río Chico, Santa Cruz, Argentina. Mastozool Neotrop 20:413–420

Quattrocchio, M.E., A. B. Borromei, , C. M. Deschamps,, S. C. Grill, and C. A. Zavala. 2008. Landscape evolution and climate changes in the late Pleistocene-Holocene, southern Pampa (Argentina): evidence from palynology, mammals and sedimentology. Quat Int 181:123–138

Quintana AC (2004) El Registro de *Ctenomys talarum* durante el Pleistoceno Tardío-Holoceno de las Sierras de Tandilia Oriental. J Neotrop Mammal 11(1):45–53

Rabassa JA, Coronato G, Bujalesky C, Salemme M, Roig C, Meglioli A, Heusser C, Gordillo S, Roig F, Borromei A, Quattrocchio M (2000) Quaternary of Tierra del Fuego, southernmost South America: an updated review. Quat Int 68-71:217–240

Rabassa JA, Coronato A, Martínez O (2011) Late Cenozoic glaciations in Patagonia and Tierra del Fuego: an updated review. Biol J Linn Soc 103:316–335

Reguero M, Candela A, Alonso R (2007) Biochronology and biostratigraphy of the Uquía formation (Pliocene–early Pleistocene, NW Argentina) and its significance in the great American biotic interchange. J S Am Earth Sci 23:1–16

Reig OA, Kiblisky P (1969) Chromosome multiformity in the genus *Ctenomys* (Rodentia, Octodontidae). Chromosoma 28:211–244

Reig O, Busch C, Ortells M, Contreras J (1990) An overview of evolution, systematics, population biology, cytogenetics, molecular biology and speciation in *Ctenomys*. Prog Clin Biol Res 335:71–96

Richards CL, Carstens BC, Knowles L (2007) Distribution modelling and statistical phylogeography: an integrative framework for generating and testing alternative biogeographical hypotheses. J Biogeogr 34:1833–1845

Roratto PA, Bartholomei-Santos ML, Freitas TRO (2011) Tetranucleotide microsatellite markers in *Ctenomys torquatus* (Rodentia). Conserv Genet Resour 3(4):725–727

Roratto PA, Fernandes FA, Freitas TRO (2014) Phylogeography of the subterranean rodent *Ctenomys torquatus*: an evaluation of the riverine barrier hypothesis. J Biogeogr 42:694–705

Slamovits CH, Cook JA, Lessa EP, Rossi MS (2001) Recurrent amplifications and deletions of satellite DNA accompanied chromosomal diversification in south American tuco-tucos (genus *Ctenomys*, Rodentia: Octodontidae): a phylogenetic approach. Mol Biol Evol 18:1708–1719

Slatkin M (1993) Isolation by distance in equilibrium and non-equilibrium populations. Evolution 47:264–279

Slatkin M, Barton NH (1989) Comparison of three indirect methods for estimating average levels of gene flow. Evolution 43:1349–1368

Stolz, J. F. B. 2006. Dinâmica populacional e relações espaciais do tuco-tuco-das-dunas (*Ctenomys flamarioni* – Rodentia – Ctenomyidae) Na Estação Ecológica do Taim – RS/Brasil. Ph.D. dissertation, Universidade Federal do Rio Grande do Sul. Porto Alegre, Brasil

Fernández-Stolz, G. P, J. F. B. Stolz, and T. R. O. Freitas. 2007. Bottlenecks and dispersal in the tuco-tuco das dunas, *Ctenomys flamarioni* (Rodentia: Ctenomyidae): in southern Brazil. J Mammal 88(4):935–945

Storfer A, Murphy MA, Evans JS, Goldberg CS, Robinson S, Spear SF, Dezzani R, Delmelle E, Vierling L, Waits LP (2007) Putting the 'landscape' in landscape genetics. Heredity 98:128–142

Swaegers J, Mergeay J, Therry L, Larmuseau MH, Bonte D, Stoks R (2013) Rapid range expansion increases genetic differentiation while causing limited reduction in genetic diversity in a damselfly. Heredity 111(5):422–429

Tammone MN, Pardiñas UFJ, Lacey EA (2018a) Contrasting patterns of Holocene genetic variation in two parapatric species of Ctenomys from Northern Patagonia, Argentina. Biol J Linn Soc 123:96–112

Tammone MN, Pardiñas UFJ, Lacey EA (2018b) Identifying drivers of historical genetic decline in an endemic Patagonian rodent, the colonial tuco-tuco, Ctenomys sociabilis (Rodentia: Ctenomyidae). Biol J Linn Soc 125(3):625–639

Teta P, D'Elía G (2020) Uncovering the species diversity of subterranean rodents at the end of the world: three new species of Patagonian tuco-tucos (Rodentia, Hystricomorpha, Ctenomys). PeerJ 8:1–35

Teta P, D'Elía G, Opazo JC (2020) Integrative taxonomy of the southernmost tucu-tucus in the world: differentiation of the nominal forms associated with Ctenomys magellanicus Bennett, 1836 (Rodentia, Hystricomorpha, Ctenomyidae). Mamm Biol 100:125–139

Tomasco IH, EP Lessa (2007) Phylogeography of the tuco-tuco Ctenomys pearsoni: mtDNA Variation and its Implication for Chromosomal Differentiation. pp. 859–882 In D Kelt, EP Lessa, J Salazar-Bravo, JL Patton (eds) The Quintessential Naturalist: Honoring the Life and Legacy of Oliver P. Pearson. University of California Publications in Zoology 134:1–981

Tomazelli LJ, Dillenburg SR, Villwock JA (2000) Late quaternary geological history of Rio Grande do Sul coastal plain, southern Brazil. Rev Bras Geosci 30(3):474–476

Tonni EP, Cione AL, Figini AJ (1999) Predominance of arid climates indicated by mammals in the pampas of Argentina during the late Pleistocene and Holocene. Palaeogeogr Palaeoclimatol Palaeoecol 147:257–281

Torgashevaa AA, Bashevaa EA, Gómez Fernández MJ, Mirol P, Borodin PM (2017) Chromosomes and speciation in Tuco-Tuco (Ctenomys, Hystricognathi, Rodentia). Russian J Genet Appl Res 7(4):350–357

Vassallo AI (1998) Functional morphology, comparative behaviour, and adaptation in two sympatric subterranean rodents genus Ctenomys (Caviomorpha: Octodontidae). J Zool 244:415–427

Vassallo AI, Kittlein MJ, Busch C (1994) Owl predation on two sympatric species of tuco-tucos (Rodentia: Octodontidae). J Mammal 75:725–732

Vekemans X, Hardy O (2004) New insights from fine-scale spatial genetic structure analysis in plant population. Mol Ecol 13:912–935

Verzi DH, Olivares AI, Morgan CC (2010) The oldest South American tuco-tuco (late Pliocene, northwestern Argentina) and the boundaries of the genus Ctenomys (Rodentia, Ctenomyidae). Mamm Biol 75:243–252

Verzi DH, Olivares AI, Morgan CC (2014) Phylogeny and evolutionary patterns of south American octodontoid rodents. Acta Palaeontol Pol 59(4):757–769

Waits LP, Cushman SA, Spear SF (2016) Applications of landscape genetics to connectivity research in terrestrial animals. In: Balkenhol N, Cushman SA, Storfer AT, Waits LP (eds) Landscape genetics: concepts, methods, applications, 1st edn. Wiley, Chichester, pp 199–219

Waples R, Gaggiotti OE (2006) What is a population? An empirical evaluation of some genetic methods for identifying the number of gene pools and their degree of connectivity. Mol Ecol 15:1419–1439

Wlasiuk G, Garza JC, Lessa EP (2003) Genetic and geographic differentiation in the Rio Negro tuco-tuco (Ctenomys rionegrensis): inferring the roles of migration and drift from multiple genetic markers. Evolution 57:913–926

Zenuto RR, Busch C (1995) Influence of the subterranean rodent Ctenomys australis (tuco-tuco) in a sand-dune grassland. Zeitschrift für Säugetierkunde 60:277–285

Zenuto RR, Busch C (1998) Population biology of the subterranean rodent Ctenomys australis (tuco-tuco) in a coastal dune-field in Argentina. Zeitschrift für Säugetierkunde 63:357–367

Part III
Organismal Biology

Chapter 6
Skull Shape and Size Diversification in the Genus *Ctenomys* (Rodentia: Ctenomyidae)

Rodrigo Fornel, Renan Maestri, Pedro Cordeiro-Estrela, and Thales Renato Ochotorena de Freitas

6.1   Introduction

The genus *Ctenomys* (Blainville 1826) is the sole member of the family Ctenomyidae, the most speciose genus of subterranean rodents. It has a large geographic distribution with species occurring in South America, from the Andes to the Atlantic coast, and from southern Peru to Tierra del Fuego in Argentina (Reig et al. 1990). Also known as tuco-tucos, it has been used as an example of the explosive cladogenesis that occurred during the Pleistocene and currently comprises more than 70 recognized valid species (Verzi et al. 2010a; Gardner et al. 2014; Bidau 2015; Freitas 2016; Teta and D'Elía 2020). The members of *Ctenomys* also have high karyotypic diversity, with diploid numbers varying from 2n = 10 (*C. steinbachi*) to 2n = 70 (*C. dorbignyi* and *C. pearsoni*), which may be the highest rate of chromosomal

R. Fornel (✉)
Programa de Pós-Graduação em Ecologia, Departamento de Ciências Biológicas,
Universidade Regional Integrada do Alto Uruguai e das Missões – Campus de Erechim,
Erechim, RS, Brazil

R. Maestri
Departamento de Ecologia, Universidade Federal do Rio Grande do Sul,
Porto Alegre, RS, Brazil
e-mail: renan.maestri@ufrgs.br

P. Cordeiro-Estrela
Departamento de Sistemática e Ecologia, Centro de Ciências Exatas e da Natureza – Campus
I, Universidade Federal da Paraíba, Jardim Universitário s/n, João Pessoa, PB, Brazil
e-mail: estrela@dse.ufpb.br

T. R. O. de Freitas
Department of Genetics, Federal University of Rio Grande do Sul, Porto Alegre,
Rio Grande do Su, Brazil
e-mail: thales.freitas@ufrgs.br

© Springer Nature Switzerland AG 2021
T. R. O. de. Freitas et al. (eds.), *Tuco-Tucos*,
https://doi.org/10.1007/978-3-030-61679-3_6

evolution among mammals (Reig and Kiblisky 1969; Reig et al. 1990; Bidau et al. 1996; Cook and Lessa 1998; Cook and Salazar-Bravo 2004). The high speciation rate and rapid evolution of morphological and cytogenetic characteristics in *Ctenomys* have been explained by their patchy distribution, limited vagility, territoriality, small effective numbers, and high karyotypic polymorphism, which are also probably associated with their subterranean lifestyle and environmental changes at the end of the Pliocene (Reig et al. 1990).

The species also vary greatly in size (from the small *C. pundti* with 100 g of body weight and 220 mm of total length to the large *C. conoveri* with 1100 g and 430 mm), color (from pale yellow-grayish to black), and in the angle of incisor procumbency (Reig et al. 1990; Vassallo 1998; Mora et al. 2003). Many species of *Ctenomys* have been characterized as scratch (claws) and chisel-tooth (incisors) diggers to build their tunnel systems; they are considered to be an interesting model to investigate functional morphological adaptations to the subterranean niche (Vassallo 1998; Lessa et al. 2008; Steiner-Souza et al. 2010; Morgan and Álvarez 2013). They possess reduced tails, strongly built anterior limbs, and comb-like hairy fringes along the manus, from which they gained the name *Cteno* (= comb) *mys* (= mice). They also show reduced ear pavilion, eyes and ears located on top of the head, and labial located posteriorly to the incisors, allowing chisel-tooth digging without ingesting soil. All of these characteristics show a high degree of fossoriality where the skull integrates different selective pressure to this habit.

Among several morphological traits that play important roles in a vertebrate, the skull is a very important anatomical structure because it houses the brain and sense organs, and it has regions of origin and insertion of muscles (Hanken and Thorogood 1993). Moreover, for most mammalian taxa, well-preserved crania and mandibles can be found in scientific collections. Many studies have explored the morphological skull variations in *Ctenomys*, at both intraspecific and interspecific levels (i.e., Vassallo 1998; Marinho and Freitas 2000; Mora et al. 2003; Freitas 2005; D'Anatro and Lessa 2006; Verzi and Olivares 2006; Fernandes et al. 2009; Fornel et al. 2010; Fernandes et al. 2012; Borges et al. 2017; Fornel et al. 2018; Kubiak et al. 2018). At the intraspecific level, a large amount of variation has been detected between allopatric and parapatric populations and among chromosomal groups (Fernandes et al. 2009; Fornel et al. 2010; Kubiak et al. 2018). Interestingly, this amount is often of the same order of magnitude as interspecific differences. Large-scale interspecific studies are few. Among them, in a study including 23 species of *Ctenomys*, Mora et al. (2003) showed significant morphological differences in skull size and in the angle of incisor procumbency, which is very likely an adaptation to digging. In the same fashion, a study of 24 species by Borges et al. (2017) found differences in bite force among *Ctenomys* resulting from different excavation strategies. However, a broad view of the morphological variations in the genus is still lacking (Lessa et al. 2008).

In recent decades, studies based on chromosomal polymorphism, biogeography, morphological similarity, and molecular data have contributed to the formulation of groupings of species within the genus *Ctenomys*. Six species groups have been proposed: *mendocinus*, Boliviano-Matogrossense, Boliviano-Paraguaio, Chaco, *opimus-fulvus*, and *pundti-talarum* (Massarini et al. 1991; Freitas 1994; Lessa and Cook 1998; Contreras and Bidau 1999; D'Elía et al. 1999; Mascheretti et al. 2000; Slamovits et al. 2001; Castillo et al. 2005; Tiranti et al. 2005). Nonetheless, Parada et al. (2011), based on molecular phylogenetic analysis, confirmed the six species groups as clades suggested two additional clades; thus, they renamed the eight groups with the oldest species in each group. Henceforth, we use these groups: (1) *boliviensis*, (2) *frater*, (3) *opimus*, (4) *tucumanus*, (5) *torquatus*, (6) *talarum*, (7) *mendocinus*, and (8) *magellanicus* (Parada et al. 2011). Freitas et al. (2012) described a new species *C. ibicuiensis*; based on cytochrome-*b* divergence, they added it and *C. dorbignyi* to the *torquatus* group. The geographic distribution of each *Ctenomys* clade and type localities of the species may be viewed in Fig. 6.1 based on data from Woods and Kilpatrick (2005), Freitas et al. (2012) and Gardner et al. (2014).

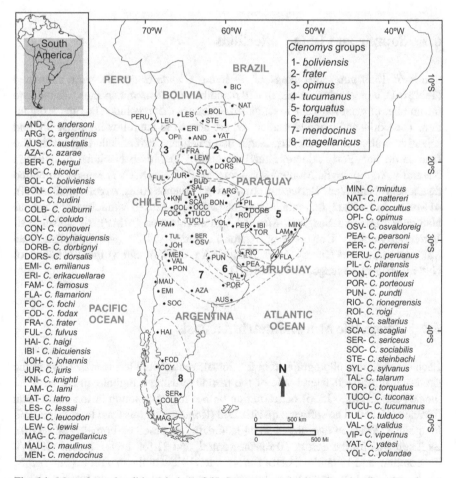

Fig. 6.1 Map of type localities (circles) of 65 *Ctenomys* species based on data from Woods and Kilpatrick (2005), Freitas et al. (2012), and Gardner et al. (2014). The meaning of abbreviations are scientific names given in the proper figure. Dashed lines and numbers indicate the eight main *Ctenomys* clades proposed by Parada et al. (2011)

Given the rapid evolution in the explosive cladogenesis of tuco-tucos in the subterranean niche (Reig et al. 1990; Castillo et al. 2005) and the lack of information about its large-scale morphological diversification, we conducted the most comprehensive survey of tuco-tucos morphological variation by including 49 species of the genus. We explored skull morphological evolution in *Ctenomys* and the morphological relationship among main clades to understand how morphological diversification occurs within subterranean radiation. The aims of this study were as follows: (1) to understand the morphological relationships among *Ctenomys* species and to test the hypothesis that the eight groups proposed for *Ctenomys* (Parada et al. 2011) based in a neutral molecular marker are supported by a quantitative analysis of the morphology of the cranium and the mandible; and (2) to test if the changes in skull shape are related to changes in size (i.e., allometry) and geographical distribution.

6.2   Sample and Data Collections

Skulls of 1359 adult specimens of 49 living *Ctenomys* species were examined (Table 6.1), along with a sample of 820 mandibles of *Ctenomys* species. We reduced the number of mandibles due to damaged structures that precluded landmark digitation. The skulls and mandibles analyzed are housed in the following museums and scientific collections: 1) Departamento de Genética, Universidade Federal do Rio Grande do Sul, Porto Alegre, Brazil (UFRGS); 2) Museo Nacional de Historia Natural y Antropología, Montevideo, Uruguay (MUNHINA); 3) Museo Argentino de Ciencias Naturales "Bernardino Rivadavia", Buenos Aires, Argentina (MACN); 4) Museo de La Plata, La Plata, Argentina (MLP); 5) Museo Municipal de Ciencias Naturales "Lorenzo Scaglia", Mar del Plata, Argentina (MMP); 6) Museum of Vertebrate Zoology, University of California, Berkeley, USA (MVZ); 7) American Museum of Natural History, New York, USA (AMNH); and 8) Field Museum of Natural History, Chicago, USA (FMNH).

6.3   Geometric Morphometric Analysis

Each cranium was photographed in the dorsal, ventral, and left lateral views of the skull and on the left lateral side of the mandible using a digital camera with 3.1 megapixels (2048 × 1536) of resolution with a macro function at the same focal distance (15 cm). The software tpsUtil 1.40 (Rohlf 2008) was used to manage skull image files. We used two-dimensional landmarks proposed by Fornel et al. (2010): 29 landmarks for the dorsal, 30 for the ventral, and 21 for the left lateral views of the cranium, and 13 landmarks for the left lateral view of the mandible (Fig. 6.2; SI 1). Anatomical landmarks were digitized for each specimen using tpsDig version

Table 6.1 Skull sample sizes of 49 species examined from the genus *Ctenomys*

Species	Clade	N	Species	Clade	N
C. argentinus	Tucumanus	3	C. magellanicus	Magellanicus	23
C. australis	Mendocinus	35	C. maulinus	–	34
C. azarae	Mendocinus	32	C. mendocinus	Mendocinus	24
C. bicolor	–	1	C. minutus	Torquatus	197
C. boliviensis	Boliviensis	59	C. nattereri	Boliviensis	1
C. bonettoi	–	2	C. occultus	Tucumanus	6
C. budini	–	1	C. opimus	Opimus	80
C. coludo	–	2	C. pearsoni	Torquatus	77
C. conoveri	Frater	4	C. perrensi	Torquatus	9
C. Coulburni	Magellanicus	30	C. peruanus	–	14
C. coyhaiquensis	Magellanicus	1	C. porteousi	Mendocinus	30
C. dorbignyi	Torquatus	13	C. pundti	Talarum	5
C. dorsalis	–	6	C. rionegrensis	Mendocinus	2
C. flamarioni	Mendocinus	34	C. roigi	Torquatus	7
C. fochi	–	3	C. scagliai	Opimus	1
C. fodax	Magellanicus	1	C. sericeus	Magellanicus	2
C. frater	Frater	11	C. sociabilis	–	15
C. fulvus	Opimus	26	C. steinbachi	Boliviensis	12
C. haigi	Magellanicus	74	C. sylvanus	–	6
C. ibicuiensis	Torquatus	16	C. talarum	Talarum	76
C. knighti	–	2	C. torquatus	Torquatus	222
C. lami	Torquatus	89	C. tuconax	–	17
C. latro	Tucumanus	8	C. tucumanus	Tucumanus	23
C. leucodon	–	8	C. yolandae	–	2
C. lewisi	Frater	13			
Total					1359

1.40 software (Rohlf 2004), and all landmarks were taken by the same person (R.F.). Coordinates were superimposed using a generalized Procrustes analysis (GPA) algorithm (Dryden and Mardia 1998). GPA removes differences unrelated to the shape: scale, position, and orientation (Rohlf and Slice 1990; Rohlf and Marcus 1993; Bookstein 1996a, b; Adams et al. 2004; Adams et al. 2013). We symmetrized both sides (left and right) of the landmarks in the dorsal and ventral views of the skull to avoid redundancy, and only the symmetric part of the variation was analyzed (Kent and Mardia 2001; Klingenberg et al. 2002; Evin et al. 2008). The size of each skull was estimated using its centroid size, the square root of the sum of the squares of the distances of each landmark from the centroid (mean of all coordinates) of the configuration (Bookstein 1991).

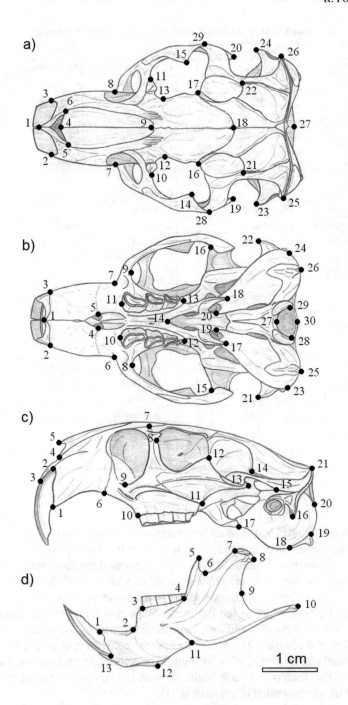

Fig. 6.2 Morphological landmark locations on skull of *Ctenomys*. a) dorsal, b) ventral, and c) lateral views of the cranium, and d) lateral view of the mandible (Fornel et al. 2010)

6.4 Shape and Size Analysis

For skull size, we summed the skull centroid size (logarithm transformed) for each view of the skull (dorsal, ventral, and lateral) to generate one value for cranium size. Centroid size variation among species was visualized through boxplots. The size difference was tested between sexes, clades, and species with a three-way ANOVA (nested: sex*clades: species). For multiple comparisons, we used Tukey's test.

Shape differences between sexes, among groups, and among species, as well as their interactions, were tested through multivariate analysis of variance (MANOVA). The Bonferroni correction for multiple comparisons was applied when needed to adjust the significance level (Wright 1992). Due to the small sample size of mandibles, we used them solely in sexual dimorphism and group comparisons, not for species analysis.

For skull shape, an exploratory analysis was carried with principal components analysis (PCA) using the variance-covariance matrix of generalized least-squares superimposition residuals. Principal components (PCs) of the covariance matrix of superimposition residuals were used as new shape variables to reduce the dimensionality of the data set as well as to work on independent variables (Baylac and Friess 2005; Evin et al. 2008). To choose the number of PCs to be included in the linear discriminant analysis (LDA), we computed correct classification percentages with each combination of PCs (Baylac and Friess 2005). We selected the subset of PCs giving the highest overall correct classification percentage. We used a leave-one-out cross-validation procedure that allowed an unbiased estimate of classification percentages (Ripley 1996; Baylac and Friess 2005). Cross-validation was used to evaluate the performance of classification by LDA. LDA explored differences in shapes between species. We performed an LDA on PCs in combined with and without the sum of the logarithms of dorsal, ventral, and lateral centroid sizes (Cordeiro-Estrela et al. 2006).

Mahalanobis distances were used to compute trees with a neighbor-joining algorithm to visualize the morphological relationships among groups and among species, with and without size information. The visualization of shape differences for the skull views were obtained through multivariate regression of shape variables on discriminant axes and the consensus configuration.

6.5 Geographical Structure of Skull Shape

We tested the association between skull morphology and geographical distribution of *Ctenomys* species following Fornel et al. (2010). We used a morphological distance matrix based on Mahalanobis distances, calculated from skull morphometric data, and a geographic distance matrix based on geodesic distances of each species calculated with the software Geographic Distance Matrix Generator, version 1.2.3 (Ersts 2009). The correlation between the two distance matrices was tested with an

RV coefficient randomization test (Heo and Gabriel 1997), using 10,000 random permutations as implemented in the *ade4* library. We also tested for the degree of association of cranial shape with spatial distribution with a two-block partial least-squares analysis (2B-PLS) between shape data and geographical coordinates of centroids of specimens for each species as implemented in the *geomorph* library. Finally, we tested for global spatial autocorrelation of centroid size and mean shape of species at different spatial scales with a Moran's I correlogram for size and by the centered Mantel static for shape (Bjørnstad et al. 1999). The number of distance classes was chosen according to Legendre and Legendre (1998).

For all statistical analyses and to generate the graphics, we used the R language and environment for statistical computing version 2.14.1 for Windows (R Development Core Team, http://www.R-project.org), as well as the following libraries: *MASS* (Venables and Ripley 2002), *ape* version 1.8–2 (Paradis et al. 2004), *stats* (R Development Core Team 2009), *ade4* (Dray and Dufour 2007), and *geomorph* (Adams and Otarola-Castillo 2013). Geometric morphometric procedures were carried out with the *Rmorph* package: a geometric and multivariate morphometrics library for R (Baylac 2008).

6.6   Variation in Cranium and Mandible Size

The centroid size of the cranium and mandible was significantly different between sexes (generally males are larger than females), among clades, and among species (Table 6.2). The interaction term between species and sex indicates that sexual dimorphism varies significantly among species of the genus *Ctenomys* (Table 6.2). The variance of the three factors tested, represented by mean squares value, shows that most of the variance in skull and mandible size is found among the eight clades, then among species and finally between sexes (Table 6.2).

Table 6.2 ANOVA of logarithms of centroid size (log CS) of the craniums (dorsal, ventral, and lateral views pooled = three views) and mandibles of *Ctenomys* between sexes and among clades, species, and the sex-versus-species interaction (n.s., not significant)

Sincranium	Mean of squares	d.f.	F	P
SKULL $_{\text{log CS 3 views}}$				
Sex	5.68	1	190.304	< 0.001
Clade	60.54	7	289.742	< 0.001
Species	32.09	25	42.999	< 0.001
Sex × species	0.73	20	1.223	n.s.
MANDIBLE $_{\text{log CS}}$				
Sex	0.55	1	110.38	< 0.001
Clade	7.29	7	206.51	< 0.001
Species	4.34	17	50.63	< 0.001
Sex × species	0.11	17	1.34	< 0.001

Among species, we found a large variation in centroid size, with five significant different groupings for Tukey's test ($P < 0.05$) (Fig. 6.3a). A partial superimposition only among few species by the size distribution is unimodal with extreme values for *C. conoveri* with a skull about 2.4 times larger than the smallest species, *C. latro* (Fig. 6.3a).

Among the eight *Ctenomys* clades, we also found five significant different groupings for skull centroid size (Fig. 6.3b). The *bolivienesis* and *frater* clades have the larger centroid size, and the *magellanicus* clade is the smaller one. There are large variations in skull centroid size within some clades, such as in *frater*, *mendocinus*,

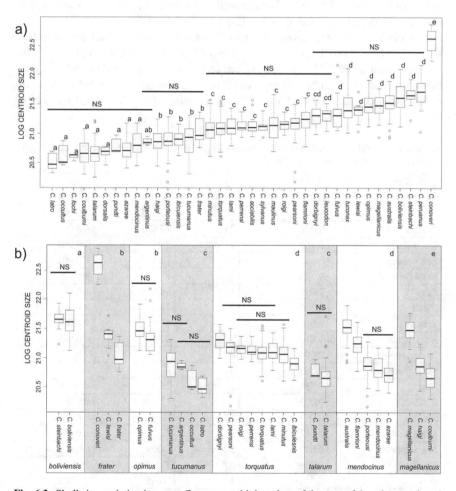

Fig. 6.3 Skull size variation in genus *Ctenomys*, with boxplots of the sum of dorsal, ventral, and lateral log centroid sizes. (**a**) Among species in increasing order, with sample size larger than two skulls for each species. (**b**) Skull size variation within and among the eight *Ctenomys* clades. Different letters above boxes represent significant differences among groups (Tukey's test). NS, nonsignificant pairwise comparison at the 5% level

and *magellanicus*. We found a smaller variation among species within *boliviensis*, *opimus*, *talarum*, *torquatus*, and *tucumanus* clades (Fig. 6.3b).

6.7 Variation in Cranium and Mandible Shape

The MANOVA showed a highly significant sexual dimorphism in shape for the dorsal, ventral, and lateral views of the cranium, as well as for the mandible (Table 6.3). The interaction between sex and clade effects was also significant for shape for all views (Table 6.3). The interaction between sex and species was significant just for dorsal and ventral views of the skull (Table 6.3).

For clades, Table 6.3 shows the significant difference in skull and mandible shape. For MANOVA comparisons between different effects, the highest F value was found for groups. Using the F values as a differentiation index, there is three times more differentiation among clades than among species or between sexes, except for the mandible (1.2 times). The pairwise MANOVA among the eight

Table 6.3 MANOVA of cranium and mandible shapes of *Ctenomys* between sexes and among clades, and species (n.s., not significant)

Skull (Dorsal)	Wilks' λ	d.f.	F	P
Sex	0.678	1	26.276	< 0.001
Clade	0.001	7	82.292	< 0.001
Species	0.007	25	12.926	< 0.001
Sex × clade	0.762	7	2.2	< 0.001
Sex × species	0.641	20	1.264	< 0.001
Skull (ventral)	Wilks' λ	d.f.	F	P
Sex	0.651	1	29.81	< 0.001
Clade	0.002	7	78.854	< 0.001
Species	0.01	25	11.854	< 0.001
Sex × clade	0.816	7	1.64	< 0.001
Sex × species	0.67	20	1.132	< 0.05
Skull (lateral)	Wilks' λ	d.f.	F	P
Sex	0.734	1	20.058	< 0.001
Clade	0.003	7	71.524	< 0.001
Species	0.009	25	12.01	< 0.001
Sex × clade	0.842	7	1.38	< 0.01
Sex × species	0.687	20	1.059	n.s.
Mandible	Wilks' λ	d.f.	F	P
Sex	0.641	1	20.65	< 0.001
Clade	0.028	7	24.911	< 0.001
Species	0.033	17	8.5	< 0.001
Sex × clade	0.706	7	1.881	< 0.001
Sex × species	0.601	17	1.13	n.s.

Ctenomys clades using three views of the skull pooled shows that all comparisons were significant (Table 6.4), as well as for the mandible (Table 6.5). Moreover, the percentage of correct classification using LDA shows higher values for the *torquatus* group for three views of the skull (100% pooled and separated) and smaller values for the *tucumanus* group for the mandible (56%) (Table 6.6). The morphological structure with the higher percentage of correct classification on average was the dorsal view of the skull (91.07%), whereas the mandible had a smaller percentage (73.86%) (Table 6.6).

The PCA did not show a clear ordination in the skull or mandible shape scores for different axes. However, in the discriminant analysis for the eight *Ctenomys* groups, we found an ordination in the two first Canonical Variate (CV) axes. For the dorsal view of the skull, CV1 shows an ordination from east to west and CV2 shows a north-to-south ordination, with *boliviensis* having positive scores and *magellanicus* having negative ones (Fig. 6.4a). These two axes explain 57.9% of the variation. In CV1, the negative scores are represented by a proportionally enlarged auditory meatus and the positive scores by a proportionally elongated rostrum (Fig. 6.4c). The negative scores on CV2 represent a proportionally shorter and narrow rostrum and the positive scores represent a proportionally enlarged zygomatic arch (Fig. 6.4b). For the ventral view of the skull, CV1 showed a small ordination west to east, and CV2 showed a small ordination north to south (Fig. 6.4e). These two axes explain 54.5% of the variation. In CV1, the positive scores are represented by a proportionally enlarged tympanic bullae (Fig. 6.4g). CV2 showed negative scores with a narrower skull compared with positive scores (Fig. 6.4f). For the lateral view of the skull, the CV1 showed east-to-west ordination, and CV2 showed north-to-south ordination (Fig. 6.5a). These two axes explain 59.4% of the variation. In CV1, the negative scores are represented by a proportionally flat skull and enlarged tympanic bullae, whereas the positive scores show a deep rostrum (Fig. 6.5c). CV2 showed negative scores with a proportionally enlarged tympanic bulla, whereas the positive scores showed a deep and elongated rostrum and enlarged zygomatic arch (Fig. 6.5b). For the mandible lateral view, the two first CVs explain 50.7% of the variation, and there is no clear ordination in these data (Fig. 6.5e). CV1 showed negative scores for proportionally shorter coronoid processes and elongated

Table 6.4 Pairwise MANOVA of cranium shape. Summary F values and significance among eight *Ctenomys* clades (results for pooled dorsal, ventral, and lateral datasets)

	boliviensis	*frater*	*opimus*	*tucumanus*	*torquats*	*talarum*	*mendocinus*
Frater	28.71*	–					
Opimus	124.39*	45.34*	–				
Tucumanus	77.36*	24.09*	38.09*	–			
Torquatus	277.92*	105.82*	195.99*	84.63*	–		
Talarum	113.33*	30.39*	25.79*	23.01*	68.45*	–	
Mendocinus	149.33*	73.51*	35.79*	46.93*	214.37*	13.77*	–
Magellanicus	247.59*	87.81*	77.61*	5.46*	212.92*	28.98*	33.87*

*$P < 0.001$; after Bonferroni correction

Table 6.5 Pairwise MANOVA of mandible shape. Summary of F values and significance among eight *Ctenomys* clades

	boliviensis	frater	opimus	tucumanus	torquats	talarum	mendocinus
Frater	12.13*	–					
Opimus	13.88*	18.04*	–				
Tucumanus	9.22*	9.13*	19.43*	–			
Torquatus	18.74*	26.02*	57.36*	10.26*	–		
Talarum	18.25*	12.54*	26.31*	12.08*	29.63*	–	
Mendocinus	15.06*	11.18*	19.35*	17.23*	61.42*	8.87*	–
Magellanicus	12.79*	9.77*	12.11*	15.31*	33.81*	16.24*	19.29*

$^{*}P < 0.001$; after Bonferroni correction

Table 6.6 Percentage of correct classification from linear discriminant analysis (LDA) of shape of dorsal, ventral, and lateral views of the cranium of the three pooled datasets and for the lateral view of the mandible for *Ctenomys* clades

Clade	Skull$_{Dorsal}$	Skull$_{Ventral}$	Skull$_{lateral}$	Skull$_{3 views}$	Mandible
1- *boliviensis*	94.44	97.22	100	97.22	66.66
2- *frater*	92.86	92.85	82.14	89.28	69.23
3- *opimus*	86.92	85.98	90.65	87.85	83.33
4- *tucumanus*	95	90	95	92.5	56
5- *torquatus*	100	100	100	100	93.29
6- *talarum*	74.63	77.61	77.61	76.11	70.21
7- *mendocinus*	87.84	83.11	83.78	86.48	82.05
8- *magellanicus*	96.87	96.09	85.15	96.09	70.12
Average	91.07	90.35	89.29	90.69	73.86

condylar processes relative to the positive scores (Fig. 6.5g). For CV2, the main difference is a proportionally deep mandibular body negative score (Fig. 6.5f). The phenogram for Mahalanobis distances for clades shows a geographic structure for the skull (Figs. 6.4d, h, and 6.5d) but not for the mandible (Fig. 6.5h). For the skull, there is a clear pattern with *frater* and *boliviensis* clades in one side of the tree (north of the geographic distribution) and *mendocinus* and *magellanicus* in the other side of the tree (south of the geographic distribution) (Figs. 6.4d, h and 6.5d).

As previously mentioned, we found significant differences in shape among all clades for all morphological structures (Tables 6.4 and 6.5). Figure 6.6 presents the consensus shape of each clade for each view of the cranium and also for the lateral view of the mandible. To facilitate the interpretation, the clades are arranged from the north (*boliviensis*) to the south (*magellanicus*) of the distribution, and the arrows indicate the main shape differences. Basically, for skull shape, the northern shapes are more robust, with larger rostrum and larger zygomatic arches, whereas the southern shapes are more gracile, with delicate rostrum and jugal with thinner bones than northern groups (Fig. 6.6). For the mandible, despite significant differences among all clades, visualizing these differences is less obvious than for the skull shape (Fig. 6.6).

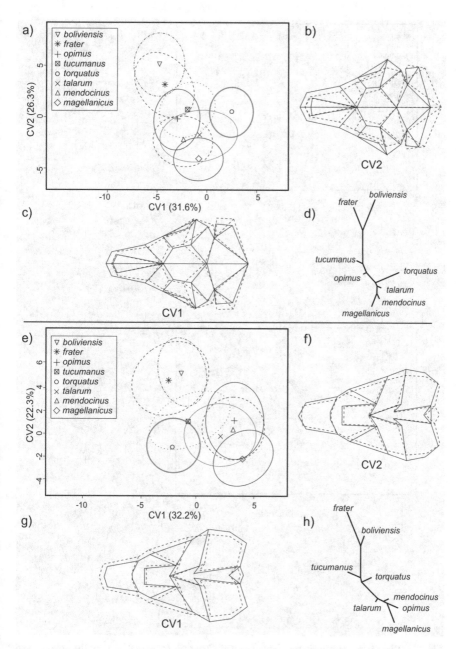

Fig. 6.4 Discriminant analysis for eight clades of *Ctenomys* for dorsal and ventral views of the cranium. (**a**) Scatter plot of canonical variate axes (CV1 and CV2) for dorsal view of the skull. (**b**) Skull shape variation for second canonical variate axis (CV2). (**c**) Skull shape variation for first canonical variate axis (CV1). (**d**) Phenogram of Mahalanobis distance for dorsal view of the skull. (**e**) Scatter plot of canonical variate axes (CV1 and CV2) for ventral view of the skull with cranium. (**f**) Skull shape variation for second canonical axis (CV2). (**g**) Skull shape variation for first canonical axis (CV1). (**h**) Phenogram of Mahalanobis distance for the ventral view of the skull. Symbols represent group mean and ellipses represent 95% confidence interval. The percentage of variance explained by each axis is given in parenthesis. In grids of shape variation for each canonical axis, solid lines indicate positive scores and dashed lines indicate negative scores

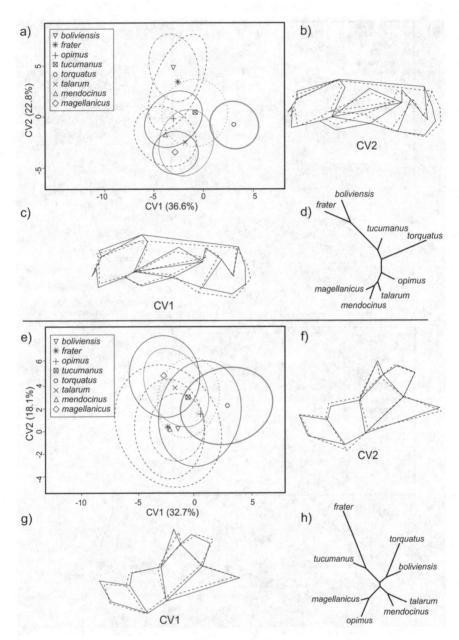

Fig. 6.5 Discriminant analysis for eight clades of *Ctenomys* for lateral view of the cranium and lateral view of the mandible. (**a**) Scatter plot of the canonical variate axes (CV1 and CV2) for lateral view of the skull. (**b**) Skull shape variation for second canonical axis (CV2). (**c**) Skull shape variation for first canonical axis (CV1). (**d**) Phenogram of Mahalanobis distance for lateral view of the skull. (**e**) Scatter plot of the canonical variate axes (CV1 and CV2) for the mandible. (**f**) Mandible shape variation for second canonical axis (CV2). (**g**) Mandible shape variation for first canonical axis (CV1). (**h**) Phenogram of Mahalanobis distance for the lateral view of the mandible. Symbols represent mean plot and ellipses represent 95% confidence interval. The percentage of variance explained by each axis is given in parenthesis. In grids of shape variation for each canonical axis, solid lines indicate positive scores and dashed lines the negative ones

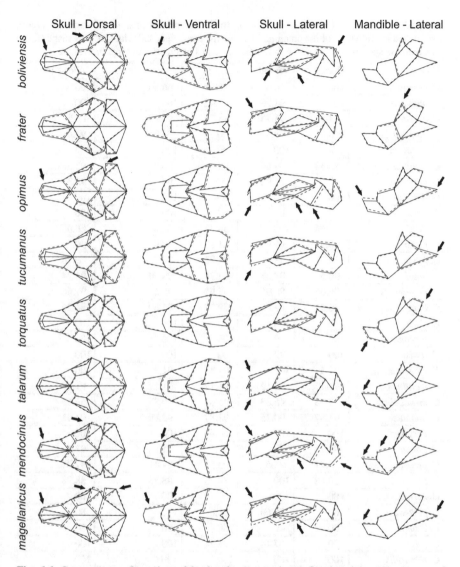

Fig. 6.6 Consensus configuration of landmarks (mean shape) for the eight main groups of *Ctenomys* for dorsal, ventral, and lateral views of the cranium, and lateral view of the mandible. Consensus configuration for all groups are given in dashed lines. The arrows indicate the main differences among groups. See the text for more detailed description

Species-level comparisons indicated highly significant differences among *Ctenomys* species for skull and mandible shapes (Table 6.3). The percentage of correct classifications by the LDA function for each species with skull sample size equal to or superior to three individuals (total of 37 species) is given in Table 6.7. The majority of species (33 of 37 species) were correctly classified, with percentages higher than 75% for shape for three views together. Nineteen species were

Table 6.7 Percentage of correct classification from discriminant analysis of shape for dorsal, ventral, and lateral views of the cranium, individually, combined, and with size plus shape (form) for *Ctenomys* species. All species with sample size equal or superior to three individuals

Species	Dorsal	Ventral	Lateral	3-views shape	3-views form
C. argentinus	100	66.66	66.66	66.66	100
C. australis	100	94.28	91.42	100	100
C. azarae	84.37	93.75	93.75	90.62	96.87
C. boliviensis	96.61	98.3	100	96.61	96.61
C. conoveri	100	100	100	100	100
C. Coulburni	93.33	86.66	80	93.33	93.33
C. dorbignyi	100	76.92	100	100	92.3
C. dorsalis	100	100	100	100	100
C. flamarioni	100	100	100	100	100
C. fochi	66.66	100	33.33	66.66	66.66
C. frater	90.9	90.9	81.81	90.9	90.9
C. fulvus	79.92	38.46	76.92	80.76	80.76
C. haigi	90.54	87.83	83.78	90.54	95.94
C. ibicuiensis	96.66	86.95	95.65	95.65	97.66
C. lami	86.51	82.02	79.77	85.39	84.26
C. latro	100	100	75	100	100
C. leucodon	100	100	100	100	100
C. lewisi	100	92.3	100	100	100
C. magellanicus	91.3	100	91.3	91.3	100
C. maulinus	91.17	94.12	91.17	91.17	91.17
C. mendocinus	70.83	45.83	45.83	70.83	66.66
C. minutus	91.37	90.35	91.87	92.89	94.41
C. occultus	100	100	66.66	100	100
C. opimus	93.75	95	97.5	95	97.5
C. pearsoni	84.41	93.5	93.5	84.41	87.01
C. perrensi	88.88	100	88.88	88.88	88.88
C. peruanus	100	100	100	100	100
C. porteousi	60	76.66	66.66	70	80
C. Pundit	80	40	80	80	80
C. roigi	100	57.14	85.71	100	100
C. sociabilis	100	100	100	100	100
C. steinbachi	100	100	100	100	100
C. sylvanus	100	83.33	100	100	100
C. talarum	86.84	82.89	82.89	86.84	86.84
C. torquatus	96.84	94.59	94.59	97.29	96.84
C. tuconax	94.12	100	88.23	100	100
C. tucumanus	91.3	78.26	91.3	91.3	91.3
Average	91.85	87.21	86.62	91.7	93.28

correctly classified in at least one view of the skull with 100%. Of these 19 species, 7 reached 100% of classification for all views of the skull. The worst classification was for *C. Fochi,* with 33.33% for the lateral view of the skull. The size contribution to form (shape plus size) increases correct classification on average, with 1.58% in reclassification (Table 6.7).

The skull shape phenogram for 49 *Ctenomys* species shows two phonetic groups highly congruent with the phylogenetic topology (the *torquatus* and *boliviensis* groups) and one group partially congruent (the *tucumanus* group; Fig. 6.7a). For the tree using shape plus size (form), the groups of *torquatus, boliviensis,* and *tucumanus* were congruent with the phylogenetic hypothesis; *opimus, mendocinus,* and *magellanicus* groups were partially congruent (Fig. 6.7b). In both trees, the largest branches were found for *C. conoveri,* which appear closer to *C. peruanus.* Moreover, in the two phenograms, *C. sociabilis* is closer to the *mendocinus* group species (*C. australis* and *C. flamarioni*) (Fig. 6.7).

6.8   Geographical Structure of Cranial Shape

For the association between variations in skull morphology and geographic distances among species populations, the RV test showed significant correlation for the dorsal (r = 0.45, $P < 0.001$), ventral (r = 0.37, $P < 0.001$), and lateral views of the skull (r = 0.25, $P < 0.05$). Two block partial least squares also showed that shape covaries significantly with latitude and longitude for dorsal (r = 0.82, $P = 0.001$), ventral (r = 0.67, $P = 0.002$), and lateral views of the skull (r = 0.695, $P < 0.001$). Autocorrelograms showed no spatial autocorrelation for size. Shape showed significant spatial autocorrelation at different spatial scales, especially at a 245-km interval (dorsal $P = 0.012$ at a 245-km interval, $P = 0.02$ at a 1073-km interval; ventral $P = 0.014$ at a 245-km interval, $P = 0.009$ at a 1528-km interval and $P = 0.02$ at a 3835-km interval; lateral $P = 0.037$ at a 245-km interval).

6.9   Morphological Variation in *Ctenomys*: An Overview

The morphological variation of skull shape in *Ctenomys* is spatially structured, which suggests homogeneity of the subterranean niche across geographical space. Especially at a 245-km scale, the strong spatially structured populations (migration/ mating patterns) are mainly responsible for the cranial evolution in *Ctenomys.* The mechanisms of geographic isolation coupled with the spatial gradient of variation in the subterranean niche are likely to be the more important selective agents in shaping the ctenomyid cranium.

We found a high variation in size within *Ctenomys* groups, except the *torquatus* group (Fig. 6.3a and b) as observed in Fornel et al. (2018). Apparently, size shows a similar range of variation within groups but is constrained to intermediate values in

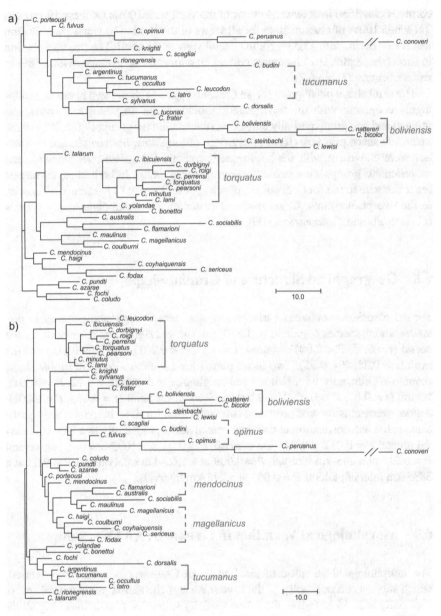

Fig. 6.7 Neighbor-joining trees of Mahalanobis distances among 49 *Ctenomys* species for dorsal, ventral, and lateral views of the skull pooled. (**a**) Phenogram for *Ctenomys* skull shape. (**b**) Phenogram for skull form (shape plus size). The solid keys indicate monophyletic groups congruent with phylogenetic hypothesis, and dashed keys indicate partially congruent groups with phylogenetic hypothesis

the *torquatus* group. *C. conoveri* is the single species with extreme values that fall outside the range of other species. This variation within clades in skull size could result from adaptive radiation with size like a line of least evolutionary resistance (Marroig and Cheverud 2005). Medina et al. (2007) found that body size follows the converse Bergmann's rule at the interspecific level in *Ctenomys*. These authors suggested that this pattern could be related to seasonality, ambient energy, primary productivity, or predation pressure. Our data show significant differences in skull size; apparently, this variation does not follow a clear latitudinal pattern. In general, cranium size by clades was ordered from north to south seem to decrease in size (Fig. 6.3b).

The Mahalanobis trees from the different views, except of the mandible, show that morphology follows a geographic structure. The *boliviensis* and *frater* groups at one extreme and the *magellanicus* group are on a north-south gradient (Figs. 6.4 and 6.5). These results are congruent with an isolation-by-distance pattern proposed for *Ctenomys* at the intraspecific level with molecular data (Mora et al. 2006, 2007). Moreover, our results show a large difference between northern and southern and between eastern and western species of *Ctenomys*. We found a gradient from north to south with more robust species in the north and more delicate species in the south—a skull morphological cline for shape but not completely for size. This pattern of isolation-by-distance in skull morphometric data was also was observed at the intraspecific level in *C. minutus* (Fornel et al. 2010). The *boliviensis* and *frater* groups in the north of *Ctenomys* distribution have a skull with strong and enlarged zygomatic arches with long processes and a deep and enlarged rostrum (the "robust" shape). In contrast, the *magellanicus* group have a more delicate skull with a thin zygomatic arch and a narrow rostrum (the "gracile" shape) (Fig. 6.6).

The phenogram for cranium shape among *Ctenomys* groups (Figs. 6.4 and 6.5) showed a more geographic signal than mandible shape. The skull is more complex than mandible in genetic and anatomical complexity, as well as a number of developmental modules (Atchley and Hall 1991; Caumul and Polly 2005). Therefore, we suggest that mandible shape in *Ctenomys* is a more functionally and developmentally constrained structure than the skull.

6.10 Congruence Between Morphological and Molecular Data

One question of this study was whether morphological phenograms are congruent with phylogenetic trees from molecular data. The answer is partly that the skull morphology trees are not completely congruent with molecular phylogeny (Fig. 6.7a, b). The only completely congruent groups were *torquatus, boliviensis,* and *tucumanus*. Moreover, our results showed that *C. dorbignyi* is associated with the *torquatus* group, as also found by Mascheretti et al. (2000) and Freitas et al. (2012), who both used molecular data. Therefore, we propose the inclusion of

. *C. dorbignyi* in the *torquatus* group as well as *C. ibicuiensis*, as suggested by Freitas et al. (2012), based on molecular and skull morphological similarities.

The *tucumanus* clade (*C. argentinus, C. latro, C. occultus,* and *C. tucumanus*) is partially congruent for shape data (Fig. 6.7a) and congruent for form data (Fig. 6.7b). The *magellanicus* group is only partially congruent because *C. magellanicus* is not associated with this group's other species. We propose that *C. coulburni* be integrated into the *magellanicus* group because the skull form and geographical distribution are very similar to the species from this clade. The *opimus* and *talarum* groups are not congruent. The *mendocinus* clade is not congruent, but there is a close association between *C. flamarioni* and *C. australis* (Fig. 6.7b). Moreover, the strong relationship between *C. flamarioni* and *C. australis* was already described by Massarini and Freitas (2005).

The most singular *Ctenomys* skull belongs to *C. conoveri*. Besides the strong morphological difference observed in the phenograms, *C. conoveri* is associated with *C. peruanus* (Fig. 6.7a and b). This high morphological divergence in *C. conoveri* was noted by Osgood (1946), who proposed a new subgenus *Chacomys* to accommodate only *C. conoveri* species (Lessa and Cook 1998). The subgenus *Haptomys* was proposed by Thomas (1916) to comprise only *C. leucodon*, which differs from other tuco-tucos because the incisors are extremely proodont. Our results for skull morphology do not support the *Haptomys* group because, in the phenograms, *C. leucodon* was associated with other groups.

We found a significant congruence at the interspecific level and support for some groups. Cardini (2003) enumerated several factors that might explain the lack of correspondence between phenotypic and phylogenetic trees, such as sampling error, retention of plesiomorphic traits, genetic drift, morphological convergence, and misrepresentation of the phylogenetic analysis. This last factor occurs in *Ctenomys*, which lacks a complete gene tree representing all species of the genus (Lessa and Cook 1998; Mascheretti et al. 2000).

6.11   Size Versus Shape

We found a weak but significant correlation between the size and shape of the skull. The percentages of correct classification for species increased only 1.58% on average when size was added. The skull size is highly variable within groups (the *torquatus* group is an exception). Thus, the relationship between size and shape in the genus *Ctenomys* seems considerable in the phylogenetic context, and the interspecific allometries are not negligible. The robust shape in the north and the gracile shape in the southern *Ctenomys* species are not completely related to skull size. Our results support the observations made by Verzi et al. (2010b) for postnatal ontogenetic series for five *Ctenomys* species, which suggested that interspecific skull shape differences found in *Ctenomys* are not associated with differences in size alone. Our results suggest that skull size shows a large variation within groups while skull shape shows a large variation among *Ctenomys* groups. Therefore, the skull shape is more similar within groups, but the size is more variable within groups.

6.12   Morphological Diversification

In the genus *Ctenomys*, Cook and Lessa (1998) and Parada et al. (2011) found an increase in the diversification rate at the base of *Ctenomys* clade. These authors suggested an increase in diversification of approximately ~3 mya. However, Rowe et al. (2011), using multilocus molecular phylogenetic, suggested that the genus *Rattus*, despite rapid diversification, displayed little ecomorphological divergence among species and did not fit the model of adaptive radiation.

Nevertheless, we found a remarkable variation in skull form in genus *Ctenomys*, and our results support a great morphological diversification in shape and size. However, the cause of this diversification remains unclear (Parada et al. 2011). Despite this, our data show that *Ctenomys* varies in body size, coat color, karyotype, sperm morphology, angle of incisor procumbency, and skull form.

The mandible shape was less variable than cranium. This characteristic is consistent with conserved morphology by biomechanical constraints. For skull shape, we have low support for an ecological speciation (Gavrilets and Losos 2009). Despite *Ctenomys* occupying large latitudinal and longitudinal gradients and several environments, their life history in burrow systems buffers above-ground conditions (Medina et al. 2007). However, we found a continuum in variation in skull shape among the eight main groups, a strong signal of geographical variation, and support for genetic drift as being the more likely cause generating the observed pattern. Simultaneously, the selection for underground life (digging, communicating, etc.) is likely strong and should drive constraints on cranial shape. This pattern suggests that the adaptive landscape of *Ctenomys* is holey (sensu Gavrilets 2003), with low fitness genotypes not fit for underground life and highly fit genotypes that form continuously within the underground niche. This scenario suggests a model of geographical parapatric speciation without strong and/or obvious barriers to gene flow. Current evidence suggests this model to be likely. Reproductive isolation has been investigated in sister species by Lopes and Freitas (2012) and Freitas (2006), showing the presence of hybrid zones at the intraspecific and interspecific levels between sister taxa *Ctenomys lami* and *C. minutus*. Few other hybrids have been detected either because population/species differentiation is rapid as predicted analytically (Gavrilets 2003) or because of limited taxon sampling. Genetic differentiation compatible with isolation by distance model has been found in most species analyzed except for *C. australis* (Mora et al. 2010), coupled with the fine-scale geographic structure for *C. talarum* (Mora et al. 2007), *C. rionegrensis* (Wlasiuk et al. 2003), *C. minutus* (Lopes and Freitas 2012), *C. flamarioni* (Fernandez-Stolz et al. 2007), *C. pearsoni* (Tomasco and Lessa 2007), *C. torquatus* (Roratto et al. 2015), and even *C. lami* (Lopes and Freitas 2012) with a distribution area of 78 × 12 km. Isolation by distance differentiation signal in the skull was also found in *C. minutus* (Fornel et al. 2010), *C. rionegrensis* (D'Anatro and Lessa 2006), and likely in *C. torquatus* (Fernandes et al. 2009). Ecological differentiation has been rarely tackled within the niche of tuco-tucos but metagenomic results indicate differentiation in diet between distantly related sympatric/parapatric species or between populations (Lopes et al. 2015), despite significant differences in skull morphology between habitats (Fornel et al. 2010).

In summary, we found remarkable skull shape variation in *Ctenomys* forming a continuum with a strong geographical signal and little evidence of distinct selection regimes between species, which agrees with patterns observed by Lessa et al. (2008), indicating that morphological adaptations in ctenomyids are more variable and complex than supposed anteriorly. In this interspecific scenario, parapatric speciation in a holey adaptive landscape is likely. This skull shape variation could be the result of small populations, limited dispersal, and weak selection in *Ctenomys*.

Acknowledgements We are very grateful to Fabiano A. Fernandes, Daniza Molina-Schiller, and Gisele S. Rebelato for their help with photographing skulls. We thank Michel Baylac for the Rmorph package. Thanks for all curators and collection managers that provided access to *Ctenomys* specimens: Enrique González (MUNHINA), Olga B. Vacaro and Esperança A. Varela (MACN), Diego H. Verzi and A. Itatí Olivares (MLP), A. Damián Romero (MMP), James L. Patton, Eileen A. Lacey, and Christopher Conroy (MVZ), Eileen Westwig (AMNH), and Bruce D. Patterson (FMNH). This work was supported by *Conselho Nacional de Desenvolvimento Científico e Tecnológico* (CNPq); *Coordenação de Aperfeiçoamento de Pessoal de Nível Superior* (CAPES); *Fundação de Amparo à Pesquisa do Rio Grande do Sul* (FAPERGS); *Departamento de Genética –* UFRGS; and *Projeto Tuco-tuco*. P. C-E. was supported by the CNPq/CAPES PROTAX Program for Taxonomy. R. F. was supported for this project by a doctoral fellowship from *Conselho Nacional de Desenvolvimento Científico e Tecnológico* (CNPq) [grant proc. no. 142953/2005-9].

Supporting Information

SI 1.
Definitions of morphological landmarks with numbers and localizations for each view of the cranium and lower jaw of *Ctenomys* (represented in Fig. 6.2). The same definitions were used by Fornel et al. (2010).

Dorsal View of the Cranium

1, anterior tip of the suture between premaxillas; 2–3, anterolateral extremity of incisor alveolus; 4, anterior extremity of the suture between nasals; 5–6, anteriormost point of the suture between nasal and premaxilla; 7–8, anteriormost point of the root of zygomatic arch; 9, suture between nasals and frontals; 10–11, anterolateral extremity of lacrimal bone; 12–13, point of least width between frontals; 14–15, tip of extremity of superior jugal process; 16–17, anterolateral extremity of suture between frontal and squamosal; 18, suture between frontals and parietals; 19–20, tip of posterior process of jugal; 21–22, anterolateral extremity of suture between parietal and squamosal; 23–24, anterior tip of external auditory meatus; 25–26, point of maximum curvature on mastoid apophysis; 27, posteriormost point of occipital along the midsagittal plane; 28–29, lateral extremity of suture between jugal and squamosal.

Ventral View of the Cranium

1, anterior tip of suture between premaxillas; 2–3, anterolateral extremity of incisor alveolus; 4–5, lateral edge of incisive foramen in suture between premaxilla and maxilla; 6–7, anteriormost point of root of zygomatic arch; 8–9, anteriormost point of orbit in inferior zygomatic root; 10–11, anteriormost point of premolar alveolus; 12–13, posterior extremity of III molar alveolus; 14, posterior extremity of suture between palatines; 15–16, anteriormost point of intersection between jugal and squamosal; 17–18, posteriormost point of pterygoid processes; 19–20, anterior extremity of tympanic bulla; 21–22, anterior tip of external auditory meatus; 23–24, posterior extremity of mastoid apophysis; 25–26, posterior extremity of paraoccipital apophysis; 27, anteriormost point of foramen magnum; 28–29, posterior extremity of occipital condyle in foramen magnum; 30, posteriormost point of foramen magnum along midsagittal plane.

Lateral View of the Cranium

1, inferiormost point of incisor alveolus; 2, posteriormost point of incisor alveolus; 3, anteriormost point of premaxilla; 4, anteriormost point of the suture between nasal and premaxilla; 5, anterior tip of nasal; 6, inferiormost point of suture between premaxilla and maxilla; 7, suture between premaxilla, maxilla and frontal in superior zygomatic root; 8, inferiormost point of suture between lacrimal and maxilla; 9, inferiormost point of infraorbital foramen in inferior zygomatic root; 10, anteriormost point of premolar alveolus; 11, extremity of inferior jugal process; 12, extremity of superior jugal process; 13, tip of posterior jugal process; 14, medial point of suture between parietal and squamosal; 15, posterior extremity of postglenoid fossa; 16, inferior extremity of mastoid apophysis; 17, inferior extremety in suture between pterygoid and tympanic bulla; 18, anteriormost margin of paraoccipital apophysis; 19, posteriormost margin of paraoccipital apophysis; 20, posterior extremity of intersection between occipital and tympanic bulla; 21, superior extremity of lambdoidal crest.

Lateral View of the Mandible

1, upper extreme anterior border of incisor alveolus; 2, extreme of the diastema invagination; 3, anterior edge of the premolar alveolus; 4, intersection between molar alveolus and coronoid process; 5, tip of the coronoid process; 6, maximum of curvature between the coronoid and condylar processes; 7, anterior edge of the articular surface of the condylar process; 8, tip of the postcondyloid process; 9, maximum curvature between condylar and angular processes; 10, tip of the angular process; 11, intersection between mandibular body and masseteric crest; 12, posterior extremity of the mandibular symphysis; 13, posterior extremity border of incisor alveolus.

Literature Cited

Adams DC, Otarola-Castillo E (2013) Geomorph: an R package for the collection and analysis of geometric morphometric shape data. Methods Ecol Evol 4:393–399

Adams DC, Rohlf FJ, Slice DE (2004) Geometric Morphometrics: ten years of progress following the "revolution". Ital J Zool 71:5–16

Adams DC, Rohlf FJ, Slice DE (2013) A field comes of age: geometric morphometrics in the 21st century. Hystrix 24:7–14

Atchley WR, Hall BK (1991) A model for development and evolution of complex morphological structures. Biol Rev 66:101–157

Baylac M (2008) Rmorph: a R geometric and multivariate morphometrics library

Baylac M, Friess M (2005) Fourier descriptors, Procrustes superimposition and data dimensionality: an example of cranial shape analysis in modern human populations. In: Slice DE (ed) Modern morphometrics in physical anthropology. Springer, New York, NY, pp 145–166

Bidau CJ (2015) Family Ctenomyidae. In: Patton JL, Pardiñas UFJ, D'Elía G (eds) Mammals of South America, Vol 2. Rodents The University of Chicago Press, pp 818–876

Bidau CJ, Gimenez MD, Contreras JR (1996) Especiación cromosómica y la conservacion de la variabilidad genética: El caso del género Ctenomys (Rodentia, Caviomorpha, Ctenomyidae). Mendeliana 12:25–37

Bjørnstad ON, Rohlf AI, Lambin X (1999) Spatial population dynamics: analyzing patterns and processes of population synchrony. TREE 14:427–432

Blainville HMD (1826) Sur une nouvelle espèce de Rongeur fouisseur du Brésil. Bulletin de la Societe philomathique de Paris 3:62–64

Bookstein FL (1991) Morphometric tools for landmark data: geometry and biology. Cambridge University Press, London

Bookstein FL (1996a) Biometrics, biomathematics and the morphometric synthesis. Bull Math Biol 58:313–365

Bookstein FL (1996b) Combining the tools of geometric morphometrics. In: Marcus LF, Corti M, Loy A, Naylor G, Slice DE (eds) Advances in morphometrics. Plenum Publishing Corporation, New York, NY, pp 131–151

Borges LR, Maestri R, Kubiak BB, Galiano D, Fornel R, Freitas TRO (2017) The role of soil features in shaping the bite force and related skull and mandible morphology in the subterranean rodents of genus Ctenomys (Hystricognathi: Ctenomyidae). J Zool. https://doi.org/10.1111/jzo.12398

Cardini A (2003) The geometry of the marmot (Rodentia: Sciuridae) mandible: phylogeny and patterns of morphological evolution. Syst Biol 52:186–205

Castillo AH, Cortinas MN, Lessa EP (2005) Rapid diversification of south American Tuco-tucos (*Ctenomys*: Rodentia, Ctenomyidae): contrasting mitochondrial and nuclear intron sequences. J Mammal 86:170–179

Caumul R, Polly PD (2005) Phylogenetic and environmental components of morphological variation: skull, mandible, and molar shape in marmots (*Marmota*, Rodentia). Evolution 59:2460–2472

Contreras JR, Bidau CJ (1999) Líneas generales del panorama evolutivo de los roedores excavadores sudamericanos del género *Ctenomys* (Mammalia, Rodentia, Caviomorpha, Ctenomyidae). Ciencia Siglo XXI, Fundación Bartolomé Hidalgo, Buenos Aires

Cook JA, Lessa EP (1998) Are rates of diversification in subterranean south American tuco-tucos (genus *Ctenomys*, Rodentia: Octodontidae) ususlly high? Evolution 52:1521–1527

Cook JA, Salazar-Bravo J (2004) Heterochromatin variation among the chromosomally diverse tuco-tucos (Rodentia–Ctenomyidae) from Bolivia. In: Sánchez-Cordero V, Medellín RA (eds) Contribuciones Mastozoológicas en Homenaje a Bernardo Villa. Instituto de Biología y Instituto de Ecología, UNAM, pp 120–142

Cordeiro-Estrela P, Baylac M, Denys C, Marinho-Filho J (2006) Interspecific patterns of skull variation between sympatric Brazilian vesper mice: geometric morphometrics assessment. J Mammal 87:1270–1279

D'Anatro A, Lessa EP (2006) Geometric morphometric analysis of geographic variation in the Río negro tuco-tuco, *Ctenomys rionegrensis* (Rodentia: Ctenomyidae). Mamm Biol 71:288–298

D'Elía G, Lessa EP, Cook JA (1999) Molecular phylogeny of Tuco-Tucos, genus *Ctenomys* (Rodentia: Octodontidae): evaluation of the *mendocinus* species group and the evolution of asymmetric sperm. J Mamm Evol 6:19–38

Dray S, Dufour AB (2007) The ade4 package: implementing the duality diagram for ecologists. J Stat Softw 22:1–20

Dryden IL, Mardia KV (1998) Statistical shape analysis. John Wiley & Sons, New York, NY

Ersts PJ (2009) Geographic distance matrix generator (version 1.2.3). American Museum of Natural History, Center for Biodiversity and Conservation: URL http:/biodiversity.informatics. amnh.org/open_source/gdmg, last Accessed August 16, 2009

Evin A, Baylac M, Ruedi M, Mucedda M, Pons J-M (2008) Taxonomy, skull diversity and evolution in a species complex of *Myotis* (Chiroptera: Vespertilionidae): a geometric morphometric appraisal. Biol J Linn Soc 95:529–538

Fernandes FA, Fornel R, Cordeiro-Estrela P, Freitas TRO (2009) Intra- and interspecific skull variation in two sister species of the subterranean genus *Ctenomys* (Rodentia, Ctenomyidae): coupling geometric morphometrics and chromosomal polymorphism. Zool J Linn Soc 155:220–237

Fernandez-Stolz GP, Stolz JFB, Freitas TRO (2007) Bottlenecks and dispersal in the tuco-tuco das dunas, *Ctenomys flamarioni* (Rodentia: Ctenomyidae), in southern Brazil. J Mammal 88:935–945

Fernandes FA, Fornel R, Freitas TRO (2012) Ctenomys brasiliensis Blainville (Rodentia: Ctenomyidae): clarifying the geographic placement of the type species of the genus Ctenomys. Zootaxa 3272:57–68

Fornel R, Cordeiro-Estrela P, Freitas TRO (2010) Skull shape and size variation in *Ctenomys minutus* (Rodentia: Ctenomyidae) in geographical, chromosomal polymorphism, and environmental contexts. Biol J Linn Soc 101:705–720

Fornel R, Cordeiro-Estrela P, Freitas TRO (2018) Skull shape and size variation within and between *mendocinus* and *torquatus* groups in the genus *Ctenomys* (Rodentia: Ctenomyidae) in chromosomal polymorphism context. Genet Mol Biol 41:263–272

Freitas TRO (1994) Geographical variation of heterochromatin in *Ctenomys flamarioni* (Rodentia-Octodontidae) and its cytogenetic relationships with other species of the genus. Cytogenet Cell Genet 67:193–198

Freitas TRO (2005) Analysis of skull morphology in 15 species of the genus *Ctenomys*, including seven Karyologically distinct forms of *Ctenomys minutus* (Rodentia: Ctenomyidae). In:

Lacey EA, Myers P (eds) Mammalian diversification from chromosomes to Phylogeography (a Celebration of the career of James L. Patton). University of California Publications in Zoology, Berkeley, pp 131–154

Freitas TRO (2006) Cytogenetic status of four *Ctenomys* species in the south of Brazil. Genetica 126:227–235

Freitas TRO (2016) Family Ctenomyidae (Tuco-tucos). In: Wilson DE, Lcher Jr TE, Mittermeier RA, org. (eds) Handbook of the mammals of the world. Lagomorphs and rodents. I. 6th ed. Lynx Edicions Publications, Barcelona, vol 6, pp 1–900

Freitas TRO, Fernandes FA, Fornel R, Roratto PA (2012) An endemic new species of tuco-tuco, genus *Ctenomys* (Rodentia: Ctenomyidae), with a restricted geographic distribution in southern Brazil. J Mammal 93:1355–1367

Gardner SL, Salazar-Bravo J, Cook JA (2014) New species of *Ctenomys* Blainville 1826 (Rodentia: Ctenomyidae) from the Lowlands and Central Valley of Bolivia. Faculty Publications from the Harold W. Manter Laboratory of Parasitology, p 722

Gavrilets S (2003) Models of speciation: what have we learned in 40 years? Evolution 57:2197–2215

Gavrilets S, Losos JB (2009) Adaptive radiation: contrasting theory with data. Science 323:732–733

Hanken J, Thorogood P (1993) Evolution and development of the vertebrate skull–the role of pattern formation. Trends Ecol Evol 8:9–15

Heo M, Gabriel KR (1997) A permutation test of association between configurations by means of RV coefficient. Commun Stat Simul Comput 27:843–856

Kent JT, Mardia K (2001) Shape, Procrustes tangent projections and bilateral symmetry. Biometrika 88:469–485

Klingenberg CP, Barluenga M, Meyer A (2002) Shape analysis of symmetric structures: quantifying variation among individuals and asymmetry. Evolution 56:1909–1920

Kubiak BB, Maestri R, de Almeida TS, Borges LR, Galiano D, Fornel R, Freitas TRO (2018) Evolution in action: soil hardness influences morphology in a subterranean rodent (Rodentia: Ctenomyidae). Biol J Linn Soc 4:766–776

Legendre P, Legendre L (1998) Numerical ecology. Elsevier, Amsterdam. 853p

Lessa EP, Cook JA (1998) The molecular phylogenetics of tuco-tucos (genus *Ctenomys*, Rodentia: Octodontidae) suggests as early burst of speciation. Mol Phylogenet Evol 9:88–99

Lessa EP, Vassallo AI, Verzi DH, Mora MS (2008) Evolution of morphological adaptations for digging in living and extinct Ctenomyid and Octodontid rodents. Biol J Linn Soc 95:267–283

Lopes CM, Freitas TRO (2012) Human impact in naturally patched small populations: genetic structure and conservation of the burrowing rodent, tuco-tuco (*Ctenomys lami*). J Hered 103:672–681

Lopes CM, Barba M, Boyer F, Mercier C, Filho PJSS, Heidtmann LM, Galiano D, Kubiak BB, Langone PQ, Garcias FM, Giely L, Coissac E, Freitas TRO, Taberlet P (2015) DNA metabarcoding diet analysis for species with parapatric vs sympatric distribution: a case study on subterranean rodents. Heredity 114:525–536

Marinho JR, Freitas TRO (2000) Intraspecific craniometric variation in a chromosome hybrid zone of *Ctenomys minutus* (Rodentia, Hystricognathi). Mamm Biol 65:226–231

Marroig G, Cheverud JM (2005) Size as a line of least evolutionary resistance: diet and adaptive morphological radiation in New World monkeys. Evolution 59:1128–1142

Mascheretti S, Mirol PM, Giménez MD, Bidau CJ, Contreras JR, Searle JB (2000) Phylogenetics of the speciose and chromosomally variable rodent genus *Ctenomys* (Ctenomyidae: Octodontidae), based on mitochondrial cytochrome b sequences. Biol J Linn Soc 70:361–376

Massarini AI, Freitas TRO (2005) Morphological and cytogenetics comparison in species of the *mendocinus*-group (genus *Ctenomys*) with emphasis in *C. australis* and *C. flamarioni* (Rodentia-ctenomyidae). Caryologia 58:21–27

Massarini A, Barros MA, Ortells M (1991) Evolutionary biology of fossorial Ctenomyinae rodents (Caviomorpha: Octodontidae). I. Cromosomal polymorphism and small karyotypic differentiation in central Argentinian populations of tuco-tucos. Genetica 83:131–144

Medina AI, Martí DA, Bidau CJ (2007) Subterranean rodents of the genus *Ctenomys* (Caviomorpha, Ctenomyidae) follow the converse to Bergmann's rule. J Biogeogr 34:1439–1454

Mora MS, Olivares AI, Vassallo AI (2003) Size, shape and structural versatility of the skull of the subterranean rodent *Ctenomys* (Rodentia, Caviomorpha): functional and morphological analysis. Biol J Linn Soc 78:85–96

Mora MS, Lessa EP, Kittlein MJ, Vassallo AI (2006) Phylogeography of the subterranean rodent *Ctenomys australis* in sand-dune habitats: evidence of population expansion. J Mammal 87:1192–1203

Mora MS, Lessa EP, Cutrera AP, Kittlein MJ, Vassallo AI (2007) Phylogeographical structure in the subterranean tuco-tuco *Ctenomys talarum* (Rodentia: Ctenomyidae): contrasting the demographic consequences of regional and habitat-specific histories. Mol Ecol 16:3453–3456

Mora SM, Mapelli FJ, Gacciotti OE, Kittlein MJ, Lessa EP (2010) Dispersal and population structure at different spatial scales in the subterranean rodent *Ctenomys australis*. BMC Genet 11:9

Morgan CC, Álvarez A (2013) The humerus of south American caviomorph rodents: shape, function and size in a phylogenetic context. J Zool 290:107–116

Osgood WH (1946) A new octodont rodent from the Paraguayan Chaco. Fieldiana Zool 31:47–49

Parada A, D'Elía G, Bidau CJ, Lessa EP (2011) Species groups and the evolutionary diversification of tuco-tucos, genus *Ctenomys* (Rodentia: Ctenomyidae). J Mammal 92:671–682

Paradis E, Strimmer K, Claude J, Jobb G, Open-Rhein R, Dultheil J, Bolker NB (2004) APE: analyses of phylogenetics and evolution in R. R package version 1. pp 8–2

R Development Core Team (2009) Stats – R: a language and environmental for statistical computing. R Development Core Team, Vienna. Available at: http://www.r-project.org

Reig OA, Kiblisky P (1969) Chromosome multiformity in the genus *Ctenomys* (Rodentia: Octodontidae). Chromosoma 28:211–244

Reig OA, Busch C, Ortells MO, Contreras JR (1990) An overview of evolution, systematics, population biology, cytogenetics, molecular biology and speciation in *Ctenomys*. In: Nevo E, Reig OA (eds) Evolution of the subterranean mammals at the organismal and molecular levels. Progress in clinical and biological research. Wiley-Liss, New York, NY, pp 71–96

Ripley BD (1996) Pattern recognition and neural networks. Cambridge University Press, Cambridge

Rohlf FJ (2004) TPSDig, Version 1.40 Stony Brook, NY: Department of Ecology and Evolution, State University of New York. Available at: http://life.bio.sunysb.edu/morph/

Rohlf FJ (2008) TPSUtil, Version 1.40 Stony Brook, NY: Department of Ecology and Evolution, State University of New York. Available at: http://life.bio.sunysb.edu/morph/

Rohlf FJ, Marcus LF (1993) A revolution in morphometrics. TREE 8:129–132

Rohlf FJ, Slice D (1990) Extensions of the Procrustes method for the optimal superimposition of landmarks. Syst Zool 39:40–59

Roratto PA, Fernandes FA, Freitas TRO (2015) Phylogeography of the subterranean rodent *Ctenomys torquatus*: an evaluation of the riverine barrier hypothesis. J Biogeogr 42:694–705

Rowe KC, Aplin KP, Baverstock PR, Moritz C (2011) Recent and rapid speciation with limited morphological Disparity in the genus *Rattus*. Syst Biol 60:188–203

Slamovits CH, Cook JA, Lessa EP, Rossi MS (2001) Recurrent amplifications and deletions of satellite DNA accompanied chromosomal diversification in south American tuco-tucos (genus *Ctenomys*, Rodentia: Octodontidae): a phylogenetic approach. Mol Biol Evol 18:1708–1719

Steiner-Souza F, Freitas TRO, Cordeiro-Estrela P (2010) Inferring adaptation within shape diversity of the humerus of subterranean rodent *Ctenomys*. Biol J Linn Soc 100:353–367

Teta P, D'Elía G (2020) Uncovering the species diversity of subterranean rodents at the end of the world: three new species of Patagonian tuco-tucos (Rodentia, Hystricomorpha, *Ctenomys*). Peer J:1–35

Thomas O (1916) Two new argentine rodents, with a new subgenus of *Ctenomys*. Ann Mag Nat Hist 18:303–306

Tomasco I, Lessa EP (2007) Phylogeography of the tuco-tuco Ctenomys pearsoni: mtDNA varia-
    tion and its implication for chromosomal differentiation. In: Kelt DA, Lessa EP, Salazar-Bravo
    JA, Patton JL (eds). The Quintessential Naturalist: Honoring the Life and Legacy of Oliver
    P. Perason. University of California Publication in Zoology Series, Berkeley, California
Tiranti SI, Dyzenchauz FJ, Hasson ER, Massarini AI (2005) Evolutionary and systematic relation-
    ships among tuco-tucos of the *Ctenomys pundti* complex (Rodentia: Octodontidae): a cytoge-
    netic and morphological approach. Mammalia 69:69–80
Vassallo AI (1998) Functional morphology, comparative behaviour, and adaptation in two sym-
    patric subterranean rodents genus *Ctenomys* (Caviomorpha: Octodontidae). J Zool (Lond)
    244:415–427
Venables WN, Ripley BD (2002) MASS: modern applied statistics with S, 4th edn. Springer,
    New York
Verzi DH, Olivares AI (2006) Craniomandibular joint in south American burrowing rodents
    (Ctenomyidae): adaptations and constraints related to a specialized mandibular position in dig-
    ging. J Zool (Lond) 270:488–501
Verzi DH, Olivares AI, Morgan CC (2010a) The oldest south American tuco-tuco (late Pliocene,
    northwestern Argentina) and the boundaries of the genus *Ctenomys* (Rodentia, Ctenomyidae).
    Mamm Biol 75:243–252
Verzi DH, Álvarez A, Olivares AI, Morgan CC, Vassallo AI (2010b) Ontogenetic trajectories of key
    morphofunctional cranial traits in south American subterranean ctenomyid rodents. J Mammal
    91:1508–1516
Wlasiuk G, Garza JC, Lessa EP (2003) Genetic and geographic differentiation in the Rio Negro
    tuco-tuco (*Ctenomys rionegrensis*): inferring the roles of migration and drift from multiple
    genetic markers. Evolution 57:913–926
Woods CA, Kilpatrick CW (2005) Infraorder Hystricognathi Brandt, 1855. In: Wilson DE, Reeder
    DM (eds) Mammal species of the world. 3, vol 2. Smithsonian Institution Press, Washington,
    pp 1538–1600
Wright SP (1992) Adjusted P-values for simultaneous inference. Biometrics 48:1005–1013

Chapter 7
Biomechanics and Strategies of Digging

Aldo I. Vassallo, Federico Becerra, Alejandra I. Echeverría, Guido N. Buezas, Alcira O. Díaz, M. Victoria Longo, and Mariana Cohen

7.1 Introduction

For small rodents, running and hiding are the main strategies to protect themselves from predators, and burrows offer an excellent shelter against most of them (Reichman and Smith 1990; Andino et al. 2014). Particularly, in arid and semiarid ecosystems, digging (i.e., to break up and remove the soil) and burrowing (i.e., to hide in burrows, to construct by tunneling, or to progress by or as if by digging) are common behaviors in many mammals (Nevo 1999; Whitford and Kay 1999; Lacey et al. 2000; Cutrera et al. 2006). Numerous terrestrial mammals have evolved fossorial adaptations, and rodents, in particular, have repeatedly diversified into underground habitats (Hopkins 2005; McIntosh and Cox 2019). Across the globe, in all continents but Australia and Antarctica, at least 250 extant rodent species (38 genera, 6 families – according to the classification applied) spend most of their lives in self-constructed burrows (Begall et al. 2007). Life underground is achieved through several distinct methods of burrow excavation. From studies performed in terrariums (Giannoni et al. 1996), as well as in natural soils (Vassallo 1998), it is known that tuco-tucos break up the soil through the use of both the foreclaws of the manus (scratch-digging) and the incisors (upper and lower) in a chewing motion

A. I. Vassallo (✉) · F. Becerra · A. I. Echeverría · G. N. Buezas
Grupo Morfología Funcional y Comportamiento, Instituto de Investigaciones Marinas y Costeras (IIMyC, UNMDP-CONICET), Universidad Nacional de Mar del Plata (UNMdP), Consejo Nacional de Investigaciones Científicas y Técnicas (CONICET),
Mar del Plata, Buenos Aires, Argentina
e-mail: avassall@mdp.edu.ar; fbecerra@mdp.edu.ar; aiechever@mdp.edu.ar

A. O. Díaz · M. V. Longo · M. Cohen
Grupo Histología e Histoquímica, Instituto de Investigaciones Marinas y Costeras (IIMyC, UNMDP-CONICET), Universidad Nacional de Mar del Plata (UNMdP), Consejo Nacional de Investigaciones Científicas y Técnicas (CONICET),
Mar del Plata, Buenos Aires, Argentina
e-mail: adiaz@mdp.edu.ar; mvlongo@mdp.edu.ar

© Springer Nature Switzerland AG 2021
T. R. O. de. Freitas et al. (eds.), *Tuco-Tucos*,
https://doi.org/10.1007/978-3-030-61679-3_7

(chisel-tooth digging) according to soil requirements. However, the use of these different digging tools varies between species and habitats occupied. Scratch-digging is the predominant mode of digging among rodents. A final mode of digging, and the least common among rodents, is head-lift digging (Hopkins 2005), which is performed through the use of the nose in a spade-like manner, and sometimes assisted by the lower incisors.

Tuco-tucos are proficient scratch-diggers that dig primarily by means of vigorous scraping movements, which comprise rapid alternating strokes of their forefeet (e.g., Vassallo 1998). These movements allow them to loosen the soil through the use of the hands, especially the foreclaws. Recently, Echeverría et al. (2019) found that the palms of tuco-tucos' hands show a pad pattern similar to that observed in other fossorial rodents (all Spalacinae occurring at least in Europe; several bathyergids: *Georychus, Cryptomys,* and *Bathyergus*; *Pedetes capensis* Pedetidae; Geomyidae and the dipodine *Jaculus*; for references see Ade and Ziekur 1999). This pattern is defined by a lack of distinct distal pads accompanied by strongly developed proximal pads (the hypothenar and thenar pads, this latter has been now redescribed as a false thumb by Echeverría et al. 2019). According to Ade and Ziekur (1999), this condition indicates that the animals can dig by using their paws in a hoe or scraper-like manner, since digits II-III-IV-V project ventrally and the flexed palm forms a kind of scraper or hoe.

According to the requirements of the substrate, tuco-tucos complementarily use their large and procumbent incisors (i.e., incisors with a protrusion angle greater than 90°) to assist in loosening and breaking obstacles such as rocks, nodules of $CaCO_3$ ("tosca") or hard soil, and fibrous roots (see Ubilla and Altuna 1990; De Santis et al. 1998; Vassallo 1998; Stein 2000; Becerra et al. 2013). For instance, Vassallo (1998) found that when confronted with sandy and friable soils, *C. talarum* (Los Talas' tuco-tuco) exclusively use their forelimbs to break down the soil. Conversely, when confronted with harder and clayey soils, they behave as scratch- and chisel-tooth diggers, also using the incisors to dig. The digging strategy used is perhaps related to the external forces that the muscles involved can exert, probably being greater in those muscles that act during tooth-digging (Tables 7.1, and 7.2). Although the actual force values at scratch-digging are still unknown, anatomy-based estimations of forces exerted by selected muscles done at the level of the foreclaw's tip and the incisors support this statement (Table 7.2).

The bite force of several *Ctenomys* species was measured *in vivo* (Becerra 2015). For instance, in *C. talarum,* bite force ranges from 32 to 27 N (in adult males and females, respectively). Considering that soil hardness of *C. talarum*'s typical habitat averages 100 N/cm^2, and taking into account the incisor's cross-section, it was assessed that the pressure exerted by jaw adductor muscles at the level of the incisors is three times higher than that required for soil penetration (Becerra et al. 2011).

The following sections will introduce the basic biomechanical principles underlying force production and transmission during digging. The burrowing system of tuco-tucos is described, emphasizing its dynamic attributes and dependence on the characteristics of the environment. Finally, we addressed the functional morphology

Table 7.1 Miology of skull and pectoral appendicular regions in *Ctenomys*[a]

Structure	Muscle	Origin	Insertion	Action
Jaw	Masseter superficialis	By a round and strong tendon, from a small area on the anteroventral surface of the zygomatic arch	Medial surface of the angle of the mandible, it inserts high up the deep medial fossa, near and posterior	Jaw adductor
	Masseter profundus	From the anteroventral surface of the zygomatic arch	Lateral surface of the mandible, on the masseteric crest	
	Posterior masseter	It originates from the posteroventral surface of the zygomatic arch	Lateral surface of the mandible ventral to the condyle, on the postcondyloid process	
	Zygomaticomandibularis	From the medial surface of the zygomatic arch	On the lateral surface of the mandible	
	Zygomaticomandibularis infraorbitalis	Represented by fibers of the zygomaticomandibularis that arise from the bony walls of the infraorbital foramen. It originates from the fossa on the lateral side of the rostrum, the superior zygomatic root, and the medial surface of the maxillary and jugal parts of the zygomatic arch	Lateral surface of the mandible. Along the dorsal aspect of the masseteric fossa	
	Temporalis	According to Woods (1972), the muscle is composed of these three parts. It originates from the surface of cranium, may include the temporalis fascia and the surface of the cranium superior and posterior to the zygomatic arch	On the coronoid process of the mandible. The main part inserts on the tip, the orbital part inserts mostly on the medial surface, the posterior part inserts on the lateral surface	
	Pterygoideus	Externus. From the edge of the lateral pterygoid plate, the surface of the alisphenoid bone, and the adjacent edge of the maxillary bone Internus. From the margin and deep inside the pterygoid fossa. Two parts	Externus. Below and posterior to the condyle on the medial surface of the condyloid process. Internus. Medial side of the mandible, on the dorsal surface of the flattened angular process	

(continued)

Table 7.1 (continued)

Structure	Muscle		Origin	Insertion	Action
Shoulder		Subscapularis	From the entire subscapular fossa of the scapula	On the lesser tuberosity of the humerus	Medial rotator of the arm. It provides shoulder joint stability
	Rotator cuff	Supraspinatus	Two parts. From the anterior surface of the proximal half of the scapular spine, the ventral surface of the spine in the region of the great scapular notch, and the surface of the septum separating this muscle from the infraspinatus. The remaining and larger part is from the supraspinous fossa, the superior border of the scapula, and the anteromedial surface of subscapularis muscle	The muscle parts are moderately fused. The part from the scapular spine is more superficial than the second part, and inserts onto its surface. The combined parts narrow to a thin tendon which inserts on the dorsocranial surface of the greater tuberosity of the humerus	External rotation of the shoulder joint. Shoulder joint stabilizer
		Infraspinatus	From the surface of the infraspinous fossa, the dorsal margin of the axillary border under the teres major muscle, the vertebral border, and the spine of the scapula. The part of the muscle from the spine originates on the posterior margin of the spine proximal to the great scapular notch and on the ventral surface of the spine in the region of the notch	The muscle narrows to a tendon and inserts on the greater tuberosity of the humerus. The muscle passes anteroventrally beneath the spine of the scapula (great scapular notch). The notch extends along the distal two-thirds of the length of the scapular spine	External rotation of the shoulder joint. Shoulder joint stabilizer
		Teres minor	From the distal two-fifths of the axillary border of the scapula, and from the anterior surface of the well-developed aponeurotic envelope. The muscle is small, and largely covered by the infraspinatus muscle	On the distal edge of the greater tuberosity. The insertion is via a tendon and is distal to and separate from the insertion of the infraspinatus muscle	External rotation of the shoulder joint. Adduction and extension of the shoulder. When the humerus is stabilized, abducts the inferior angle of the scapula. Shoulder joint stabilizer.

(continued)

Table 7.1 (continued)

Structure	Muscle	Origin	Insertion	Action
	Teres major	From the most posterior margin of the vertebral border and the dorsal third of the axillary border of the scapula. Some fibers also originate from the fascial surface of the infraspinatus the subscapularis muscles	Distally, on the shaft of the humerus	Its main action is flexing the shoulder according to anatomical descriptions. Acting alone from a position with the shoulder semi-flexed, it would thus be expected to raise the elbow, and its activity may relate to initial paw lift. However, due to its insertion medially on the humeral neck, it is possible that in some positions of the humerus it could exert a force tending to medially rotate the humerus, and thus contribute to medial movement of the paw, which occurs at the same time and to which it was also highly correlated
	Latissimus dorsi	From the last five thoracic vertebrae and the lumbodorsal fascia	On the medial side of the humerus below the lesser tuberosity. The muscle joins with the teres major muscle to insert via a common tendon	Its action is to pull the arm backward and upward
	Dorsoepitrochlearis	Dorsolateral border of the m. latissimus dorsi	Posteroventral aspect of the olecranon	Forelimb extensor
	Triceps longus	Axilar border of the scapula	Posterior surface of the olecranon	Forelimb extensor

(continued)

Table 7.1 (continued)

Structure	Muscle	Origin	Insertion	Action
	Triceps medialis	Distal four-fifths of the shaft of the humerus	Posterodorsal surface of the olecranon. The deep fibers are continuous with the anconeus muscle and insert on the lateral aspect of the olecranon	Forelimb extensor
	Triceps brachii (=triceps lateralis)	Ventrolateral surface of the greater tuberosity of the humerus	On the ventrolateral surface of the olecranon and onto the surface of the triceps longus muscle	Forelimb extensor
	Biceps brachii	The short head, from the tip of the coracoid process. Long head, from the base of the coracoid process and lip of the glenoid fosa	The two heads merge into one and inserts on the brachial ridge of the ulna. Distally, inserts secondary on the radius	Forelimb flexor
	Supinator	On the proximal 2/3 of the humerus. From the surface of the radial collateral ligament and from the deep surface of the distal end of the lateral epicondyle	Fleshy. On the proximal half of the lateral surface of the radius	Forearm supinator
Forearm	Pronator teres	Medial epicondyle of the humerus. Fleshy	Middle 2/3 of the radius	Forearm pronator
	Palmaris longus	Medial epicondyle of the humerus	Thenar and hypothenar pads	Hand flexor
	Flexor carpi ulnaris	Ventral surface of the olecranon and proximal end of the ulna	Base of the pisiform bone	Hand flexor
	Flexor carpi radialis	Medial epicondyle of the humerus	On the base of the second and third metacarpals	Hand flexor
	Extensor carpi radialis	Humerus. From the proximal end of the lateral epicondylar ridge	Half of the second and third metacarpals	Hand extensor
	Extensor carpi ulnaris	Lateral epicondyle of the humerus and proximal half of the ulna (distal to the middle of the semilunar notch)	On the unciform bone	Hand extensor

(continued)

Table 7.1 (continued)

Structure	Muscle	Origin	Insertion	Action
	Flexor digitorum profundus	Via four heads. The first and the second, from the medial epicondyle of the humerus; the third from the proximal 2/3 of the radius; the fourth, from the proximal 2/3 of the ulna	The tendons join in one that passes through the wrist and splits into four tendons. They insert onto the sesamoids of the terminal phalanxs of digits II to V. From the tendon of digit II, a side tendon develops and passes out onto the thumb	Digits flexor
	Sublimis (=flexor digitorum superficialis)	Medial epicondyle of the humerus	Splits in three tendons which pass out onto the base of digits II, III, and IV. Beyond the metacarpophalangeal junction, each tendon splits into two small branches which pass out on either side of the large tendon of the flexor digitorum profundus and insert on the base of the second phalanx	Digits flexor
	Extensor digitorum	Lateral epicondyle of the humerus	Insertion tendon passes through the wrist (between the ulna and radio epiphyses) and splits in 4 tendons that insert on the terminal phalanges of the second through fifth digits	Digits extensor
	Abductor pollicis longus	Ulna and radius	Point of bifurcation between the thumb and the thenar pad	False thumb and thumb abductor
Hand	Flexores breves profundi	Distal carpal corresponding to each digit	Onto the sesamoids at the metacarpophalangeal joints of each digit	Digits flexors
	Abductor pollicis brevis	Proximal border of the radial sesamoid (skeletal support of the false thumb)	Onto the radial side of the metacarpal of the thumb	Thumb abductor
	Flexor pollicis brevis	Base of the radial sesamoid	Base of the metacarpal of the thumb	Thumb flexor
	Adductor pollicis brevis	Common deep palmar ligament	Incorporates to the muscle complex connecting the false thumb, the thumb, and the digit II	Digits adductor

[a]Taken from Woods (1972); Druzinsky et al. (2011); Echeverría et al. (2017, 2019)

Table 7.2 Mechanical advantage and forces exerted by selected forelimb and jaw adductor muscles involved in digging in *Ctenomys talarum*

	Triceps longus	Biceps brachii	Masseter superficialis	Masseter profundus
Muscle mass (g)	0.78	0.10	0.61	0.60
% muscle mass to body mass	0.53	0.04	0.42	0.41
PCSA (mm^2)	27	6	72	94
Li/Lo, mechanical advantage	0.21	0.16	0.20	0.30
Fi, muscle internal force (N)	8.23	1.77	18.03	23.55
Fo, external force (N) at claws (triceps, biceps) and incisors (masseter)	1.73	0.28	3.61	7.06

of the so-called digging tools, i.e., incisors, claws, and functionally related structures, including issues related to its structural strength.

7.2 A Biomechanical Approach to Digging Behavior

Briefly, we will introduce why a biomechanical approach allows us to understand the tuco-tucos' digging behavior. Classical mechanics deals with the behavior of physical entities when they are affected by forces or displacements. Biomechanics, as a discipline, integrates anatomical attributes and principles of classical mechanics to understand the functioning and adaptations of animals at different levels of organization, from the cellular level, organs or structures, to the whole organism (e.g., Vogel 2013). Locomotion and feeding are two of the main functions on which biomechanics focuses on. Terrestrial locomotion requires the production of forces to sustain the body by counteracting the force of gravity, as well as to propel the animal and produce its displacement (Biewener 2003). On the other hand, feeding requires the production of forces to disaggregate a food item (Ungar 2010). In subterranean rodents, such as *Ctenomys*, part of the anatomical structures used for the basic functions of locomotion and feeding (which they possess because they are terrestrial mammals) were co-opted to produce, in addition, a new function, digging. These now partially modified structures are called "digging tools," but it is important to note that they have not lost their original function. For instance, when a tuco-tucos moves inside its burrow or moves over the surface to access to aboveground plant parts, it does so according to a typical mammal locomotor pattern (Lobo Ribeiro 2017). Of course, digging tools involve more anatomical elements and relationships between them than specifically the incisors and foreclaws. For example, for the incisors to properly perform as digging tools, it is necessary that the cervical musculature act stabilizing the head, so that these muscles are well developed in chisel-tooth digging rodents, as well as the jaw muscles (Hildebrand 1985; Van Daele et al. 2009).

One of the most important structural modifications concerning digging is the increase of the mechanical advantage (MA) of the involved muscles. In musculoskeletal systems (e.g., limbs; jaw and associated muscles), the mechanical advantage is a measure of the amplification of force exerted by a certain arrangement of muscles and bones, given by the equation:

$$MA = Fo / Fi = Li / Lo; Fo = Li / Lo \times Fi.$$

Where Fo is the resultant external force exerted (or applied) by the system, Fi is the contributive fraction of the internal force – produced by muscle contraction – to the Fo (effective force), and Li and Lo are determined by the moving bones that rotate on a joint, that is, the lever arms. This formalism of classical mechanics, known as the static equilibrium equation, allows us to analyze the anatomical characters involved in digging in two complementary approaches. On the one hand, it allows us to interpret the function of a complex feature consisting of different bones, joints, muscles, and sites of origin and insertion, i.e., understanding the relationship between structure and function (Fig. 7.1). On the other hand, it allows us to predict on which anatomical attributes natural selection could act on, leading to adaptations related to specialized behaviors such as digging and subterranean habits.

7.3 The Burrow System in Tuco-Tucos

For mammals that evolved to live underground, the burrow represents an exosomatic structure which constitutes, in fact, a materialization of the relationship between the individuals and the environment. The nexus established by the tuco-tuco's burrow with its habitat is so strong and complex that it is difficult, or even arbitrary, to establish the boundaries between the species' phenotype and the surrounding environment. Burrows serve a number of purposes such as thermoregulation (they provide a stable environment), shelter against predators, procurement, and storage of food, facilitation of social interactions and mating (e.g., Fleming and Brown 1975; Ellison 1995; Ebensperger and Blumstein 2006; Antinuchi et al. 2007; Zelová et al. 2010). The tuco-tuco's burrow structure was well studied by Antinuchi and Busch (1992). These authors found that tuco-tucos are capable of building extensive and elaborated tunnel systems with many feeding openings, which are characterized by a branching structure; i.e., a main axial tunnel (representing 48% of the total length), a single nest, and a variable number of lateral blind and foraging tunnels. The burrow openings are of two types: some serve to take out the soil removed by the animal (product of the maintenance of the burrow or the construction of new tunnels), which are usually surrounded by loose soil that takes the form of lobed or fan-shaped mounds, and others are used as an exit to the outside to collect plant material, or eventually interact with other tuco-tucos, for example, during the excursions made by the males to visit the female's burrow during the reproductive season (Zenuto et al. 2002). The studied species by Antinuchi and Busch (1992),

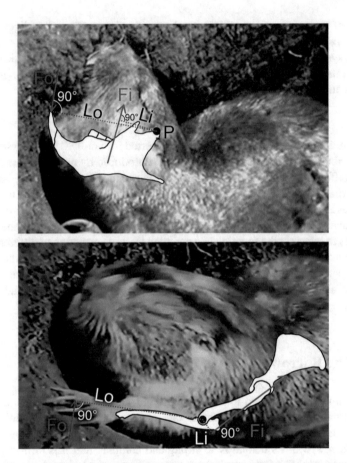

Fig. 7.1 Biomechanical analysis of chisel-tooth and scratch-digging in tuco-tucos. *Above*: The mandible is in static equilibrium between two forces acting in opposite directions (third-order lever system). The force produced by the mandibular adductor musculature (Fi) tends to rotate the jaw by closing it, while (Fo) is the reaction force that the substrate exerts on the lower incisors which tends to rotate the jaw in the opposite direction. *Below*: The forearm (ulna and radius) is in static equilibrium between two forces acting in the same direction (first-order lever system). The force produced by the arm extensor musculature (Fi) tends to rotate it by extending the elbow, while (Fo) is the reaction force that the substrate exerts on the claws which tends to rotate the forearm in the opposite direction. Each force acts on its respective lever arm, Li and Lo. This static equilibrium can be thought of as the situation prior to the breakage of the ground either by the incisors (above) or claws (below), so Fi x Li = Fo x Lo. P represents the joint's position (pivot). Photographs taken from a video film of a *C. talarum* adult male (body mass 150 g) digging in soil typical of its habitat; *above*: modified from Becerra et al. (2011). In white, over imposition of mandible (above) and ulna, humerus, and scapula (below)

C. talarum, lives in well-vegetated grassy and highly variable habitats, ranging from sand dunes to inland grasslands with sandy and friable to hard, clayey soils (Malizia et al. 1991). The soil hardness at which this species is found is 15–25 kg/cm^2 for 5 cm of penetration (Vassallo 1998). It should be noted that this tuco-tuco species is not among the largest species of *Ctenomys*, but despite their small body size (♀ = 122 g, and ♂ = 154.4 g; Malizia and Busch 1991), it can build tunnel systems with a burrow area of 8 ± 6 m^2, and a burrow length of 14 ± 8 m in average (Antinuchi and Busch 1992).

Burrow systems are dynamic structures. The animals prolong the galleries daily toward patches with abundant vegetation, while other sections are abandoned and obliterated with sediment to prevent the transit of predators (Dentzien-Dias and Figueiredo 2015). The existence of areas surrounding *C. mendocinus* burrow systems showing past signs of burrowing activity and a relatively lower vegetation cover (Rosi et al. 2000) agrees with this observation. These facts are further evidence that burrow systems are dynamic structures, where grazed areas with partly depleted food are temporally abandoned, and feeding galleries are extended through microhabitats with relatively higher food availability, as suggested by Rosi et al. (2000). This burrowing dynamic allows tuco-tucos to access plant resources distributed over a relatively large area, by means of short and fleeting excursions to the surface from several burrow openings. There is evidence of intraspecific differences in digging dynamics: populations of *C. mendocinus* that occupy different habitats in terms of plant abundance and diversity differ in burrow attributes such as the total length of the burrow and number of gallery bifurcations (Rosi et al. 2000). A radio-telemetry interspecific study by Cutrera et al. (2010) found that the home-range size of the sand dune tuco-tuco *C. australis* was ~19 times larger than that of Los Talas' tuco-tuco *C. talarum*, in spite of *C. australis* being only 2 to 3 times larger and having a BMR 1.3–2.6 higher than C. *talarum*. These authors suggested that the differences in the composition of the plant community of the different habitats of these species could affect the turnover rate of the consumable plants, and ultimately habitat food availability through time, leading the sand dune tuco-tuco's burrow to cover greater areas to meet their energy requirements.

In line with this, Kubiak et al. (2017) argued that *C. minutus* inhabiting sand dunes probably need longer galleries covering larger areas to access sufficient food resources because this habitat has relatively less plant biomass than other habitats occupied by the species. In sum, and contrary to what happens with those species that use the burrow only as shelter and/or breeding, subterranean rodents carry out daily routine digging activities to build their burrow in such a way that its structure and dynamics are in line with the variable attributes of the occupied habitat and the energy requirements of the species.

Contrary to what one might think in the case of a subterranean rodent, burrowing does not seem an activity to which tuco-tucos devote much to the daily time budget. In fact, in *C. talarum* it has been estimated that the daily lengthening of the burrow for accessing new patches of vegetation is ~1 m/day (Vassallo 2006), and ~3 m in larger species such as *C. australis* (AIV, pers. obs. based on the daily appearance of new mounds). Based on a digging speed of 1.5–5 m/hour, which varies depending on the hardness of the soil (Vassallo 1998; Luna and Antinuchi 2007), it can be inferred that daily lengthening can demand 12–40 net minutes of digging. Pioneering studies by Vleck (1979) showed that progressing by digging requires energy expenditure 360–3400 times greater than moving the same distance aboveground.

However, looking at the daily energy expenditure, digging represents only 12% of the thermoregulation cost, and 10% of the maintenance cost, as estimated by Antinuchi et al. (2007). Even daily displacements through the tunnel system (patrolling; access to foraging openings) would consume more energy than digging. Hence, it is intriguing why a relatively limited activity concerning daily time and energy budgets is associated with major changes in morphology. These include those shown by tuco-tucos, especially in its biomechanical and functional morphology attributes. Probably, this is partly due to the relatively high forces required to disaggregate and transport the substrate.

7.4 Functional Morphology of Digging Strategies

7.4.1 Scratch-Digging

The morphological adaptations related to scratch-digging are well-known in mammals, and they are mainly found in the forelimbs. For example, the superstructures (i.e., bone eminences such as the teres major, deltoid, and olecranon processes, and the distal, lateral, and medial epicondyles of the humerus) of long bones are strongly developed in scratch-diggers. One of the main functions of bone superstructures is to provide an attachment point for tendons and ligaments, which transmit force from the contracting muscles to the skeleton (Mc Henry and Corruccini 1975; Polly 2007; Lessa et al. 2008). Thus, in scratch-diggers, the more developed bone superstructures on forelimbs allow an increased strength in flexing the large digits and the wrist, extending the elbow, flexing the humerus on the scapula, and stabilizing the shoulder (Hildebrand 1985; see Table 7.1). For example, a study of limb morphology and function in caviomorph rodents not including *Ctenomys* (Elissamburu and Vizcaíno 2004) showed that, in general terms, caviomorph diggers are characterized by morphofunctional indices (ratios) that represent higher humeral and ulnar robustness, higher deltoid and epicondylar development, and increased mechanical advantage of elbow extensor muscles. In *Ctenomys*, Echeverría et al. (2014) investigated the postnatal ontogeny of limb proportions and functional indices in *C. talarum*, and found that during the early development of this species, several morphological traits that are typically associated with scratch-digging are already present (suggesting that they are prenatally shaped), and other traits develop progressively. For instance, when compared with juveniles and adults, young pups show relatively more robust humeri, ulnae, and femora (being robustness the anteroposterior diameter at diaphysis divided by length), wider humeral epicondyles, longer zeugopodial elements, a proportionally shorter olecranon, as well as a poorly developed teres major processes and an incipient deltoid crest (Echeverría et al. 2014). Bone robustness can be related to the need to support the body mass during locomotion or of developing the forces required for more specific functions of the limb, such as digging activity (Elissamburu and De Santis 2011). Thus, the greater bone robustness observed in young *C. talarum* may indicate a possible compensation for lower bone stiffness, while the wider epicondyles may be associated with improved effective forces in those muscles that originate onto them (see Table 7.1), compensating the lower muscular development. Echeverría et al. (2014) also found a gradual increase in the relative olecranal length (i.e., the index of fossorial ability: IFA = olecranal length/functional ulnar length), suggesting a gradual enhancement in the scratch-digging performance due to an improvement in the mechanical advantage of forearm extensors (i.e., m. triceps and m. dorsoepitrochlearis). Middle limb indices, which indicate the extent to which the forelimb is apt for fast movement (brachial index) and how well the hindlimbs are apt for speed (crural index), is higher in pups than in juveniles-adults, reflecting relatively more gracile limbs in their middle segments, which is in accordance with their incipient fossorial ability.

In terms of behavioral development, and in accordance with what was previously observed at the morphological level by Echeverría et al. (2014), Echeverría et al. (2015) observed that in *C. talarum* scratch-digging and burrowing express early during postnatal ontogeny, particularly, during lactancy age (Fig. 7.2). In this study, it was found that the digging of a "true burrow" (i.e., a completely closed tunnel of at least the animal's body length) clearly preceded the dispersal age, and it happened around 18 (lactation) and 47 (post-weaning) postnatal days (Fig. 7.2). Also, it was observed that young pups are capable of loosening the soil using their foreclaws, and remove the accumulated substrate using their hindfeet as adults do (i.e., via an *inchworm* locomotion in reverse, see Hickman 1985). Prior to weaning age, young are able to construct simple burrows to shelter, and they dig for the first time their own burrow in the absence of either a burrowing demonstrator or an early subterranean environment (i.e., a natal burrow). Thus, in this species, immature digging behavior gets expressed early during ontogeny and develops progressively. Taking into account that (a) some morphological adaptations related to scratch-digging are already present in very young individuals and others develop progressively (Echeverría et al. 2014), (b) subterranean lifestyle is assumed to be highly

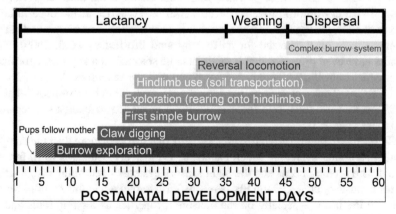

Fig. 7.2 Postnatal ontogeny of salient behaviors related to digging in *C. talarum*. (Modified from Echeverría 2011)

costly due to the high energetic demand of tunnel extension (Vleck 1979; Luna et al. 2002), and (c) *C. talarum* displays a long maternal care period (~2 months; Zenuto et al. 2002), the early occurrence of scratch-digging and burrow construction would provide of enough time to reach a proper musculoskeletal and behavioral development, mainly to deal with energetic and biomechanical demands. Epigenetic effects of the early biomechanical environment have been documented in several groups of mammals and, therefore, it is widely accepted that both genetic and epigenetic factors determine the final shape and strength of the skeleton (Carter et al. 1998; Nowlan and Prendergast 2005). For example, physical activity may promote normal development of muscles and bones (e.g., Herring and Lakars 1981), and patterns of muscle contraction can cause a differential growth in those bone areas that are closest to peak strains (Young et al. 2009; and references therein). Thus, functional demands would affect relatively more the position of bone structures associated with muscle and ligament attachment (e.g., deltoid process) than the attainment of the fundamental form of the skeleton, as well as the presence of bone superstructures (e.g., condyles, articular surfaces, tuberosities, grooves) (Murray 1936; Hall 1978).

7.4.2   Tooth-Digging

Ctenomys species have spread throughout a remarkable portion of South America, having to confront a huge variety of environmental conditions when digging is performed (Bidau 2015). For many of those species, scratch-digging would not be enough to break down the clayey rocks, roots, and all sorts of obstacles in the soil, and a less flexible – but still more robust and powerful – apparatus needs to show up in the scene: the jaws (see, for instance, Vassallo 1998; Lessa et al. 2008; Becerra et al. 2013). In order to do so, these animals use their procumbent incisors to apply powerful forces, much more powerful even than the expected ones for a small-to-medium size caviomorph rodent (Becerra et al. 2011, 2013, 2014; Becerra 2015; Borges et al. 2016; Fig. 7.3). Even though this strong biting – relative to body size – seems to have been selected for an underground life (Lacey et al. 2000), some authors have even proposed that it might also be selected as a key factor for males success at controlling a certain habitat range and the resources therein (Becerra et al. 2012a). This fact may be the reason why, even though tuco-tucos inhabit a wide spectrum of soils, bite force is not directly related to habitat's mechanical constrains (Fig. 7.3).

Chisel-tooth digging is a burrowing strategy that involves more than the complex craniomandibular system – which cannot be exclusively dedicated to it, but it is shared with feeding needs and the main sensory systems – recruiting also much of the neck musculature (AIE and FB, pers. obs.). Nevertheless, in this section, we will focus on the head, especially on the incisors' properties as digging tools and the overall biomechanics of biting.

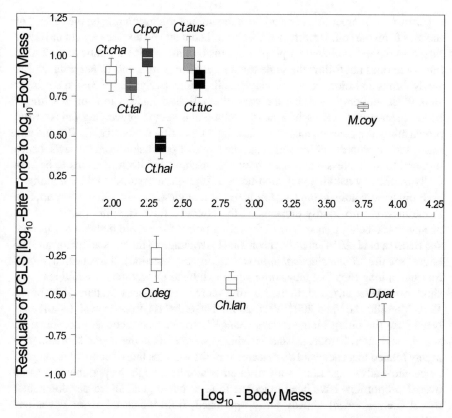

Fig. 7.3 Scatter-plot of residual values of a phylogenetic generalized least squared regression (PGLS) between the log-transformed bite force and log-transformed body mass – to avoid the potential distortive effect of either body size or evolutionary affinity – in six *Ctenomys* species and four other caviomorph rodents as step-wise out-groups. Phylogeny was built upon cytochrome-b sequences available on GenBank. Species used with GenBank vouchers in brackets: *Ct.cha, C. "chasiquensis"* (MF770015.1); *Ct.por, C. porteousi* (AF370682.1); *Ct.tal, C. talarum* (AF370698.1); *Ct.aus, C. australis* (AF370697.1); *Ct.tuc, C. tuconax* (AF370684.1); *Ct.hai, C. haigi* (HM777476.1); *O.deg., Octodon degus* (Fam. Octodontidae; AF007058.1); *M.coy, Myocastor coypus* (Fam. Echimyidae; AF422919.1); *Ch.lan, Chinchilla lanigera* (Fam. Chinchillidae; JX312692.1); *D.pat, Dolichotis patagonum* (Fam. Caviidae; GU136724.1). Gray scale within *Ctenomys* species (except for *C. "chasiquensis"*) represents the soil hardness measured in situ. Taking into account that the phylogeny was constructed at the species level, standard deviations were rescaled based on the mean values after PGLS. Data from Becerra (2015)

It has been previously observed that tuco-tucos attack most hard obstacles by anchoring the upper incisors and applying the force at the lower incisors tip, by repetitive firing movements of the mandible (Vassallo 1998; Becerra et al. 2013). Therefore, incisors' resistance toward abrasion is highly important. Justo et al. (1995), De Santis et al. (2001), and Vieytes et al. (2007) studied the incisors' *schmelzmuster* (i.e., enamel pattern) in ctenomyids. They found out that the thicker external (radial) enamel would be present when a more intense tooth-digging habit is performed, in order to withstand the concomitant microgrinding on the surface of incisors from the soil. Nevertheless, while soil content and texture seem to be highly linked to the external enamel properties (Justo et al. 1995; De Santis et al. 2001), bite force does not follow the same trend (e.g., *C. australis* and *C. haigi* inhabiting sandy dunes and more compact, clayey soils, respectively; Fig. 7.3). On the inner side, those authors found that the enamel's bulk is organized in Hunter-Schreger bands (decussated prisms in a nearly orthogonal setup), preventing cracks from propagating as a consequence of surface fracture during tooth-digging. Beyond any macroscopic property of incisors (e.g., the angle of protrusion from the jaws, bending, and torsion stress resistance), their microscopic architecture seems to be specially molded by nature giving tuco-tucos a "high-performance" tool for digging.

From a more holistic (and anatomical) point of view, the amount of force applied at the incisors' tip strictly depends on the effective force (i.e., the fraction of the muscle force being transmitted to the biting point; for a more detailed discussion, see Becerra et al. 2014) and the mechanical advantage of the jaws adductor musculature (see the "*Biomechanical approach of digging*" section). It has been assumed for quite a long time that the shorter snout exhibited by tuco-tucos would lead to a shorter out-lever arm and, therefore, to a more advantageous system (Hildebrand 1988; Vassallo and Verzi 2001; Verzi 2002). Nevertheless, Becerra et al. (2014) analyzed how the biting biomechanics change between tuco-tucos and non-digging, social, caviomorph rodents, and strikingly showed that the snout shortening is highly related to a backward displacement of the muscles line of action, keeping the mechanical advantage fairly similar. In other words, it might be possible that these overall proportions have been inherited from a rodent generalized developmental model (i.e., *bauplan*) already fairly optimized, from which all lesser deviations radiated.

On a mesoscale, it would be logical to think that the angle at which incisors protrude from the jaws slightly but directly affects the out-lever arm, and inversely do so to the mechanical advantage and bite force (on equivalent muscle input setups). Yet, previous studies have found some positive correlation between procumbency angle and a burrowing lifestyle, beyond phylogenetic or body size differences (Lessa 1990; Lessa et al. 2008; Becerra et al. 2012b). These results might be indicating that this "mechanically disadvantageous" condition would be negligible compared to the great improvement of having front-forwarded incisors for better anchoring to the soil. Within the genus *Ctenomys*, on the other side, the story seems to differ in one step further showing no clear pattern of procumbency variation with regard to skull's gross morphology or environment (see Mora et al. 2003; Verzi et al. 2010; Echeverría et al. 2017). This fact leads us to consider a putative scenario, at which tuco-tucos as a group would have already achieved a state of the character above what is strictly needed for tooth-digging. Therefore, at the highly specialized ctenomyid *bauplan* – much more than the generalized caviomorph one – incisors would protrude at an angle wide enough to withstand the mechanical demands of soil break down, and deviations from this model would be unrelated to these physical needs.

On the flip side, and being negligible the above-mentioned differences in proportion of bony elements (i.e., mechanical advantage), differences in bite force should be a consequence of mere behavioral dissimilarities, or of an uneven degree of development for the adductor musculature. In this regard, Becerra et al. (2014) showed that when both superficial and deep masseter muscles and both heads of the zygomaticomandibularis muscle (*sensu* Druzinsky et al. 2011; Table 7.1) pull in concert, they account for over 90% of the bite force. For all of them, except for the head of the latter coming through the infraorbital foramen (plesiomorphic condition for all hystricomorph rodents), a significantly greater degree of development was coincidently shown for tuco-tucos than for other non-subterranean caviomorph rodents (Becerra et al. 2011, 2013, 2014). A closer look at all jaw adductor muscles across ctenomyids, based on data from these last references, showed that they do not depend on body size (all *p-values* > 0.05). We could, then, assume that the strong biting that tuco-tucos apply during tooth-digging has not been evolutionarily achieved by rearranging the skull morphology, but by hypertrophying their musculature. Likewise, preliminary studies that analyze the masseteric musculature of *C. talarum* and *Cavia aperea*, a non-digging and gregarious caviomorph rodent, showed an increase in the heterogeneity of fiber types in tuco-tucos (Fig. 7.4), with higher proportions of fibers capable of generating more powerful forces (Longo et al. 2017, 2018).

Summarizing, ctenomyid species have evolved associated to an underground lifestyle, developing some anatomical specializations which help them accomplish tooth-digging tasks. Although gross head morphology does not enhance their digging performance, their hypertrophied adductor jaw musculature lets them apply biting forces stronger than needed to easily break down the soil and obstacles while burrowing. Moreover, incisors macro- and microstructure allow tuco-tucos not only to attack the soil at an advantageous angle but also to withstand the high stress and abrasion produced by specific clayey, silty, or sandy soils. Overall, the fact that specializations observed within this genus (e.g., bite force, adductor musculature development, and procumbency angle) clearly exceed the particular needs for tooth-digging could respond to a highly derived *bauplan*, or to cover other (behavioral?) needs; e.g., mating competition. Being that for one or the other, it could also explain the lack of association between any of those specializations and soil conditions.

7.5 Structural Strength of Bones

When a subterranean rodent like the tuco-tuco digs using the so-called digging tools, it must exert forces strong enough to break down the soil, which is allowed by their powerful forelimb and mandibular muscles. In turn, when these forces are exerted against the substrate, the digging tools receive reaction forces of equal magnitude and opposite direction (i.e., the Newtonian principle of action and reaction, which also applies to subterranean rodents! See vector Fo in Fig. 7.1). These reaction forces are transmitted to the rest of the body, in particular to the bones, which

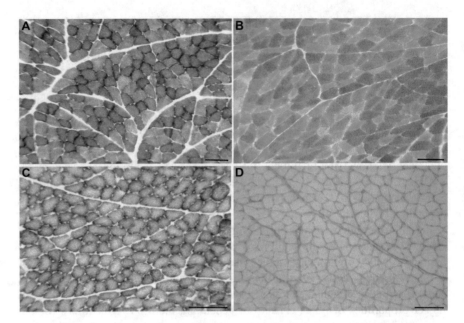

Fig. 7.4 Fiber types of superficial masseter muscles in *Ctenomys talarum* (A, B) and *Cavia aperea* (C, D). (A, C) Succinate dehydrogenase activity, indicative of the oxidative capacity of fibers. (B, D) Periodic acid Schiff, to detect the glycolitic activity of fibers. Different staining intensities indicate different fiber types. Scale bars: 100 μm

must be able to resist them without structural failure, i.e., without breakage or permanent deformation. Since these forces are concentrated in places such as sites of muscle origin and insertion, sites of contact between bones (e.g., joints), or the tip of digging tools, this force per unit area which can be expressed as stress and measured in pascals (i.e., newton per square meter). As an adaptation to these efforts and their concomitant stresses, not only the claws and incisors but also the limb long bones and skull of subterranean rodents are relatively more robust than those of their non-digging species counterparts (Becerra et al. 2012b; Echeverría et al. 2014; McIntosh and Cox 2016). One method to estimate the stress that these forces produce is by means of the development of a virtual model (Fig. 7.5A, B) with the same geometry as real bones, both internally and externally, from axial tomographic images (Fig. 7.5C) and specific software for image integration (see it in detail in Buezas et al. 2019). The mechanical properties of bone are assigned to this model, a procedure known as finite element analysis (FEA; Rayfield 2007). In a second stage, the model is subjected to in silico experiments, at which the application and reception of forces are simulated. Ideally, these forces should be consistent with the real forces that animals exert in nature. To measure these forces, experimental values are obtained from live animals with force transducers, while biomechanical estimations are based upon muscle dissections and lever arm measurements (e.g., Becerra et al. 2011, 2013, 2014). By using a FEA approach, it was quantified that the stress to which *C. talarum* skull is subjected to during tooth-digging can vary in

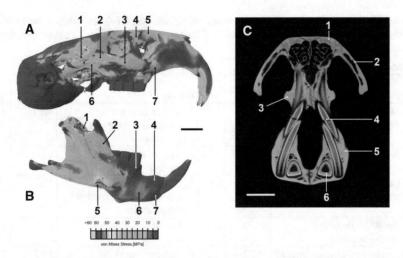

Fig. 7.5 Structural bone resistance in *C. talarum*'s skull. Finite element analysis (FEA) showing the stress to which the cranium (**A**) and mandible (**B**) are subjected during incisor biting. The FE model was loaded with the corresponding jaw adductor muscle forces, which resulted in a bite force similar to the one measured in live specimens using a force transducer. Areas subjected to relatively high stresses are towards the red side of the spectrum (von Mises stress >25 MPa). Note that, within the cranium, these areas spread throughout the posterior portion of the zygomatic arch (zones 1, 2, 6), dorsal origin of the anterorbital bar (4), dorsal- and ventral-most regions within the diastema (5, 7), and the lingual side of incisors. In the mandible, areas of high stress mainly spread throughout the ascending ramus and anterior end of masseteric crest (1, 5). Descriptions, stress values and safety factors for numbered zones are provided in Table 7.3. Coronal microCT slice at the level of molar 1 (**C**), highlighting some features responsible of the architecture-based mechanically enhanced skull in this species (1, thick frontal bones; 2, anterorbital bar; 3, enlarged maxillar bones; 4, long-rooted molariform teeth; 5, thick mandible; 6, incisor roots moving back to the ascending ramus). Scale bars: 5 mm.

more than one order of magnitude depending on the anatomical region being considered (Buezas et al. 2019).

Another question related to structural strength is: what is the relationship between the (theoretical) maximum stress that the bone can withstand, and the corresponding values achieved during the real performance in nature? A proper way to answer this question is by calculating the safety factor (SF), a mechanical estimation of how below the stress at a certain point is from that one at which bone begins to experience a nonreversible change of shape in response to applied forces (e.g. Currey 2006). Therefore, these values are computed as the quotient between the critical yield stress and the local von Mises stress. It is shown that stress and SF values are more heterogeneous within the mandible than on the cranium in *C. talarum* (Table 7.3). This is because of the natural variation of mandibular bone thickness: thick at the diastema and the mandibular corpus, where incisor and molar roots are hosted (regions 3 and 4, respectively), and relatively thin in the ascending ramus (regions 2 and 5).

Table 7.3 Anatomical descriptions, von Mises stress values (in MPa), and safety factors (SF) in 14 regions across the mandible and cranium in *Ctenomys talarum* (see Fig. 7.5), corresponding to an incisor bite force of 32 N

Cranium			Mandible		
Region	Stress (MPa)	SF	Region	Stress (MPa)	SF
1 Most posterior end of the zygoma	84.19	1.94	1 Region between the coronoid and the condyloid processes	66.7	2.4
2 Most dorsal region of the zygoma	26.87	6.07	2 Base of the coronoid process	21.2	7.7
3 Central region of the zygoma	18.21	8.95	3 Dorsal region of the mandible, at the level of the premolar	3.3	49.4
4 Dorsal origin of the anterorbital bar	25.72	6.34	4 Dorsal region of the diastema	9.7	16.8
5 Frontal region at the level of the diastema's posterior end	19.31	8.44	5 Anterior end of the masseteric crest	52.8	3.1
6 Postero-ventral region of the zygoma	19.23	8.95	6 Lateral region of the diastema	6.2	26.3
7 Posterior end of the diastema – anterior end of the zygoma	23.73	6.87	7 Antero-ventral region of the diastema	6.5	25.1

Summarizing, biting forces received at incisors tips during chisel-tooth digging are directly transmitted to the skull, a structure which provides anchorage and protection to sensory organs and the central nervous system. Bone distribution and skull geometry in subterranean rodents like tuco-tucos can have significant effects on the adequate dissipation of biting forces and stress dampening. Together with the complexity of the cranial sutures (Buezas et al. 2017), these traits seem to be important structural adaptations related to the burrowing behavior.

7.6 Conclusions

As reviewed above, the predominant mode of digging among subterranean rodents is scratch-digging, followed by chisel-tooth digging when soils or underground obstacles become tougher to disaggregate or break down. Tuco-tucos – as other subterranean mammals – are characterized by several physiological, morphological, and behavioral adaptations related to their subterranean lifestyle (Nevo and Reig 1990; Nevo 1999) and, strikingly, they are among the few subterranean rodents capable of performing those two digging methods on an efficient manner (see other remarkable examples in Jarvis and Sale 1971; Hickman and Brown 1973; Lessa and Thaeler Jr. 1989). This functionally dual capacity – coupled with the highly optimized biomechanics of both musculoskeletal systems – allows tuco-tucos to inhabit a wide range of habitats and soils, from the Andean Puna above 4000 m to the coastal sandy dunes of the Atlantic, and from the mesic and humid Pampas to the dry Chaco and Monte desert (Reig et al. 1990; Bidau 2015; Fig. 7.1). Particularly, the soil types where they are found include very friable and soft sandy soils; gravelly soils; hard, clayey soils, hard soils rich in humus, and with a low percentage of silt or clay; sandy rock-free soils; well-compacted, humus-rich, or loamy soils; deep humid soils; black soil but extremely light; compact sandy soils, among others (see, for example, Table 7.1 in Echeverría et al. 2017). However, in spite of occupying a wide geographical and environmental range, the ecological niche they occupy and the behaviors they exhibit are broadly homogeneous (Mora et al. 2003). In general terms, subterranean rodents classified as scratch-diggers according to their morphological specializations are often found in sandy and friable soils, whereas chisel-tooth diggers are found in a broader range of soils (Lessa and Thaeler Jr. 1989). Thus, species such as *C. talarum* (both chisel-tooth and scratch-digger) efficiently performs in both well-compacted, humus-rich, or sandy soils – where the fairly sympatric *C. australis* (a mainly scratch-digger) is confined (Vassallo 1998). Concluding, other factors – besides morphofunctional specializations – such as physiological adaptations or predation pressure might be constraining the presence of a particular species in specific soils, and this must be taken into account to explain habitat preferences in *Ctenomys* species.

Acknowledgments We thank the editors for the invitation to contribute to this collective volume on *Ctenomys* evolution. This study was carried out under the support of the National Council for Scientific and Technical Research of Argentina (CONICET) (grants PIP 2014–2016 N° 11220130100375 and UE N°0073) and National University of Mar del Plata (grants EXA918/18 and ING516/18). We thank colleagues from various countries and laboratories for the friendly exchange of knowledge about an underground friend in common, the Tuco-Tuco.

References

Ade M, Ziekur I (1999) The forepaws of the rodents *Cryptomys hottentotus* (Bathyergidae) and *Nannospalax ehrenbergi* (Muridae, Spalacinae): phylogenetic and functional aspects. Mitteilungen aus dem Museum für Naturkunde in Berlin/Zoologische Reihe 75:11–17

Andino N, Borghi CE, Giannoni SM (2014) Characterization and selection of microhabitat by *Microcavia australis* (Rodentia: Caviidae): first data in a rocky habitat in the Hyperarid Monte Desert of Argentina. Mammalia 80:71–81

Antinuchi CD, Busch C (1992) Burrow structure in the subterranean rodent *Ctenomys talarum*. Mamm Biol 57:163–168

Antinuchi CD, Zenuto RR, Luna F, Cutrera AP, Perissinotti PP, Busch C (2007) Energy budget in subterranean rodents: insights from the tuco tuco *Ctenomys talarum* (Rodentia: Ctenomyidae). In: Kelt DA, Lessa EP, Salazar-Bravo JA, Patton JL (eds) The quintessential naturalist: honoring the life and legacy of Oliver P. Pearson. The University of California Press, Berkeley, pp 111–140

Becerra F (2015) Aparato masticatorio en roedores caviomorfos (Rodentia, Hystricognathi): Análisis morfofuncional, con énfasis en el género *Ctenomys* (Ctenomyidae). Ph.D. dissertation, Universidad Nacional de Mar del Plata. Mar del Plata, Buenos Aires, Argentina

Becerra F, Echeverría AI, Vassallo AI, Casinos A (2011) Bite force and jaw biomechanics in the subterranean rodent Talas tuco-tuco (*Ctenomys talarum*) (Caviomorpha: Octodontoidea). Can J Zool 89:334–342

Becerra F, Echeverría AI, Marcos A, Casinos A, Vassallo AI (2012a) Sexual selection in a polygynous rodent (*Ctenomys talarum*): an analysis of fighting capacity. Zoology 115:405–410

Becerra F, Vassallo AI, Echeverría AI, Casinos A (2012b) Scaling and adaptations of incisors and cheek teeth in caviomorph rodents (Rodentia, Hystricognathi). J Morphol 273:1150–1162

Becerra F, Casinos A, Vassallo I (2013) Biting performance and skull biomechanics of a chisel tooth digging rodent (*Ctenomys tuconax*; Caviomorpha; Octodontoidea). J Exp Zool 319:74–85

Becerra F, Echeverría AI, Casinos A, Vassallo AI (2014) Another one bites the dust: bite force and ecology in three caviomorph rodents (Rodentia, Hystricognathi). J Exp Zool 321A:220–232

Begall S, Burda H, Schleich CE (2007) Subterranean rodents: News from underground. Springer-Verlag, Berlin

Bidau CJ (2015) Family Ctenomyidae Lesson, 1842. Pp. 818–877 in Mammals of South America, Volume 2 Rodents. (J.L. Patton, U.F.J. Pardiñas, and G. D'Elía, eds.). The University of Chicago Press, Chicago

Biewener AA (2003) Animal locomotion, 1st edn. Oxford University Press, Oxford

Borges LR, Maestri R, Kubiak BB, Galiano D, Fornel R, de Freitas TRO (2016) The role of soil features in shaping the bite force and related skull and mandible morphology in the subterranean rodents of the genus *Ctenomys* (Hystricognathi: Ctenomyidae). J Zool 301:108–117

Buezas GN, Becerra F, Vassallo AI (2017) Cranial suture complexity in caviomorph rodents (Rodentia;Ctenohystrica). J Morphol 278:1125–1136

Buezas GN, Becerra F, Echeverría AI, Cisilino A, Vassallo AI (2019) Mandible strength and eometry in relation to bite force: a case study in three caviomorph rodents. J Anat 234:564–575

Carter D, Mikic B, Padian K (1998) Epigenetic mechanical factors in the evolution of long bone epiphyses. Zool J Linnean Soc 123:163–178

Currey JD (2006) Bones: structure and mechanics. Princeton University Press, Princeton

Cutrera AP, Antinuchi CD, Mora MS, Vassallo AI (2006) Home-range and activity patterns of the south American subterranean rodent *Ctenomys talarum*. J Mammal 87:1183–1191

Cutrera AP, Mora MS, Antenucci CD, Vassallo AI (2010) Intra- and interspecific variation in home-range size in sympatric tuco-tucos, *Ctenomys australis* and *C. talarum*. J Mammal 91:1425–1434

De Santis LJM, Moreira GJ, Justo ER (1998) Anatomía de la musculatura branquiomerica de algunas especies de *Ctenomys* Blainville, 1826 (Rodentia, Ctenomyidae): Caracteres adaptativos. Boletín de la Sociedad de Biología de Concepcion 69:89–107

De Santis LM, Moreira GJ, García Esponda CM (2001) Microestructura del esmalte de os incisivos de *Ctenomys azarae* y *C. talarum* (Rodentia, Ctenomyidae). Mastozoología Neotropical 8:5–14

Dentzien-Dias PC, Figueiredo AEQ (2015) Burrow architecture and burrowing dynamics of *Ctenomys* in foredunes and paleoenvironmental implications. Palaeogeogr Palaeoclimatol Palaeoecol 439:166–175

Druzinsky RE, Doherty AH, De Vree FL (2011) Mammalian masticatory muscles: homology, nomenclature, and diversification. Integr Comp Biol 51:224–234

Ebensperger LA, Blumstein DT (2006) Sociality in New World hystricognath rodents is linked to predators and burrow digging. Behav Ecol 17:410–418

Echeverría AI, Becerra F, Vassallo AI (2014) Postanatal ontogeny of limb proportions and functional indices in the subterranean rodent *Ctenomys talarum* (Rodentia, Ctenomyidae). J Morphol 275(8):902–913

Echeverría AI, Biondi LM, Becerra F, Vassallo AI (2015) Postnatal development of subterranean habits in tuco-tucos *Ctenomys talarum* (Rodentia, Caviomorpha, Ctenomyidae). J Ethol 34(2):107–118

Echeverría AI, Becerra F, Buezas GN, Vassallo AI (2017) Bite it forward… bite it better? Incisor procumbency and mechanical advantage in the chisel-tooth and scratch-digger genus *Ctenomys* (Caviomorpha, Rodentia). Zoology 125:53–68

Echeverría AI, Abdala V, Longo MV, Vassallo AI (2019) Functional morphology and identity of the thenar pad in the subterranean genus *Ctenomys* (Rodentia, Caviomorpha). J Anat 235:940–952

Elissamburu A, De Santis L (2011) Forelimb proportions and fossorial adaptations in the scratch-digging rodent *Ctenomys* (Caviomorpha). J Mammal 92:683–689

Elissamburu A, Vizcaíno S (2004) Limb proportions and adaptations in caviomorph rodents (Rodentia: Caviomorpha). J Zool 262:145–159

Ellison GTH (1995) Is nest building an important component of thermoregulatory behaviour in the pouched mouse (*Saccostomus campestris*). Physiol Behav 57:693–697

Fleming TH, Brown GJ (1975) An experimental analysis of seed hoarding and burrowing behaviour in two species of Costa Rican heteromyid rodents. J Mammal 56:301–315

Giannoni SM, Borghi CE, Roig VG (1996) The burrowing behavior of *Ctenomys eremophilus* (Rodentia, Ctenomyidae) in relation with substrate hardness. Mastozoología Neotropical 3:161–170

Hall BK (1978) Developmental and cellular skeletal biology. Academic, New York

Herring SW, Lakars TC (1981) Craniofacial development in the absence of muscle contraction. J Craniofac Genet Dev Biol 1:341–357

Hickman GC (1985) Surface-mound formation by the tuco-tuco, *Ctenomys fulvus* (Rodentia: Ctenomyidae), with comments on earthpushing in other fossorial mammals. J Zool 205:385–390

Hickman GC, Brown LN (1973) Mound-building behavior of the southeartern pocket gopher (*Geomys pinetis*). J Mammal 54:786–790

Hildebrand M (1985) Digging of quadrupeds. In: Hildebrand M, Bramble M, Liem KF, Wake DB (eds) Functional vertebrate morphology. The Belknap Press of Harvard University Press, Cambridge, MA, pp 89–109

Hildebrand M (1988) Analysis of vertebrate structure, 3rd. edn. Wiley, New York

Hopkins SSB (2005) The evolution of fossoriality and the adaptive role of horns in the Mylagaulidae (Mammalia: Rodentia). Proceedings of the Royal Society of London, B. Biological Sciences 272:1705–1713

Jarvis JUM, Sale JB (1971) Burrowing and burrow patterns of east African mole-rats *Tachyoryctes*, *Heliophobius* and *Heterocephalus*. J Zool 163:451–479

Justo ER, Bozzolo LE, De Santis LM (1995) Microstructure of the enamel of the incisors of some ctenomyid and octodontid rodents (Rodentia, Caviomorpha). Mastozoología Neotropical 2:43–51

Kubiak BB, Galiano D, de Freitas TRO (2017) Can the environment influence species home-range size? A case study on *Ctenomys minutus* (Rodentia, Ctenomyidae). J Zool 302:171–177

Lacey E, Patton JL, Cameron GN (2000) Life underground: the biology of subterranean rodents. Chicago University Press, Chicago

Lessa EP (1990) Morphological evolution of subterranean mammals: integrating structural, functional, and ecological perspectives. In: Nevo E, Reig OA (eds) Evolution of subterranean mammals at the organismal and molecular levels. Wiley, New York, pp 211–230

Lessa EP, Thaeler CS Jr (1989) A reassessment of morphological specializations for digging in pocket gophers. J Mammal 70:689–700

Lessa EP, Vassallo AI, Verzi DH, Mora MS (2008) Evolution of morphological adaptations for digging in living and extinct ctenomyid and octodontid rodents. Biol J Linn Soc 95:267–283

Lobo Ribeiro L (2017) Locomoção de *Ctenomys talarum* (Rodentia: Ctenomyidae): um estudo com câmera de alta velocidade. M.S. thesis, Universidade do Estado do Rio de Janeiro. Rio de Janeiro, Rio de Janeiro, Brazil

Longo MV, Díaz AO and Vassallo AI. (2017) Morfología funcional de la musculatura masetérica del roedor subterráneo *Ctenomys talarum*: histoquímica de tipos de fibras. XXX Jornadas Argentinas de Mastozoología. Bahía Blanca, 14–17 de Noviembre

Longo MV, Díaz AO, and Vassallo AI (2018) Histoquímica del músculo masetero de *Cavia aperea* y su correlación con las demandas funcionales. XIX Congreso de Ciencias Morfológicas y 16 Jornadas de educación de la Sociedad de Ciencias Morfológicas de La Plata. La Plata, 25–26 de Octubre

Luna F, Antinuchi CD (2007) Energy and distribution in subterranean rodents: sympatry between two species of the genus *Ctenomys*. Comp Biochem Physiol 147A:948–954

Luna F, Antinuchi CD, Busch C (2002) Digging energetics in the south American rodent *Ctenomys talarum* (Rodentia, Ctenomyidae). Can J Zool 80:2144–2149

Malizia AI, Busch C (1991) Reproductive parameters and growth in the fossorial rodent *Ctenomys talarum* (Rodentia: Octodontidae). Mammalia 55:293–306

Malizia AI, Vassallo AI, Busch C (1991) Population and habitat characteristics of two sympatric species of *Ctenomys* (Rodentia: Octodontidae). Acta Theriol 36:87–94

Mc Henry HM, Corruccini RS (1975) Distal humerus in hominoid evolution. Folia Primatol 23:227–244

McIntosh AF, Cox PG (2016) The impact of gape on the performance of the skull in chisel-tooth digging and scratch digging mole-rats (Rodentia: Bathyergidae). R Soc Open Sci 3:160568

McIntosh AF, Cox PG (2019) The impact of digging on the evolution of the rodent mandible. J Morphol 280:176–183

Mora MS, Olivares AI, Vassallo AI (2003) Size, shape and structural versatility of the skull of the subterranean rodent Ctenomys (Rodentia, Caviomorpha): functional and morphological analysis. Biol J Linn Soc 78:85–96

Murray PDF (1936) Bones: a study of the development and structure of the vertebrate skeleton. Cambridge University Press, Cambridge

Nevo E (1999) Mosaic evolution of subterranean mammals. Oxford University Press, Oxford

Nevo E, Reig OA (eds) (1990) Evolution of subterranean mammals at the organismal and molecular levels. Alan R. Liss, New York

Nowlan NC, Prendergast PJ (2005) Evolution of mechanoregulation of bone growth will lead to non-optimal bone phenotypes. J Theor Biol 235:408–418

Polly PD (2007) Limbs in mammalian evolution. In: Hall BK (ed) Fins into limbs. Evolution, development and transformation. University of Chicago Press, Chicago, pp 245–268

Rayfield EJ (2007) Finite element analysis and understanding the biomechanics and evolution of living and fossil organisms. Annual Rev Earth Planetary Sci 35:541–576

Reichman OJ, Smith SC (1990) Burrows and burrowing behavior by mammals. In: Genoways HH (ed) Current mammalogy. Plenum Press, New York, pp 197–244

Reig OA, Busch C, Contreras JR, Ortells MO (1990) An overview of evolution, systematics, population biology, cytogenetics, molecular biology, and speciation in Ctenomys. In: Nevo E, Reig OA (eds) Evolution of subterranean mammals at the organismal and molecular levels. Wiley, New York, pp 71–96

Rosi MI, Cona MI, Videla F, Puig S, Roig VG (2000) Architecture of *Ctenomys mendocinus* (Rodentia) burrows from two habitats differing in abundance and complexity of vegetation. Acta Theriol 45:491–505

Stein BR (2000) Morphology of subterranean rodents. In: Lacey EA, Patton JL, Cameron GN (eds) Life underground: the biology of subterranean rodents. The University of Chicago Press, Chicago, pp 19–61

Ubilla M, Altuna C (1990) Analyse de la morphologie de la main chez des espèces de *Ctenomys* de l'Uruguay (Rodentia, Octodontidae): adaptations au fouissage et implications évolutives. Mammalia 54(1):107–118

Ungar PS (2010) Mammal teeth: origin, evolution and diversity. John Hopkins University Press, Baltimore

Van Daele P, Herrel A, Adriaens D (2009) Biting performance in teeth-digging African mole-rats (*Fukomys*, Bathyergidae, Rodentia). Physiol Biochem Zool 82:40–50

Vassallo AI (1998) Functional morphology, comparative behaviour, and adaptation in two sympatric subterranean rodents genus *Ctenomys* (Caviomorpha: Octodontidae). J Zool 244:415–427

Vassallo AI (2006) Acquisition of subterranean habits in tuco-tucos (Rodentia, Caviomorpha, *Ctenomys*): role of social transmission. J Mammal 87:939–943

Vassallo AI, Verzi DH (2001) Patrones craneanos y modalidades de masticación en roedores caviomorfos (Rodentia, Caviomorpha). Bol Soc Biol Concepc 72:139–145

Verzi DH (2002) Patrones de evolución morfológica en Ctenomyinae (Rodentia, Octodontidae). Mastozoología Neotropical 9:309–328

Verzi DH, Álvarez A, Olivares AI, Morgan CC, Vassallo AI (2010) Ontogenetic trajectories of key morphofunctional cranial traits in south American subterranean ctenomyid rodents. Journal o Mammalogy 91:1508–1516

Vieytes EC, Morgan CC, Verzi DH (2007) Adaptive diversity of incisor enamel microstructure in south American burrowing rodents (family Ctenomyidae, Caviomorpha). J Anat 211:296–302

Vleck D (1979) The energy cost of burrowing by the pocket gopher *Thomomys bottae*. Physiol Zoology 52:122–136

Vogel S (2013) Comparative biomechanics: Life's physical world, 2nd edn. Princeton University Press, Princeton

Whitford WG, Kay RF (1999) Bioperturbation by mammals in deserts: a review. J Arid Environ 41:203–230

Woods CA (1972) Comparative myology of jaw, hyoid, and pectoral appendicular regions of new and Old World hystricomorph rodents. Bull Am Mus Nat Hist 147:115–198

Young NM, Hallgrímsson B, Jr Garland T (2009) Epigenetic effects on integration of limb lengths in a mouse model: selective breeding for high voluntary locomotor activity. Evol Biol 36:88–99

Zelová J, Šumbera R, Okrouhlík J, Burda H (2010) Cost of digging is determined by intrinsic factors rather than by substrate quality in two subterranean rodent species. Physiol Behav 99(1):54–58

Zenuto RR, Vassallo AI, Busch C (2002) Comportamiento social y reproductivo del roedor subterráneo solitario *Ctenomys talarum* (Rodentia: Ctenomyidae) en condiciones de semicautiverio. Rev Chil Hist Nat 75:165–177

Chapter 8
Adaptive Pelage Coloration in *Ctenomys*

Gislene Lopes Gonçalves

8.1 Introduction

In rodents, pelage color tends to resemble background habitat coloration, suggesting an adaptive significance (Sumner 1934; Dice and Blossom 1937; Cott 1940; Endler 1978; Krupa and Geluso 2000). But how does it work in subterranean lineages, in which individuals expend most of their lifetime in burrowing systems (Lacey et al. 2000)? Surprisingly, long stand studies in pocket gophers (*Thomomys bottae* and *Geomys bursarius*) and the Israeli subterranean mole rat (*Spalax ehrenbergi*) have demonstrated similar patterns to aboveground rodents, i.e., a strong correlation between dorsal pelage and soil coloration (Ingles 1950; Kennerly 1954, 1959; Krupa and Geluso 2000; Heth et al. 1988), presumably reflecting an influence of selective pressure when they are active on the surface. This concealment coloration is also substantiated in *Ctenomys* (Langguth and Abella 1970; Vassalo et al. 1994), in which pelage color varies continuously, both inter- and intraspecies (Langguth and Abella 1970; Freitas and Lessa 1984; Wlasiuk et al. 2003; Gonçalves and Freitas 2009; Gonçalves et al. 2012). Overall, coat coloration ranges from light to dark brown in tuco-tucos (Fig. 8.1). However, brown with white patterns, grayish, and melanic phenotypes are also present. Similarly, variation is found in the background environment, as species are spread throughout South America, including a variety– and vast areas – of habitats, e.g., pampas of Puna (above 4000 m), high mountain steppes, low valleys of the west, dunes of the Atlantic coast of the east, mesic and humid plains, desert or semi-deserts, open areas among subtropical

G. L. Gonçalves (✉)
Departamento de Genética, Universidade Federal do Rio Grande do Sul,
Porto Alegre, RS, Brazil

Departamento de Recursos Ambientales, Facultad de Ciencias Agronómicas,
Universidad de Tarapacá, Arica, Chile
e-mail: lopes.goncalves@ufrgs.br

© Springer Nature Switzerland AG 2021
T. R. O. de. Freitas et al. (eds.), *Tuco-Tucos*,
https://doi.org/10.1007/978-3-030-61679-3_8

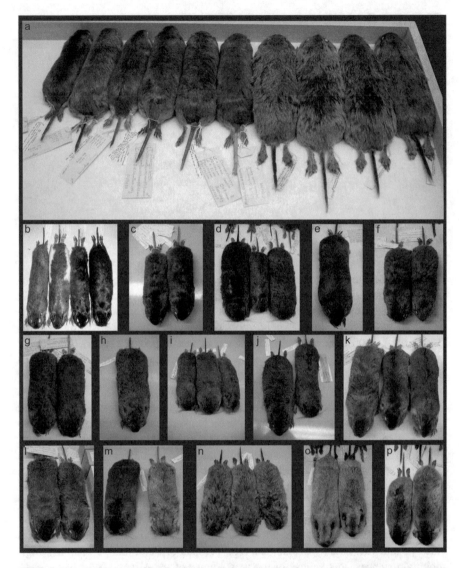

Fig. 8.1 Inter- and intraspecific variation in pelage color of *Ctenomys*. (**a**) general view of a speci-mens' drawer of tuco-tucos from the Museum of Vertebrate Zoology (MVZ), revealing the typical brown pattern found. (**b**) *C. torquatus*; (**c**) *C. yolandae*; (**d**) *C. haigi*; (**e**) *C. bonettoi*; (**f**) *C. roigi*; (**g**) *C. dorbigny*; (**h**) *C. magellanicus*; (**i**) *C. maulinus*; (**j**) *C. sociabilis*; (**k**) *C. argentinus*; (**l**) *C. perrensi*; (**m**) *C. mendocinus*; (**n**) *C. fulvus*; (**o**) *C. peruanus*; (**p**) *C. opimus*. (Photographs (except b [from G. L. Gonçalves]) by T. R. O. Freitas – courtesy of mammal collection from the Museum of Vertebrate Zoology, UC Berkeley)

forests, and steppes of Terra del Fuego (Reig et al. 1990; Lacey et al. 2000; Bidau 2015; Freitas 2016).

In particular, two pairs of species that live in the Atlantic coast catches not only evolutionary biologists but anyone's eyes for its marked differences in pelage associated with habitat background. The first is *Ctenomys australis* Rusconi, 1934 and *Ctenomys talarum* Thomas, 1989, occurring in a coastal dune region in southern Buenos Aires province of Argentina (Contreras and Reig 1965; Reig et al. 1990), and the second is *Ctenomys flamarioni* Travi, 1981 and *Ctenomys minutus* Nehring, 1887 (Freitas 1995a, b), which inhabit the southern Brazil coastal plain (Fig. 8.2a). *C. australis* and *C. flamarioni* have blonde coat color (light phenotype) and inhabit the sandy dunes, whereas *C. talarum* and *C. minutus* have brown pelage (dark phenotype) and inhabit sandy fields (Fig. 8.2b) that correspond to a continuum of coastal dunes toward the continent (Freitas 1995a, b; Busch et al. 2000) (Fig. 8.3a); these two habitats can be distinguished by soil color (Fig. 8.3b) and hardness, and plant cover (Malizia et al. 1991; Cutrera et al. 2010; Kubiak et al. 2015; Lopes et al. 2015; Kubiak et al. 2018). Phylogenetic relatedness between and within these pair of species also vary. *C. australis*, *C. talarum*, and *C. flamarioni* belong to the mendocinus species group, whereas *C. minutus* are placed in the torquatus species group (Parada et al. 2011; Chap. 2, this volume). In this context, the repeated phenotypes might represent convergence to similar habitats, in which ecological function is potentially cryptic anti-predation behavior (Langguth and Abella 1970; Vassalo et al. 1994), which has never been explored.

Two studies have investigated pelage variation in *Ctenomys* from an evolutionary genetics perspective. First, Wlasiuk et al. (2003) demonstrated that genetic drift underlies pelage forms in different populations of *Ctenomys rionegrensis* Langguth and Abella (1970) that include brown, dark-backed, and melanic phenotypes. Second, Gonçalves et al. (2012) performed a molecular approach targeting a key gene-driven of coatcolor – the Melanocortin 1 receptor (MC1R) –, including a wide range of species with distinct color pelages.

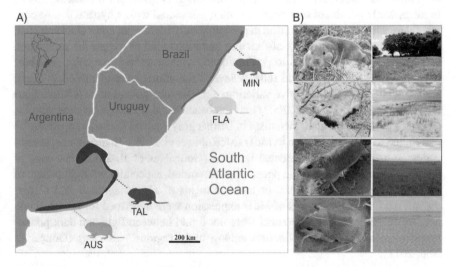

Fig. 8.2 (**a**) Geographic distribution of *Ctenomys flamarioni* (FLA), *Ctenomys minutus* (MIN), *Ctenomys australis* (AUS), and *Ctenomys talarum* (TAL) in the coastal plain of Argentina and southern Brazil with schematic shades of its pelage. (**b**) Convergence pattern of light-dark phenotypes (FLA-MIN and AUS-TAL) inhabiting contiguous habitats of sandy dunes and sandy fields

A)

B)

Fig. 8.3 (a) Habitats of *Ctenomys* in the coastal system: sandy dunes and sandy fields. (b) Soil coloration of each species' habitat. TAL, *C. talarum*; MIN, *C. minutus*; AUS, *C. australis*; FLA, *C. flamarioni*

Hair and skin color in rodents are largely determined by the amount, type, and distribution of melanin packaged in the melanosomes of epidermal cells and hair follicles (Jackson 1997). Mc1r acts as a pigmentary switch in the production of melanin: when activated by α-melanocyte stimulating hormone (α-MSH), it signals the production of black/brown pigment (eumelanin) and in the absence (or inhibition) of a-MSH, red/yellow pigment (pheomelanin) is synthesized (Jackson 1997). In mice, Mc1r dominant mutations are often associated with a hyperactive or constitutively active receptor resulting in predominantly black coat color (Jackson et al. 1994), whereas recessive loss-of-function mutations tend to trigger the production of pheomelanin, which leads to predominantly yellow or red coat color (Robbins et al. 1993). Similarly, in wild rodents several mutations were identified in Mc1r, and associated with the adaptive variation, e.g., the rock pocket mice (*Chaetodipus intermedius*; Nachman et al. 2003) and the beach mice (*Peromyscus maniculatus*; Hoekstra et al. 2006); also, melanism in British gray squirrel (*Sciurus carolinensis*) was linked to a 24-bp deletion in Mc1r (McRobie et al. 2009). In tuco-tucos, several coding substitutions were detected in Mc1r (Gonçalves et al. 2012), but none of them with plausible link to the phenotypes examined, especially the light pelage of *C. australis* and *C. flamarioni*, or melanic forms of *C. rionegrensis* and *C. torquatus*. Additionally, patterns of Mc1r expression were described for dorsal, flank, and ventral regions, but differences were not found between light and dark phenotypes; even though, the distinction among body regions was clear (Gonçalves et al. 2012).

8.2 Pelage Variation: From Genotype to Phenotype

Simple Mc1r mutations of large effect have not contributed to adaptive differences among species of tuco-tucos, thus the variation in coat-color among *Ctenomys* suggests that this trait might have a more complex or even polygenic basis. Finding the genes underlying this variation is probably a daunting task, which will require mapping and association studies involving more markers and defined populations. A suitable candidate gene is the Agouti signaling protein (Agouti), an antagonist of Mc1r; in mice, a local expression that varies both spatially and temporarily (Bultman et al. 1992; Siracusa 1994) results in suppression of synthesis of eumelanin and increased production of phaeomelanin. Agouti is the second most important gene linked with adaptive pelage color variation in rodents (e.g., beach mice (Steiner et al. 2007)), which remains to be explored, particularly in the blonde pelages of tuco-tucos, such as in *C. australis*, *C. flamarioni*, and *C. mendocinus* that also present an intraspecific variation of lighter pelage (see Fig. 8.1).

Typically, wild rodents have a pelage pattern of light ventral, which results from constitutive Agouti expression and associated production of phaeomelanin. In contrast, dorsal hairs have a banded pattern (commonly referred to as agouti hair): terminal and subterminal bands and a base. This banding derives from a pulse of Agouti expression during the intermediary phase of the hair cycle, resulting in the deposition of phaeomelanin during the middle of hair growth and deposition of eumelanin at the beginning and end of hair growth (Hoekstra and Nachman 2006). In the agouti-type pelage distinct variables may be target by the selection, as the distribution of pigment, i.e., bandwidth, and the density of pigment deposited in it, resulting in lighter or darker phenotypes. A few studies have dissected the pigment structure in the hair (e.g., *Peromyscus* (Linnen et al. 2009); *Spalax* (Singaravelan et al. 2010, 2013)) and ultimately inferred its contribution to overall appearance and convergence as well.

In this chapter, an original study on the pigmentation of *Ctenomys* is reported from a morphological perspective, hypothesizing an association of pelage and soil coloration. The hair pattern and pigment density are characterized in species of tuco-tucos from the Atlantic Coastal dune system that present repeated adaptive phenotypes, to test the existence of convergence.

8.3 Quantifying Hair, Pelage, and Soil Coloration

In vertebrates, the visible color spectrum typically ranges from 400 to 700 nm (blue to red) (Krupa and Geluso 2000). Therefore, it was used to measure the pelage and soil coloration. A total of 123 specimens of *C. talarum* (TAL = 20), *C. minutus* (MIN = 40), *C. australis* (AUS = 28), *C. flamarioni* (FLA = 35) from both field-caught and taxidermized specimens from scientific collections of the following institutions were used: Universidade Federal do Rio Grande do Sul (UFRGS),

Universidad Nacional de Mar del Plata (UNMDP), and Museo Municipal de Ciencias Naturales Lorenzo Scaglia (MCNLS) (Appendix). For each specimen, three body regions were analyzed: dorsal, flank, and ventral, determined to infer distinct selective pressures, since there are a differential influence on the individual's overall coloration (i.e., dorsal and flank are considered more relevant for evolution (see Linnen et al. 2009; Manceau et al. 2010)).

Quantification was obtained by the pixel densitometry method; the unit is defined as Gray for the RGB (red, green, and blue) system. Samples (pelage and soil) were photographed with the Munsell (X-Rite Inc.) universal color card to correct the value obtained in relation to the standardized black and white estimates, using the following formula:

$$\text{Calculated Color} = \frac{\text{Obtained Value} - \text{Black}}{\text{White} - \text{Black}}$$

The photographs were analyzed using the software AxioVision version 4.8 (Carl Zeiss Microimaging System Inc.); the central area was delimited using the outline spline tool, and the densitometry values were individually generated for red, green, and blue pixels. For each sample, three measurements were performed and averaged for each pixel. Then, the global average of RGB pixels was calculated.

Microscopic slides were prepared for the dorsal, flank, and ventral regions of each specimen, plucking 10 guard hairs per individual per region. Hairs were rinsed in 50% ethanol and immersed in colorless enamel under the coverslip. Each slide was photographed with a Sony® Cybershot DS20 camera attached to the Leica® M125 stereoscopic microscope using the 0.8X magnification for the whole hair, and 10X for the terminal and subterminal band images. The photographs were analyzed using the AxioVision, measuring hair width, and terminal and subterminal bandwidth. Also, the densitometry values of the pigment deposited in the terminal and subterminal bands were analyzed, zooming the same region analyzed (largest diameter) for all species.

For habitat characterization, soil samples were collected along an 80 m-transect, randomly delineated in each habitat. For *C. flamarioni* and *C. minutus* sampling was placed in Xangri-lá (29°47'S; 50°01'W) and Osório (29°31'S; 50°32'W) Municipalities, in southern Brazil. For *C. australis* and *C. talarum*, sampling sites were located in Necochea Municipality (38°03'S; 57°49'W and 38°02'S; 57°56'W, respectively), in Argentina. Soil samples were taken from the surface in every 10 m of transects and stored in 15 ml tubes. Additionally, eight samples were randomly taken from the burrowing system of each species for sampling comparison of underground vs aboveground. A total of 64 samples were individually placed in Petri dishes and dehydrated at 58 °C for 24 h. For plant coverage analysis, a specific area was photographed in each sampling stations of *C. minutus* and *C. flamarioni*, using a 1 m tape measure at the center of the image as a reference, in order to standardize the area (1 m^2). The percentage of plant coverage was estimated using the Braun-Blanquet method (1932). Previously published data from *C. australis* and *C. talarum* were taken from Cutrera et al. (2010).

Normal distribution of variables was tested using the Kolmogorov-Smirnov test, which is suitable for small sample sizes (Steinskog et al. 2007). Also, the heterogeneity of variance was tested with Bartlets test. Most of the data fit in a normal curve; however, significant heterogeneity variance was found. Thus, the data were treated as nonparametric. For comparisons in dorsal, flank, and ventral regions for differences in the distribution (bandwidth) and density (color) of pigment deposited in hair and pelage, the Kruskal-Wallis nonparametric test was used, followed by Dunn's multiple paired comparisons; the p-value (<0.05) was adjusted for multiple comparisons using the Bonferroni. Also, this test was used to compare microhabitat characteristics (soil coloration and plant cover) among species. To test the existence of an association between soil and pelage (dorsal, flank, and ventral) color, a simple linear regression analysis was used. Statistical analyzes were performed using the software XLSTAT (Addinsoft). Results of bandwidth/hair width, densitometry analysis, and substrate color are presented using the box-plots graphical method, including minimum and maximum values, mean, first, and third quartiles; other values are presented as mean (χ) ± standard error (SE).

8.4 Phenotypic Variation: Pigment Distribution and Density

In the dorsal pelage, tuco-tucos have the agouti hair type, presenting the banding pattern with black and yellow pigments alternately deposited (Fig. 8.4). In the flank and ventral hairs, the terminal band is absent. FLA has almost no pigment in ventral hairs; when present, it is composed only by pheomelanin. In the other three species, hairs from the flank and ventral regions have two-band patterns (subterminal and base), with pheomelanin in a lower density. Differences in the width of the terminal and subterminal bands were observed between light and dark phenotypes (Figs. 8.4, 8.5, and 8.6). TAL and MIN have the proportional widest terminal band and the shortest subterminal band in dorsal hairs; conversely, AUS and FLA present proportionally shortest terminal width and widest subterminal band in such region (Fig. 8.6). Contrary, the subterminal band in flank and ventral hairs did not vary significantly (also in proportion) between light and dark phenotypes (Fig. 8.6).

Differences in dorsal pelage coloration were identified among species of tuco-tucos, also within the light and dark phenotypes (Fig. 8.7): TAL and MIN presented significantly higher pigment density compared to AUS and FLA. Thus, TAL represents the darkest phenotype, whereas FLA the lightest. In the flank, the dark phenotypes significantly differ to the light ones; within phenotypes, differences were found only for light pelages (Fig. 8.7). In the ventral region, there were no significant differences between light and dark phenotypes (TAL, MIN, AUS). However, FLA showed marked distinction to all other species (Figs. 8.4 and 8.7). Significant differences in the pigment density within the terminal band were found between phenotypes (Fig. 8.7): dark species presented lower values compared to light ones. Similarly, dark phenotypes had distinct values for the subterminal band compared to the light ones.

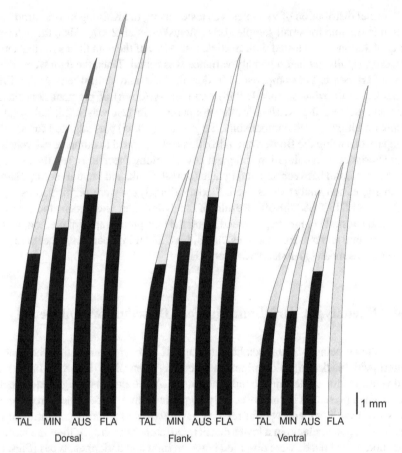

Fig. 8.4 Schematic representation of *Ctenomys* dorsal, flank, and ventral hairs, in scale. TAL, *C. talarum*; MIN, *C. minutus*; AUS, *C. australis*; FLA, *C. flamarioni*

These results suggest a remarkable influence of the density deposited in the terminal and subterminal bands on the overall coloration of an individual. Accordingly, the greatest functionality is supposed for the dorsal region in comparison to the flank (that is less intense) reinforced by the small variation. Results of pigment density in the ventral hairs corroborated this hypothesis, since no significant differences were found for the subterminal band and overall coloration between light and dark phenotypes (Fig. 8.7). Since the ventral region is relatively less exposed, the widest range of variation found might result from selective pressure relaxation. To test this assumption, the variances were estimated in several parameters analyzed (e.g., terminal and subterminal bandwidth, total hair width, pigment density within the terminal and subterminal bands in dorsal, flank, and ventral regions); eight of them presented heterogeneity among species. Not surprisingly, most occurred in parameters taken from the flank and ventral regions. In the dorsal, the terminal bandwidth shown the lowest values in light phenotypes. The dark phenotypes

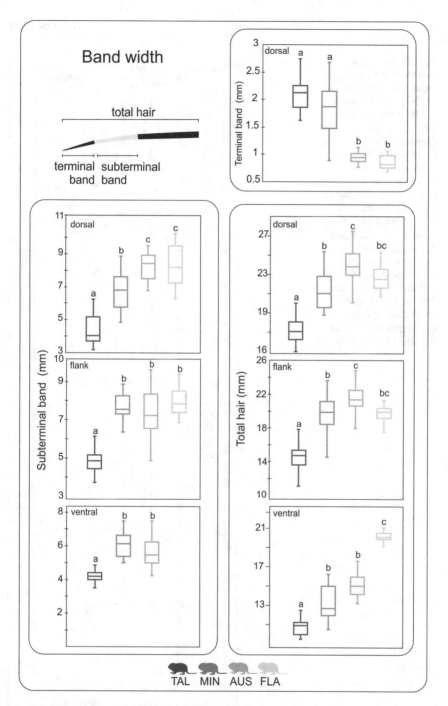

Fig. 8.5 Box-plots representing variability found in the terminal band (**a**), the subterminal band (**b**), and total hair (**c**) width in the four *Ctenomys* species: TAL, *C. talarum*; MIN, *C. minutus*; AUS, *C. australis*; FLA, *C. flamarioni*, showing the mean and first and third quartiles. Different letters over the box-plot indicate statistical significance between species, within each body region analyzed (dorsal, flank, and ventral). The colors indicate the phenotypes (see Fig. 8.2 and inlet schematic legend)

Fig. 8.6 Box-plots representing variability found in the proportional width of the terminal (**a**) and the subterminal bands (**b**) in the species of *Ctenomys*: TAL, *C. talarum*; MIN, *C. minutus*; AUS, *C. australis*; FLA, *C. flamarioni*, showing the mean, and the first and third quartiles. Different letters over the box-plot indicate statistical significance between species, within each body region analyzed (dorsal, flank, and ventral). The colors indicate distinct phenotypes (see Fig. 8.2 and inlet schematic legend)

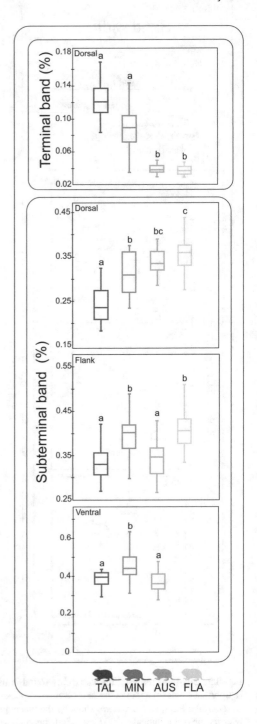

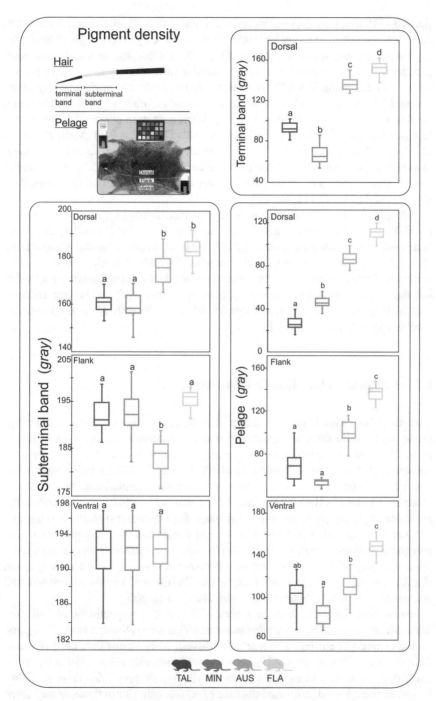

Fig. 8.7 Box-plots representing variability in hair density for the terminal (**a**) and subterminal (**b**) band and for the pelage (**c**) in the species of *Ctenomys*: TAL, *C. talarum*; MIN, *C. minutus*; AUS, *C. australis*; FLA, *C. flamarioni*, showing the mean and first and third quartiles. Different letters on the box-plot indicate statistical significance between species, within each body region analyzed (dorsal, flank, and ventral). The colors indicate distinct phenotypes (see Fig. 8.2 and inlet schematic legend)

showed 4–10 times the greatest variance compared to light for terminal bandwidth; therefore, the variation might be constrained in markedly cryptic light phenotypes, suggesting greater selective pressure (Table 8.1). Contrary, in the densitometry parameter the subterminal band showed similar variance in light and dark pheno-types, thus indicating less influence on the overall coloration comparatively to the terminal bandwidth. Significant differences in soil coloration were found between the two habitats (Fig. 8.8). No significant differences were observed between sur-face and burrow soil samples within sandy fields (TAL, $P = 0.48$; MIN, $P = 0.35$) and dunes (AUS, $P = 0.06$; FLA, $P = 0.06$), allowing sufficient representativeness of samples from transects. Similar to dorsal coat color, soil from TAL microhabitat had a higher density (i.e., lower values), whereas for FLA indicated the lowest den-sity (i.e., higher values) (Table 8.2).

Additionally, linear regression analysis indicated a strong association ($R^2 = 0.87$; $P < 0.001$) of soil with dorsal and flank coat-color ($R^2 = 0.71$; $P < 0.001$), and mod-erate association with ventral ($R^2 = 0.57$; $P < 0.001$), which is mainly influenced by FLA pelage values in relation to other species (Fig. 8.9). For plant coverage, two markedly distinct groups of values were recovered, one representing sandy fields and the other sandy dunes (Table 8.2). Estimates from sandy fields were twice as high as those in sandy dunes, and did not differ significantly ($P > 0.05$) between similar microhabitats (Table 8.2).

8.5 Cryptic Coloration in *Ctenomys*

Tuco-tucos have a predominantly fossorial habit; though they are also active aboveg-round, particularly foraging in close to burrows (Comparatore et al. 1991; Busch et al. 2000). In particular, a high frequency of mobility was described for TAL (Busch et al. 1989), AUS (Vassalo et al. 1994), and FLA (Fernández-Stolz et al. 2007; Stolz 2006). Contrary to *Spalax*, in which aboveground exposure is recog-nized as accidental (Heth 1991), the regular activity in open areas indicates that predation might be more common in *Ctenomys* than any other subterranean lineage.

Ctenomids are often preyed on by several vertebrates, for example, burrowing owl (*Athene cunicularia*), pampas fox (*Pseudalopex gymnocercus*), lesser grison (*Galictis cuja*), white-eared opossum (*Didelphis albiventris*), Molina's hog-nosed skunk (*Conepatus chinga*), small hairy armadillo (*Chaetophractus vellerosus*), and Neuwied's lancehead (*Bothrops neuwidi*) (Busch et al. 2000).

Specifically, TAL and AUS represent 16% and 2%, respectively, of owl prey items (*Athene cunicularia*, *Asio flammeus*, and *Tyto alba*) (Vassalo et al. 1994), and such difference is attributed to markedly distinct body sizes (TAL ca. 118 g and AUS ca. 360 g). Predation in AUS occurred predominantly in subadult individuals, likely due to constrain of the predator in carrying out the prey (Vassalo et al. 1994). However, there is no data on the influence of cryptic behavior in these, or any other, *Ctenomys* species preventing predation (i.e., differential survival), linking to micro-habitat selection.

Table 8.1 Analysis of significant variance among *Ctenomys talarum* (TAL), *Ctenomys minutus* (MIN), *Ctenomys australis* (AUS), and *Ctenomys flamarioni* (FLA) for different hair and pelage parameters

Parameter	Species	Var.	χ^2_{calc}	P
Terminal width – dorsal			45.72	<0.001
	TAL	8.24		
	MIN	26.16		
	AUS	1.19		
	FLA	2.83		
Total hair width – ventral			31.46	<0.001
	TAL	269.22		
	MIN	528.77		
	AUS	191.98		
	FLA	28.29		
Subterminal band width – flank			9.10	0.02
	TAL	22.25		
	MIN	28.27		
	AUS	31.85		
	FLA	76.89		
Subterminal band width – ventral			7.92	0.04
	TAL	15.31		
	MIN	15.41		
	AUS	4.68		
	FLA	8.83		
Subterminal band densitometry – dorsal			9.38	0.02
	TAL	16.85		
	MIN	57.85		
	AUS	55.32		
	FLA	25.74		
Subterminal band densitometry – flank			9.10	0.02
	TAL	22.25		
	MIN	28.27		
	AUS	31.85		
	FLA	76.89		
Subterminal band densitometry – ventral			7.92	0.04
	TAL	15.31		
	MIN	15.41		
	AUS	4.68		
	FLA	8.83		
Pelage densitometry – flank			16.58	0.001
	TAL	211.99		
	MIN	27.93		
	AUS	132.68		
	FLA	123.96		

Fig. 8.8 Association of
the color of the soil with an
appearance in the species
of *Ctenomys*: TAL,
C. talarum; MIN,
C. minutus; AUS,
C. australis, FLA,
C. flamarioni. (**a**)
Box-plots representation of
the variability found in the
color of the soil (filled
boxes) and dorsal pelage
(non-filled boxes), showing
the mean, and first and
third quartiles. Different
letters on the box-plot
indicate statistical
significance between
species, within a given
analyzed body region. The
colors indicate distinct
phenotypes (see Fig. 8.2
and inlet schematic legend)

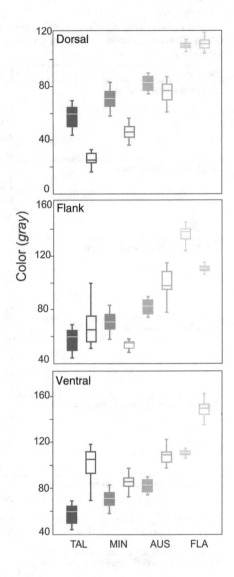

Table 8.2 Estimates of mean ± standard error of soil color and plant cover in the microhabitats of *C. talarum* (TAL), *C. minutus* (MIN), *C. australis* (AUS), and *C. flamarioni* (FLA)

	Sandy fields		Sandy dunes		Kruskal–Wallis	
	TAL	MIN	AUS	FLA	K_{obs}	P
Soil coloration	58.14 ± 2.00	70.77 ± 1.78	82.03 ± 1.37	111.06 ± 0.85	53.37	< 0.001
Plant cover	57.18 ± 16.12	60.50 ± 12.31	25.31 ± 17.36	30.50 ± 10.72	43.56	< 0.001

Fig. 8.9 Linear regression of soil color by coat color. The circles represent *Ctenomys talarum* and *Ctenomys flamarioni*, and the triangles *Ctenomys minutus* and *Ctenomys australis*. The colors indicate the phenotypes (see Fig. 8.2)

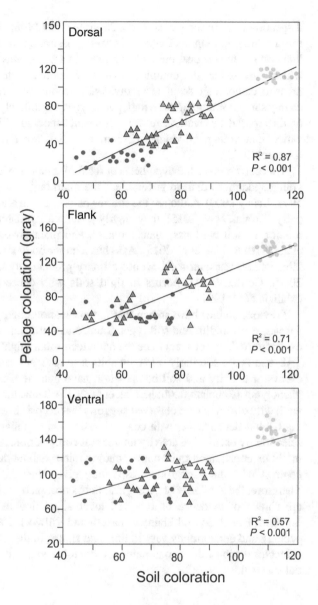

This study characterizes, for the first time, hair, pelage, and soil coloration in tuco-tucos. Specifically, data on distribution and density of pigment deposited in the terminal band of dorsal hairs highlights the biological relevance of such region also in this lineage. The smallest variances were found in the light phenotypes; therefore, dorsal coloration of AUS and FLA might be more restricted to vary. Accordingly, subtle changes in coat-color in these two species might contrast in pale dunes that present low plant cover (i.e., more exposed area), making them more susceptibility for predation capture, which remains to be tested. These species have the highest dispersion rates in the genus (Stolz 2006; Vassalo 1998; Garcias et al. 2018), rein-forcing the importance of cryptic behavior, assuming intense selective pressure in this system. In contrast, the widest range of variation was found in TAL and MIN, which may reflect the complexity of microhabitat with the highest plant cover, favoring camouflage despite an individual's overall coloration. Consequently, subtle changes in coloration of these dark phenotypes are unlikely to have an intense effect on differential survival. Moreover, a relevant aspect in MIN is the empirical obser-vation that young individuals (2–3 months old) are lighter in color than adults (Fonseca 2003).

This corroborates the hypothesis of local adaptation, whose function is to protect young specimens that, in general, are the main target of predation (Vassalo et al. 1994; Lacey 2000). Although the species pairs occur in allopatry with areas of sym-patry (Kubiak et al. 2015), they clearly present microhabitat selection, differing in relation to soil hardness, plant biomass, and plant cover (Vassalo 1998; Cutrera et al. 2010; Kubiak et al. 2015). AUS has a larger body size (Busch et al. 2000) and inhabits less resistant soils, whose primary productivity is reduced (Cutrera et al. 2010). Contrary, TAL occurs in rigid soils with dense and diverse plant cover (Malizia et al. 1991).

Previous studies have shown that the excavation energy cost is similar in these species, even in different soil types (Luna and Antinuchi 2007). Therefore, energy expenditure does not seem to be the main factor that might explain soil selection by TAL and AUS. Similarly, FLA inhabits less resistant sandy soils than MIN, and excavator activity and soil composition have non-significant differences between phenotypes (Rebelato 2006; Kubiak et al. 2015). Thus, the association in soil color-ation with dorsal pelage observed suggests that crypsis is a potential factor influenc-ing habitat dependence, with coat coloration being a significant variable, prior to selection by excavation activity and/or soil composition. Accordingly, each species might be constrained to its corresponding micro habitat due to the disadvantage of contrast with the background, especially given their high activity aboveground. Therefore, the similarity of ecological niches occupied by TAL-MIN and AUS-FLA are shreds of evidence of repeated local adaptation in dynamic habitats (e.g., Southern Brazil Coastal Plain; Tomazelli and Villwock 2000), in which population ecology and demography vary in time and space. In this context, the fixed ecologi-cal factor responsible for maintaining these local adaptations is potentially differen-tial survival.

8.6 Convergent Evolution

AUS and FLA belong to the Mendocinus species group, defined by morphological characteristics (e.g., asymmetric sperm), karyotype ($2n = 47$–48; similar G and C band patterns), and molecular data (Castillo et al. 2005; D'Elia et al. 1999; Freitas 1994; Lessa and Cook 1998; Massarini and Freitas 2005; Parada et al. 2011; Chapter 2, this volume). Such similarity of characters has raised questions such as whether these two species, recognized as phylogenetically close related, share the same most recent common ancestor (Freitas 1994, 1995a; Fernández-Stolz 2007; Malizia et al. 1991; Mora et al. 2006). Comparative assessment of skull morphology revealed significant morphological differences (Fornel 2009; Massarini and Freitas 1995, 2005; Travi and Freitas 1984). Then, it was suggested that FLA might have split from an ancestral form from Argentina, by migration, isolation, and further differentiation of AUS (Freitas 1994; Massarini and Freitas 2005). This migration would have occurred in the Pleistocene when the coastal plain was under arid conditions, approximately 100 km wider than at present; thus, the River Plate was not a relevant geographical barrier (Corrêa et al. 1992). However, current evolutionary analysis of the Mendocinus group place AUS closer to *C. mendocinus* (Parada et al. 2011). In this context, the convergence observed in AUS and FLA in terms of body size and light coloration might represent repeated evolution in the occupation of coastal environments instead of the strict retention of an ancestral character (Fig. 8.10). Interestingly, a pale pelage coloration is found in museum specimens of *C. mendocinus* deposited in the Museum of Vertebrate Zoology (see Fig. 8.1). The Mendocinus species group might have a common genetic background that underlies the quality and quantity of pigment deposited in the hair, which should be further investigated. Since the phenotypic variation of *C. mendocinus* is intraspecific, it is a candidate species to explore the hair pattern together with the Agouti gene in populations of both forms, to understand the genetic basis of this adaptive phenotype in *Ctenomys*.

The increase in the subterminal band (and consequent reduction in the terminal band) of the hair, as well as the lower pigment density, provide the dilution of the overall color of the individuals, making them paler (i.e., the blond color of AUS and

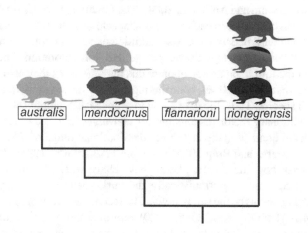

Fig. 8.10 Variation of pelage color within the mendocinus species group under a phylogenetic context (for details see D' Elia et al. Chapter 2, this volume), including *Ctenomys australis*, *Ctenomys flamarioni*, *Ctenomys mendocinus* and *Ctenomys rionegrensis*. Phenotypes observed are: blond (*australis* and *mendocinus*), brown (*mendocinus* and *rionegrensis*), pale blond (*flamarioni*), dark-backed (*rionegrensis*) and melanic (*rionegrensis*)

FLA) (Fig. 8.9). However, there are significant differences at fine-scale between light phenotypes. AUS has a dark-blond pelage compared to FLA, whose correspondence is directly reflected in its darker soil. Thus, the data generated in this study indicate convergent mechanisms of crypsis in the same ecological context. Similarly, the data showed parallel evolutionary trajectories in the generation of dark phenotypes in TAL and MIN. These species are phylogenetically distant (Parada et al. 2011), and converge in terms of body size, microhabitat, coat-color, and soil. Interestingly, mechanisms used to generate dark phenotypes are identical: increased eumelanin distribution at the tip of the hair (longer terminal width), and pheomelanin density in the subterminal band. These two small changes generate potentially advantageous phenotypes on dark soils, in which small variation is linked to the ability to turn into cryptic in a more complex environment.

8.7 Final Remarks

The determinants of color patterns in animals are still poorly understood, but three main functions are suggested: intraspecific communication, predator avoidance, and thermoregulation (Endler 1978). Tuco-tucos have a predominantly solitary habit and rarely are in direct contact with other individuals of the same species, suggesting that chemical and vocal communication rule the reproductive behavior in these animals (Francescoli 1999; Zenuto et al. 2004). Thus, it is assumed that coloration has little significant involvement in communication. Conversely, the results of this study suggest that pelage phenotypes of TAL, MIN, AUS, and FLA have an evolutionary significance of predator evasion, possibly also contributing to better thermoregulation (see Cutrera and Antinuchi 2004). The function of cripsis is reinforced by the differences in coat color in each of the four species, converging in parallel to two groups: light and dark phenotypes. Also, additional support comes from the strong association between *Ctenomys* dorsal pelage and soil coloration. Differences in plant cover of the four habitats corroborate this hypothesis, as they also show variation at the macroecological level, contributing to a fine-tuning of unique local adaptation of each species. Thus, the data allow to propose that natural selection may be the main evolutionary factor responsible for convergence in tuco-tucos. The existence of specific areas of sympatry in the distributional range of TAL-AUS (Reig et al. 1990; Contreras and Reig 1965) and MIN-FLA (Freitas 1995a; Kubiak et al. 2015) led to ask how the cryptic phenotypes behave when in contrasting habitat background, i.e., when opportunistically the dark phenotype occupy the sandy dunes, and light phenotype the sandy fields. The recent discovery of hybrids between FLA and MIN (Kubiak et al. 2015, 2020) reinforce the existence of admixture between phenotypes and habitats. Do the color phenotypes in contrasting soils have disadvantage comparatively to the cryptic ones? Will the disadvantage, if it exists, be higher in open habitats than in plant-covered fields? Quantitative studies involving controlled experiments, particularly using these natural laboratories of sympatry, are fundamental to evaluate rates of predation (i.e., prey capture) associated with cryptic behavior in these species, which will clarify the effect of coloration on the differential survival of adaptive phenotypes.

Acknowledgments I thank to M.D. Romero (Museo Municipal de Ciencias Naturales Lorenzo Scaglia) for permission to photograph and collect hair samples of *C. australis* and *C. talarum*; to C. M. Lopes for the preparation of skins of *C. minutus*, M.S. Mora for the whole support during fieldwork in Argentina to collect *C. australis* and *C. talarum* and their soil samples, T.R.O. Freitas for taking photographs of *Ctenomys* pelage in the Museum of Vertebrate Zoology, and G.R.P. Moreira for suggestions on statistical analysis and fieldwork support. Financial support: CAPES, CNPq, PPGBM-UFRGS, and FAPERGS (PRONEX 16/2551-0000485-4).

Appendix

Ctenomys specimens used:

C. talarum: UNMDP4; UNMDP5; UNMDP6; UNMDP7; UNMDP8; UNMDP9; UNMDP10; UNMDP11; UNMDP12; UNMDP13; UNMDP14; UNMDP15; UNMDP16; UNMDP17; MCNLS 93-1; MCNLS 93-3; MCNLS 93-2 UNMDP; MCNLS. *C. minutus*: TR579; TR639; TR640; TR641; TR642; TR643; TR644; TR645; TR646; TR647; TR648; TR649; TR650; TR651; TR652; TR653; TR654; TR655; TR656; TR657; TR1201; TR1202; TR1203; TR1207; TR1212; TR1219; TR1220; TR1221; TR1222; TR1225; LAMI2; TR1125; TR1126; TR1128; TR1129; TR1130; TR1132; TR1133; TR1137; TR1231; *C. australis*: UNMDP 1-1; UNMDP 1-2; UNMDP 1-3; MCNLS 81-1; MCNLS 82-22; MCNLS 82-67; MCNLS 82-68; MCNLS 82-69; MCNLS 82-71; MCNLS 82-238; MCNLS 82-239; MCNLS 82-240; MCNLS 82-241; MCNLS 82-242; MCNLS 82-243; MCNLS 82-244; MCNLS 82-245; MCNLS 84-20; MCNLS 84-23; MCNLS I-737; MCNLS I-740; MCNLS I-1044; MCNLS 1; MCNLS 2; MCNLS 4; UNMDP 37; UNMDP 38; UNMDP 39, UNMDP; MCNLS; *C. flamarioni*: PUC278; TR449; TR473; TR474; TR475; TR477; TR482; TR483; TR488; TR491; TR493; TR495; TR496; TR497; TR500; TR1152; TR1153; TR1154; DZRS01; G123; PUC408; TR476; TR478; TR479; TR480; TR484; TR485; TR489; TR490; TR494; TR498; TR499; TR1271; TR1272; TRNI1, TRNI2.

Literature Cited

Bidau CI (2015) Ctenomyidae. Ctenomys. In: Patton J, Pardiñas FU, D'Elía G (eds) Mammals of South America. Vol 2. Rodents. University of Chicago Press, Chicago, pp 818–877

Braun-Blanquet J (1932) Plant sociology: the study of plant communities. McGraw-Hill Publications in the Botanical Sciences, New York

Bultman SJ, Michaud EJ, Woychik RP (1992) Molecular characterization of the mouse agouti locus. Cell 71:1195–1204

Busch C, Malizia AI, Scaglia AO, Reig AO (1989) Spatial distribution and attributes of a population of *Ctenomys talarum* (Rodentia: Octodontidae). J Mammal 70:204–208

Busch C, Antinuchi CD, Valle CJ, Kittle MJ, Malizia AI, Vassalo AI, Zenuto R (2000) Population ecology of subterranean rodents. In: Lacey EA, Patton JL, Cameron GN (eds) Life underground, the biology of subterranean rodents. University of Chicago Press, Chicago, pp 183–226

Castillo AH, Cortinas MN, Lessa EP (2005) Rapid diversification of South American tuco-tucos (*Ctenomys*; Rodentia, Ctenomyidae): contrasting mitochondrial and nuclear intron sequences. J Mammal 86:170–179

Comparatore VM, Agnudsdei M, Busch C (1991) Habitat relations in sympatric populations of *Ctenomys australis* and *Ctenomys talarum* (Rodentia: Octodontidae) in a natural grassland. Mamm Biol 57:47–55

Contreras JR, Reig OA (1965) Datos sobre la distribución del gênero *Ctenomys* (Rodentia, Octodontidae) em la zona costera de la provincia de Buenos Aires comprendida entre Necochea y Bahía Blanca. Physis xxv(69):169–186

Correa ICS, Baitelli R, Ketzer JM, Martins R (1992) Translação horizontal e vertical do nível do mar sobre a plataforma continental do Rio Grande do Sul nos últimos 17.500 anos BP. Anais III Congresso ABEQUA, pp 225–240

Cott HB (1940) Adaptive coloration in animals. Methuen, London

Cutrera AP, Antinuchi CD (2004) Cambios en el pelaje del roedor subterráneo *Ctenomys talarum*: posible mecanismo térmico compensatório. Rev Chil Hist Nat 77:235–242

Cutrera AP, Mora MS, Antenucci CD, Vassallo AI (2010) Intra- and interspecific variation in home-range size in sympatric tuco-tucos, *Ctenomys australis* and *C. talarum*. J Mammal 91:1425–1434

D'Elia G, Lessa EP, Cook JA (1999) Molecular phylogeny of tuco-tucos, genus *Ctenomys* (Rodentia: Octodontidae): evaluation of the mendocinus species group and the evolution of asymmetric sperm. J Mamm Evol 6:19–38

Dice L, Blossom PM (1937) Studies of mammalian ecology in southwestern North America, with special attention to the colors of the desert mammals. Carnegie Inst Wash Publ 485:1–25

Endler JA (1978) A predator's view of animal color patterns. In: Hecht MK, Steere WC, Wallace (eds) Evolutionary biology, vol 11. Plenum Press, New York, pp 319–364

Fernández-Stolz GP (2007) Estudos evolutivos, filogeográficos e de conservação em uma espécie endêmica do ecossistema de dunas costeiras do sul do Brasil, Ctenomys flamarioni (Rodentia-Ctenomydae), através de marcadores moleculares microssatélites e DNA mitocondrial. Tese de doutorado, Universidade Federal do Rio Grande do Sul, Brasil. 193 pp

Fernández-Stolz GP, Stolz JFB, Freitas TRO (2007) Bottlenecks and dispersal in the tuco-tuco das dunas, *Ctenomys flamarioni* (Rodentia: Ctenomyidae), in Southern Brazil. J Mammal 88:935–945

Fonseca MB (2003) Biologia populacional e classificação etária do roedor subterrâneo tuco-tuco *Ctenomys minutus* Nehring, 1887 (Rodentia, Ctenomyidae) na planície costeira do Rio Grande do Sul, Brasil. Dissertação de mestrado, Universidade Federal do Rio Grande do Sul, Brasil. 110 pp

Fornel R (2009) Evolução na forma e tamanho do crânio no gênero *Ctenomys* (Rodentia: Ctenomydae). Tese de doutorado, Universidade Federal do Rio Grande do Sul, Brasil. 171 pp

Francescoli G (1999) A preliminary report on the acoustic communication in Uruguayan *Ctenomys* (Rodentia, Octodontidae): basic sound types. Bioacoustics 10:203–218

Freitas TRO (1994) Geographic variation of heterochromatin in *Ctenomys flamarioni* (Rodentia: Octodontidae) and its cytogenetic relationship with other species of the genus. Cytogenet Cell Genet 67:193–198

Freitas TRO (1995a) Geographic distribution and conservation of four species of the genus *Ctenomys* in southern Brazil. Stud Neotropical Fauna Environ 30:53–59

Freitas TRO (1995b) Geographic distribution if sperm forms in the genus *Ctenomys* (Rodentia: Octodontidae). Rev Bras Genét 18:43–46

Freitas TRO (2016) Family Ctenomyidae. In: Wilson DE, Lacher TEJ, Mittermeier RA (eds) Handbook of the mammals of the world: lagomorphs and rodents I. Lynx Editions, Barcelona, pp 499–534

Freitas TRO, Lessa EP (1984) Cytogenetics and morphology of *Ctenomys torquatus* (Rodentia, Octodontidae). J Mammal 65:637–642

Garcias FM, Stolz JFB, Fernández GP, Kubiak BB, Bastazini VAG, Freitas TRO (2018) Environmental predictors of demography in the tuco-tuco of the dunes (*Ctenomys flamarioni*). Mastozool Neotrop 25(2):293–304

Gonçalves GL, Freitas TRO (2009) Intraspecific variation and genetic differentiation of the collared tuco-tuco (Ctenomys torquatus) in Southern Brazil. J Mammal 90(4):1020–1031

Gonçalves GL, Hoekstra HE, Freitas TRO (2012) Striking coat colour variation in tuco-tucos (Rodentia: Ctenomyidae): a role for the melanocortin-1 receptor? Biol J Linn Soc 105:665–680

Heth G (1991) Evidence of above-ground predation and age determination of the prey in subterranean mole rats (*Spalax ehrenbergi*) in Israel. Mammalia 55:529–542

Heth G, Beiles A, Nevo E (1988) Adaptive variation of pelage color within and between species of the subterranean mole rat (*Spalax ehrenbergi*) in Israel. Oecologia 74:617–622

Hoekstra HE, Nachman M (2006) Coat color variation in Rock Pocket Mice (*Chaetodipus intermedius*): from genotype to phenotype. In: Lacey E, Myers P (eds) Mammalian diversification: from chromosomes to phylogeography (A celebration of the career of James L. Patton), vol I33. University of California Publications, Zoology, Berkeley

Hoekstra HE, Hirschmann RJ, Bundey RA, Insel PA, Crossland JP (2006) A single amino acid mutation contributes to adaptive beach mouse color pattern. Science 313:101–104

Ingles LG (1950) Pigmental variations in populations of pocket gophers. Evolution 4:353–357

Jackson IJ (1997) Homologous pigmentation mutations in human, mouse and other model organisms. Hum Mol Genet 6:1613–1624

Jackson IJ, Budd P, Horn JM, Johnson R, Raymond S, Steel K (1994) Genetics and molecular biology of mouse pigmentation. Pigment Cell Res 7:73–80

Kennerly TE Jr (1954) Local differentiation in the pocket gopher (*Geomys personatus*) in southern Texas. Tex J Sci 6:297–329

Kennerly TE Jr (1959) Contact between the ranges of two allopatric species of pocket gophers. Evolution 13:247–263

Krupa JJ, Geluso KN (2000) Matching the color of excavated soil: cryptic coloration in the plains pocket gopher (*Geomys bursarius*). J Mammal 81:86–96

Kubiak BB, Galiano D, Freitas TRO (2015) Sharing the space: distribution, habitat segregation and delimitation of a new sympatric area of subterranean rodents. PLoS One 10:e0123220

Kubiak BB, Maestri R, Almeida TS, Borges LR, Galiano D, Fornel R, Freitas TRO (2018) Evolution in action: soil hardness influences morphology in a subterranean rodent (Rodentia: Ctenomyidae). Biol J Linn Soc 125:766–776

Kubiak BB, Kretschmer R, Leipnitz LT, Maestri R, Almeida TS, Borges LR, Galiano D, Pereira JC, Oliveira EHC, Ferguson-Smith MA, Freitas TRO (2020) Hybridization between subterranean tuco-tucos (Rodentia, Ctenomyidae) with contrasting phylogenetic positions. Sci Rep 10:1502

Lacey EA (2000) Spatial and social systems of subterranean rodents. In: Lacey EA, Patton JL, Cameron GN (eds) Life underground: the biology of subterranean rodents. University of Chicago Press, Chicago/London, pp 257–293

Lacey EA, Patton JL, Cameron GN (2000) Life underground: the biology of subterranean rodents. University of Chicago Press, Chicago

Langguth A, Abella A (1970) Sobre una poblacion de tuco-tucos melanicos (Rodentia-Octodontidae). Acta Zool Lilloana 27:101–108

Lessa EP, Cook JA (1998) The molecular phylogenetics of tuco-tucos (genus *Ctenomys*, Rodentia: Octodontidae) suggests an early burst of speciation. Mol Phylogenet Evol 9:88–99

Linnen CR, Kingsley EP, Jensen JD, Hoekstra HE (2009) On the origin and spread of an adaptive allele in deer mice. Science 325:1095–1098

Lopes CM, De Barba M, Boyer F, Mercier C, da Silva Filho PJ, Heidtmann LM, Galiano D, Kubiak BB, Langone P, Garcias FM, Gielly L, Coissac E, de Freitas TRO, Taberlet P (2015) DNA metabarcoding diet analysis for species with parapatric vs sympatric distribution: a case study on subterranean rodents. Heredity 114:525–536

Luna F, Antinuchi CD (2007) Energy and distribution in subterranean rodents: sympatry between two species of the genus *Ctenomys*. Comp Biochem Physiol A Comp Physiol 147:948–954

Malizia AI, Vassallo AI, Busch C (1991) Population and habitat characteristics of two sympatric species of *Ctenomys* (Rodentia: Octodontidae). Acta Theriol 36:87–94

Manceau M, Domingues VS, Linnen CR, Rosenblum EB, Hoekstra HE (2010) Convergence in pigmentation at multiple levels: mutations, genes and function. Philos Trans R Soc B 365:2439–2450

Massarini AI, Freitas TRO (1995) Análise morfológica e citogenética de *C. flamarioni* e *C. australis* – duas espécies ecologicamente equivalentes (Rodentia: Octodontidae). Rev Bras Genet 18:487

Massarini AI, Freitas TRO (2005) Morphological and cytogenetics comparison in species of the mendocinus-group (genus *Ctenomys*) with emphasis in *C. australis* and *C. flamarioni* (Rodentia: Ctenomyidae). Caryologia 58:21–27

McRobie H, Thomas A, Kelly J (2009) The genetic basis of melanism in the Gray Squirrel (Sciurus carolinensis). J Hered 100:709–714

Mora MS, Lessa EP, Kittlein MJ, Vassallo AI (2006) Phylogeography of the subterranean rodent *Ctenomys australis* in sand-dune habitats: evidence of population expansion. J Mammal 87:1192–1203

Nachman MW, Hoekstra HE, D'Agostino SL (2003) The genetic basis of adaptive melanism in pocket mice. Proc Natl Acad Sci USA 100:5268–5273

Parada A, D'Elía G, Bidau CJ, Lessa EP (2011) Species groups and the evolutionary diversification of tuco-tucos, genus *Ctenomys* (Rodentia: Ctenomyidae). J Mammal 92:671–682

Rebelato GS (2006) Análise ecomorfológica de quatro espécies de *Ctenomys* do sul do Brasil (Ctenomyidae – Rodentia). Brasil. Dissertação de mestrado, Universidade Federal do Rio Grande do Sul, Brasil. 146 pp

Reig O, Busch C, Ortells M, Contreras J (1990) An overview of evolution, systematics, population biology, cytogenetics, molecular biology and speciation in Ctenomys. Prog Clin Biol Res 335:71–96

Robbins LS, Nadeau JH, Johnson KR, Kelly MA, Roselli-Rehfuss L et al (1993) Pigmentation phenotypes of variant extension locus alleles result from point mutations that alter MSH receptor function. Cell 72:827–834

Singaravelan N, Pavlicek T, Beharav A, Wakamatsu K, Ito S et al (2010) Spiny mice modulate eumelanin to pheomelanin ratio to achieve cryptic coloration in "evolution canyon," Israel. PLoS One 5:e8708

Singaravelan N, Raz S, Tzur S, Belifante S, Pavlicek T et al (2013) Correction: adaptation of pelage color and pigment variations in Israeli subterranean blind mole rats, *Spalax Ehrenbergi*. PLoS One 8(8):e69346. https://doi.org/10.1371/annotation/27bebc65-09c5-4c58-be6c-4f22c4fe0919

Siracusa LD (1994) The agouti gene: turned on to yellow. Trends Genet 10:423–428

Steiner CC, Weber JN, Hoekstra HE (2007) Adaptive variation in beach mice produced by two interacting pigmentation genes. PLoS Biol 5:1880–1889

Steinskog DJ, Tjøstheim DB, Kvamstø NG (2007) A cautionary note on the use of the Kolmogorov–Smirnov test for normality. Mon Weather Rev 135(3):1151–1157

Stolz JFB (2006) Dinâmica populacional e relações espaciais do tuco-tuco das dunas (*Ctenomys flamarioni*) (Rodentia-Ctenomyidae) na Estação Ecológica do Taim-RS/Brasil. Dissertação de mestrado, Universidade Federal do Rio Grande do Sul, Brasil. 71 pp

Sumner FB (1934) Does 'protective coloration' protect? Results of some experiments with fishes and birds. Proc Natl Acad Sci USA 20:559–564

Tomazelli LJ, Willwock JA (2000) O Cenozóico no Rio Grande do Sul: geologia da planície costeira. In: Holz M, de Ros LF (eds) Geologia do Rio Grande do Sul. CIGO/UFRGS, Porto Alegre, pp 375–406

Travi VH, Freitas TRO (1984) Estudos citogenéticos e craniométricos de *Ctenomys flamarioni* e *Ctenomys australis* (Rodentia: Octodontidae). Ciên Cult 36:771

Vassallo AI (1998) Functional morphology, comparative behaviour, and adaptation in two sympatric subterranean rodents genus *Ctenomys* (Caviomorpha: Octodontidae). J Zool 244:415–427

Vassalo AI, Kittlein MJ, Busch C (1994) Owl predation on two sympatric species of tuco-tucos (Rodentia: Octodontidae). J Mammal 75:725–732

Wlasiuk G, Garza JC, Lessa EP (2003) Genetic and geographic differentiation in the Rio Negro tuco-tuco (*Ctenomys rionegrensis*): inferring the roles of migration and drift from multiple genetic markers. Evolution 57:913–926

Zenuto RR, Fanjul MS, Busch C (2004) Use of chemical communication by the subterranean rodent *Ctenomys talarum* (tuco-tuco) during the breeding season. J Chem Ecol 30:2111–2126

Part IV
Environmental Relationships

Chapter 9
Environmental and Ecological Features of the Genus *Ctenomys*

Daniel Galiano and Bruno Busnello Kubiak

9.1 Introduction

Life underground imposes general constraints for most subterranean and fossorial rodent species. The habitat requirements and ecological characteristics of subterranean influence numerous aspects of their biology, including where individuals live and how they behave. Subterranean rodents of the genus *Ctenomys*, commonly referred to as "tuco-tucos," are widely distributed throughout South America. There are currently approximately 65 valid species based on morphological, karyotypic, and molecular data (D'elia, Teta and Lessa, Chap. 2 this volume). These animals are typically solitary, have low mobility, and are usually found distributed among small patches of suitable habitat. They occupy a wide range of habitat types, particularly in open areas such as grasslands, steppes, deserts, and sand dunes (Lacey et al. 2000), with a few species occurring in forest regions (Gardner et al. 2014; Leipnitz et al. 2020) (Fig. 9.1). Regional distributions of *Ctenomys* species vary substantially with soil and vegetation characteristics and resource availability. Soil features are known to influence some aspects of their biology (e.g., burrow system characteristics and excavation strategies), with temporal and spatial variation in population density largely attributed to interactions between limiting environmental factors combined with species life-history characteristics. As in other fossorial animals, tuco-tucos can increase local environmental heterogeneity at the landscape level by aiding in the formation, aeration, and mixing of soils, and may enhance the infiltration of water into the soil. This modification of vegetation and soil characteristics can in turn impact herbivore food webs. Tuco-tucos are also a significant food

D. Galiano
Laboratório de Zoologia, Universidade Federal da Fronteira Sul, Realeza, Brazil

B. B. Kubiak (✉)
Programa de Pós-Graduação em Genética e Biologia Molecular, Universidade Federal do Rio Grande do Sul, Porto Alegre, Brazil

© Springer Nature Switzerland AG 2021
T. R. O. de. Freitas et al. (eds.), *Tuco-Tucos*,
https://doi.org/10.1007/978-3-030-61679-3_9

Fig. 9.1 Mosaic of photographs of *Ctenomys* habitats, burrows, and mounds. The different types of habitat are represented, such as grasslands, steppes, sand dunes, and forest areas. Localities: (**a**) Viamão, RS, Brazil; (**b**) Xangri-lá, RS, Brazil; (**c**) Capão da Canoa, RS, Brazil; (**d**) Viamão, RS, Brazil; (**e**) Manoel Viana, RS, Brazil; (**f**) Rio Grande, RS, Brazil; (**g**) Osório, RS, Brazil; (**h**) Cáceres, MT, Brazil; (**i**) Nova Mutum, MT, Brazil; (**j**) Feliz Natal, MT, Brazil; (**k**) Feliz Natal, MT, Brazil; (**l**) Cáceres, MT, Brazil

source for many avian and mammalian predators. In this chapter, we discuss the spatial use patterns and general ecological characteristics of species interactions and social structure in tuco-tucos, as well as their impacts on the local environment and general aspects of conservation.

9.2 Spatial Utilization and Species Interactions

Tuco-tucos are usually associated with arid or semiarid habitats in open areas such as deserts, savannas, and dunes (Bidau 2015; Freitas 2016), with few exceptions of species occurring in forest regions (Gardner et al. 2014; Leipnitz et al. 2020). The tendency to occupy open habitats may be related to the ecological pressures imposed by the subterranean lifestyle. The distribution of critical resources, including suitable habitat and food supply, is thought to strongly influence the distribution of small mammal species, including subterranean rodents (Jarvis et al. 1998; Tammone et al. 2012; Galiano et al. 2014a, b). When resources such as suitable habitat or food supply are unevenly distributed, habitat use is typically nonrandom, with animals preferentially occupying patches with sufficient resource availability (Garshelis 2000).

Tuco-tucos excavate networks of tunnels in which they live and perform most of their vital activities, making life underground more energetically costly than life on the surface (Lacey et al. 2000) (see Fig. 9.1g). Above and below ground soil and vegetation features are important factors for tuco-tuco habitat choice. For example, Bongiovanni et al. (2019) demonstrated that soil depth is an important habitat feature for *Ctenomys mendocinus* activity in the Puna Desert, where climatic conditions are extreme. Individuals from this species tend to select areas with deeper soils, likely because deeper burrow systems facilitate thermoregulation in cold environments (i.e., the amplitude of temperature fluctuation diminishes with increasing soil depth) (Bennett et al. 1998; Burda et al. 2007). The presence of shrubs also has a significant positive relationship with *C. mendocinus* activity on both micro and macro scales, since shrubs can represent a seasonal food source and provide protection against aerial predators. Another study by Scheich and Zenuto (2007) demonstrated that *Ctenomys talarum* is able to identify soils with richer food supply (e.g., grasses) and discriminate quality through olfaction, corroborating the importance of vegetation as a determining factor in habitat selection for this species.

The hardness and compaction of soils can also be considered a limiting factor for habitat occupation by tuco-tucos. It is well-documented that excavating and living in harder soils is more energetically costly than in softer soils (Luna and Antinuchi 2006; Antinuchi et al. 2007). However, different species of tuco-tucos still occupy a wide range of habitat with respect to soil characteristics. This limitation seems to be overcome by some species through differential allocation of effort between the incisors and limbs during excavation (Vassallo 1998; Morgan et al. 2017) or variations in morphological structures linked to bite force (Borges et al. 2017). For example, *Ctenomys minutus* populations show variation in humerus shape and skull size shape among the habitats they occupy. Individuals in sandy field habitats with harder soils have greater bite force and use their limbs more intensively in excavation than do individuals in sand dunes with softer soils (Kubiak et al. 2018). Likewise, *C. mendocinus* exhibits variation in activity depending on habitat soil characteristics. In the Puna Desert region, soil hardness was not associated with *C. mendocinus* activity (Bongiovanni et al. 2019); however, there was a negative association between soil hardness and *C. mendocinus* activity in the Monte Desert region (Albanese et al. 2010). The differences in determining factors for habitat selection among *C. mendoncinus* populations might be explained by environmental characteristics, as the Puna Desert is a more extreme climate with lower resource availability.

Soil moisture is another factor that can influence habitat selection for *Ctenomys* species. Flooded regions were negatively correlated with the selection of macroscale habitats for *C. mendoncinus* (Bongiovanni et al. 2019) and microscale habitats for *C. minutus* (Galiano et al. 2016), although the level and duration of tolerance for higher levels of soil moisture need to be investigated further.

The degree to which habitat features influence the ecology of *Ctenomys* becomes evident for species that occupy contrasting habitat types like grasslands and sand dunes, such as *C. minutus* in the coastal region of southern Brazil (see Fig. 9.1c) and *C. talarum* in the coastal region of Argentina. Both habitat types show clear variation in soil hardness and resource availability, with sand dunes having softer soils and lower food availability compared to grasslands (Comparatore et al. 1992; Malizia et al. 1991; Galiano et al. 2016; Kubiak et al. 2018). *C. minutus* inhabiting sand dunes have a home range that is 1.75 times larger than individuals inhabiting grassland (Kubiak et al. 2017a) and a weaker bite force (Kubiak et al. 2018). In *C. talarum*, individuals from Mar de Cobo sand dunes have a comparatively larger body size than those from Necochea grasslands (Zenuto 1999), but the home range size does not differ (Cutreta et al. 2010). These species also differ in food types consumed between habitats (Del Valle et al. 2001, Lopes et al. 2015), which could be related to changes in plant species composition and subsequent differences in food availability along their geographical distribution (Lopes et al. 2020, Chap. 10 this volume). Seasonal changes in food availability can also influence *Ctenomys* feeding during the year. For example, some studies have shown that when the availability of grasses is scarce during dry seasons, consumption of shrubs increases (Madoery 1993; Comparatore et al. 1995; Altuna et al. 1999). In general, the heterogenous nature of the habitat these species occupy and environmental discontinuities along the species distribution range probably strongly influence habitat selection and resource use for *Ctenomys*.

Most tuco-tuco species are characterized by allopatric distribution. There are only three cases of sympatry officially documented to date (Contreras and Reig 1965; Reig et al. 1990; Freitas 1995; Kubiak et al. 2015) (Fig. 9.2). A fourth case describes the capture of *C. dorsalis* and *C. conoveri* at the same time and location in Colonia Fernheim, Paraguay (Londoño-Gaviria et al. 2019). However, there is no other information regarding this possible region of sympatry, and this case warrants further investigation (Londoño-Gaviria et al. 2019). One of the three confirmed areas of sympatry occur in a portion of the coastal dunes in the Necochea region in the province of Buenos Aires, Argentina, between *C. australis* and *C. talarum*. These species present well-documented habitat segregation according to soil and vegetation types (Malizia et al. 1991; Comparatore et al. 1992). This segregation is

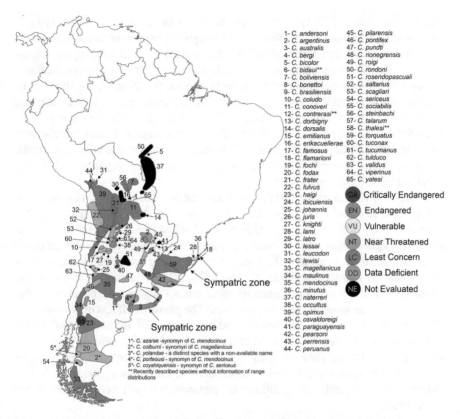

Fig. 9.2 Distribution ranges of 65 extant *Ctenomys* species in South America. Conservation status is based on IUCN (2020), and sympatric zones are indicated

likely the result of interspecific competition, as demonstrated by Vassallo (1993). On the other hand, this author also highlighted that competition plays a much smaller role in sustaining the pattern of microhabitat segregation between species compared to habitat preference associated with species differences in body size and color. There are also documented interspecific differences in home range size, with *C. australis* having a home range 19 times larger than *C. talarum* (Cutrera et al. 2010).

The other two sympatric zones occur in the coastal region of southern Brazil for the species *C. flamarioni* and *C. minutus* (Freitas 1995; Kubiak et al. 2015). One of these sympatric zones is located in the first line of sand dunes, similar to the sympatric zone in Argentina, and covers a length of approximately 15 km. The other sympatric zone occurs in sandy fields in the southern portion of their distribution and was only recently described (Kubiak et al. 2015). In the dune region of southern

Brazil, Kubiak et al. (2015) found results similar to those described for the sympatric region of Argentina, where species show segregation in microhabitat according to differences in soil hardness, plant biomass, and plant cover. Further, the authors also describe different patterns of habitat selection in allopatric and sympatric populations of *C. flamarioni*, whereas *C. minutus* selected the same habitat characteristics under both conditions. However, no interspecific differences in home range size were detected, nor were there differences between sympatric or allopatric populations for this trait (Kubiak et al. 2017b). The authors suggested that co-occurrence may not influence home range size in these species, perhaps due to environmental adaptations that facilitate coexistence (e.g., microhabitat segregation and dietary modifications) (Lopes et al. 2015).

Moreover, Kubiak et al. (2017c) pointed out that these species differ in microhabitat selection and morphological characteristics in areas of sympatry: *C. flamarioni* selects different microhabitats while *C. minutus* does not, and *C. minutus* present morphological modifications (i.e., reduction in skull length and body mass) when occurring in sympatry. This may be a result of temporal segregation of microhabitats. The displacement of morphological characteristics might reflect changes resulting from many generations of habitat segregation between *C. flamarioni* and *C. minutus* in the sympatric area. The plausible effects of both of these phenomena (i.e., spatial segregation and morphological differentiation) might point in the same direction, with the co-occurrence of species causing an ecological shift in the known zones of sympatry. In this context, interspecific interactions such as competition and competitive exclusion are likely among the main factors determining allopatric distribution in subterranean rodents (Miller 1964; Nevo 1979).

Another intriguing observation near the contact zones of *C. flamarioni* and *C. minutus* was the occurrence of hybrids (Kubiak et al. 2020). These species have extensive differences in chromosome organization (Freygang et al. 2004; Gava and Freitas 2003; Massarini and Freitas 2005), phenotype, evolutionary history (Parada et al. 2011), sperm morphology (Freitas 1995), and genetic characterization (Kubiak et al. 2020), but still can generate natural hybrids. This hybridization occurs bi-directionally (females of both species can generate hybrids). The importance of hybridization in the processes of adaptation and speciation has been rigorously debated in the literature, and the consensus that hybridization often plays an important role in evolution has recently been under debate (Taylor and Larson 2019). In the case of tuco-tucos, if hybrid individuals such as the ones mentioned above occur in higher numbers than imagined and possess some capacity to reproduce (either between hybrids themselves or with parental individuals), then interactions between tuco-tucos may be more dynamic than expected.

9.3 Social Structure

Few tuco-tuco species have been effectively characterized based on social structure and ecological features. Most known species to date are solitary, and in solitary species, individuals occupy their own burrow system and aggressively defend it. However, it is relatively common to find young individuals and pups sharing the tunnel system with their mothers for some time after the reproductive period. Mothers typically give birth to one to two pups per pregnancy (Pearson 1959; Rosi et al. 1992, 1996; Malizia and Busch 1991, 1997; Zenuto and Busch 1998; Marinho and Freitas 2006; Garcias et al. 2018). Some authors have observed that the parental care period for *C. talarum* was between 56 and 65 days (Malizia and Busch 1991; Zenuto et al. 2002) and that individuals live approximately 20–22 months (Busch et al. 1989; Malizia and Busch 1997). In this context, individuals of *C. talarum* have parental care and parent-offspring interactions that correspond to approximately 9.6% of their lifetime. However, the gestation period, extent of parental care, and number of offspring varies among tuco-tuco species (Pearson 1959; Zenuto 1999).

Male and female tuco-tucos interact at minimum during the reproductive season. Burrows are usually connected for a brief period to allow courtship and copulation (Altuna et al. 1999), and it is common to capture animals with injuries possibly resulting from conflict during mating periods (personal communication; Thales Freitas). In the solitary species *C. talarum*, laboratory studies have shown that individuals travel above ground to search for mates and copulate (frequently the male travels) (Zenuto et al. 2002). There is also evidence of below-ground access to females by males for *C. minutus* over long distances (~70 m) (Kubiak et al. 2017a).

Although most species are solitary, three patterns of sociality have been identified among tuco-tucos: solitary, facultative sociality, and social (O'Brien et al. 2020). *Ctenomys australis*, *C. flamarioni*, *C. haigi*, *C. Minutus*, and *C. talarum* have a solitary social structure in which each adult lives alone in a tunnel system and displays minimal or zero spatial overlaps with other adults (Lacey et al. 1998; Cutrera et al. 2006, 2010; Kubiak et al. 2017a, b). On the other hand, *C. sociabilis* has been shown to be social (Lacey et al. 1997), and *C. opimus* presents an intermediate pattern between these two and is considered to be facultatively social (O'brien et al. 2020). *Ctenomys rionegrensis* also shows signs of facultative sociality (Tomasco et al. 2019). The social pattern presented by *C. sociabilis* differs from most other species. Social individuals have a high overlap of home range with conspecific individuals, and burrow systems are routinely occupied by multiple adult females and, in many cases, a single adult male (Lacey et al. 1997; Lacey and Wieczorek 2003). In this context, only adult females in a group are kin, and while females of this species are philopatric, males are not (Lacey and Wieczorek 2003).

The intermediate, or facultatively social pattern, of *C. optimus* is characterized by the extensive but incomplete overlapping of home ranges (O'Brien et al. 2020). Interestingly, *C. rionegrensis* apparently has sporadic spatial overlap among adults, documented by both the capture of multiple adults in a single burrow entrance

(Lessa et al. 2005) and overlapping of home range area between females and males (Tassino et al. 2011). However, individuals of *C. rionegrensis* do not appear to regularly share the same tunnel gallery (Tassino et al. 2011; Estevan et al. 2016). According to Lacey and Wieczorek (2003), comparisons of ctenomyids over larger spatial scales are required, and interactions of subterranean rodents may be more diverse than previously expected.

9.4 The Influence of Tuco-Tucos on the Environment

Subterranean rodents excavate and inhabit extensive burrow systems, and changes in plant diversity, abundance, and community composition are typical consequences of high activity level of these animals (Andersen 1987; Contreras and Gutiérrez 1991; Huntly and Reichman 1994; Malizia et al. 2000; Campos et al. 2001; Reichman and Seabloom 2002; Kerley et al. 2004; Lara et al. 2007; Hagenah and Bennett 2013; Galiano et al. 2014a; Miranda et al. 2019). They also alter soil conditions, such as soil granulometry, aeration, and chemical and physical characteristics (Schauer 1987; Cox and Roig 1986; Borghi et al. 1990; Malizia et al. 2000; Lara et al. 2007; Šklíba et al. 2009; Hagenah and Bennett 2013). These effects of subterranean rodents on vegetation and soil can arise from burrow dynamics, diet selection, or foraging behavior (Huntly and Reichman 1994). Because of the great impact they have on entire ecosystems, including soil, water, and air content, decomposition processes of plant material, nutrient cycling, and composition of local biota (Hole 1981), subterranean rodents are regarded as ecosystem engineers (Cameron 2000; Reichman and Seabloom 2002; Reichman 2007). Their actions constitute a major factor in soil and vegetation dynamics.

Several studies have focused on the effects of *Ctenomys* on vegetation (Malizia et al. 2000; Campos et al. 2001; Tort et al. 2004; Lara et al. 2007; Galiano et al. 2014a) and soil conditions (Malizia et al. 2000; Lara et al. 2007; Galiano et al. 2014a). Most of these studies found that *Ctenomys* species alter vegetation and nutrient content of soil, although there is some variation in magnitude of these effects. For example, Galiano et al. (2014a) found significant effects of *Ctenomys minutus* activity on vegetation and soil conditions in patches of sandy fields. The authors observed a 37% reduction in plant biomass areas where individuals of the species were present versus absent. This reduction is within the range of other herbivorous subterranean rodents (e.g., plant biomass reduced by 25–50%; Reichman and Smith 1985). Additionally, reduction in plant biomass was observed for other species of *Ctenomys*. For example, *C. mendocinus* was associated with a 44% reduction in plant community biomass in the southern Puna Desert (Lara et al. 2007), while *C. talarum* was associated with a 31% reduction in biomass in grasslands of Buenos Aires Province (Malizia et al. 2000). This reduction could be a direct result of activities related to feeding and burrow construction.

Albanese et al. (2010) found that grasses were the dominant plant type consumed by *C. mendocinus* (79%), and grass leaves were the most representative item among consumed plant parts (89.5%). Decreased densities of grasses in areas with *Ctenomys* were reported by Campos et al. (2001) and Lara et al. (2007). This pattern suggests an effect of *Ctenomys* foraging on the abundance of grasses, which may in turn promote further effects on community dynamics. This modification of vegetation patterns by *Ctenomys* species suggests that they may act as keystone species. Moreover, tuco-tucos are herbivorous predators that can facilitate the propagation and reproduction of several plant species.

Burrowing activities of tuco-tucos can also affect mycorrhiza. Miranda et al. (2019) detected changes in the abundance of arbuscular mycorrhizae and dark septate endophytes as a consequence of burrowing activity by *Ctenomys* aff. *Knighti*. These authors observed that soil patches from burrows contain soil enriched with organic matter, soil microorganisms that decompose it into nutrients available for plants, fungal propagules that can establish mutualistic relationships with their roots, and greater water retention in relation to the surrounding landscape. Soil mixing by burrowing mammals is an important pedogenic process (Johnson 1990) and increases soil porosity.

During burrowing, subterranean rodents redistribute soil among different horizons which contributes to aeration, irrigation, and fertilization of soils (Reichman and Smith 1990). According to Malizia et al. (2000), the formation of new mounds by tuco-tucos (*C. talarum*) increased the levels of N, P, Na, K, and Mg; however, higher levels of Ca and pH were found in undisturbed areas. Lara et al. (2007) also found that the activity of *C. mendocinus* increased nutrient concentration (N, K, P) in bare soil compared to bare soil in undisturbed areas, and Galiano et al. (2014a) observed high concentrations of P and K just below the soil surface where *C. minutus* were present. Even though the effects on different nutrients vary between studies, it is well known that tuco-tucos and other subterranean mammals (e.g., pocket gophers, Geomyidae) modify the distribution of nutrients (Abaturov 1972; Mielke 1977; Grant and Mc Bryer 1981; Hole 1981; Spencer et al. 1985; Inouye et al. 1987; Reichman and Seabloom 2002) as a result of moving, mixing, or bringing soil to the surface from lower levels as well as the incorporation of nutrients into the soil. Moreover, the extensive excavations and their associated impacts might generate a dynamic mosaic of nutrients and soil conditions that promote diversity and maintain disturbance-dependent components of plant communities.

These animals have a significant effect on vegetation composition and dynamics as well as soil properties. While this obviously will affect ecosystem processes such as decomposition or productivity, the consequence of all the effects that tuco-tucos have on ecosystems are difficult to predict. In fact, tuco-tucos seem to be an important physical force in the ecosystems in which they occur, and their combined ecological effects may be a significant force in the maintenance and conservation of these ecosystems. To date, few studies have focused on the environmental effects of tuco-tucos in the Neotropics. While not all effects can be considered positive, the fact that tuco-tucos are a major source of heterogeneity in the ecosystems they inhabit is unquestionable.

9.5   Species Conservation

Conservation of small mammals requires knowledge of the genetically and ecologically meaningful spatial scales at which species respond to habitat modifications, and effects on small mammals may have cascading effects across the environments they inhabit (Manning and Edge 2004). The geographical distribution of species is also an important trait for biodiversity conservation and management policies because it provides relevant and meaningful information regarding the requirements for ecological success (Margules and Pressey 2000). As subterranean small mammals, tuco-tucos shows low vagility levels and high levels of population subdivision across landscapes, and consequently, low genetic variability (Reig et al. 1990; Lacey et al. 2000). These characteristics together with factors such as habitat fragmentation and degradation, urbanization, climatic changes, and limited knowledge concerning the areas of occurrence and life history of these small mammals enhance the risk of extinction of species (Gallardo et al. 1996; Freitas et al. 2012; Gómez Fernández et al. 2016; Caraballo et al. 2020). Additionally, the life history of these small mammals has contributed to difficulties in developing conservation initiatives (Fernandes et al. 2007).

Tuco-tucos have been considered agricultural pests for decades because of their fossorial habits and, consequently, are exposed to risk of predation by humans (Massoia 1970; Pearson et al. 1968; Freitas et al. 2012). The conservation status of all *Ctenomys* species is an important issue to be addressed since most species have a small geographic range that is restricted to sandy-soil grassland habitats. Such habitat specificity combined with anthropogenic threats described above and typical patchy distribution suggests higher vulnerability of tuco-tucos than presently supposed, and that conservation effort should be based on consistent and detailed studies of habitat occupation (Fernandes et al. 2007; Galiano et al. 2014b). In this context, the development of habitat conservation plans is urgently needed for many fragmented landscapes, and the first step in this process should be to better understand the responses of animal populations to different scenarios of fragmentation. For example, Mapelli and Kittlein (2009) analyzed the influence of patch and landscape characteristics on patch occupancy by *Ctenomys porteousi*. They observed that species distributions were affected not only by characteristics of the habitat patches but also by those of the surrounding landscape matrix. This implies that ongoing modification of the matrix landscape through anthropogenic activities might have important effects on the conservation of these species.

The lack of protection for areas occupied by subterranean species is also a major problem for the conservation of tuco-tucos. Caraballo et al. (2020) studied the overlap between protected areas and the 67 extant *Ctenomys* species, pointing out that only 34 of these species have significant overlapping distributions with protected areas. According to the authors, the most concerning species are those at risk (vulnerable (VU), endangered (EN), or critical endangered (CR), according to the

International Union for Nature Conservation, IUCN), and nonevaluated ones (i.e., not included in the IUCN Red List or data deficient) which have little or no overlap with protected areas. In other words, half of the extant tuco-tucos correspond to nonevaluated species, most of which are known only from the type locality or its surroundings. Under this scenario, we can assume that nonprotected areas have a unique importance for *Ctenomys* conservation, and that data deficiency and regional habitat transformation pose a serious threat to these species.

Regarding the conservation status, 53 out of 65 species (D'elía, Teta and Lessa, Chap. 2 this volume) were evaluated according to the IUCN Red List. Of these, approximately 24.5% were considered to be under some degree of threat (one classified as VU, nine as EN, and three as CR), three (5.7%) are considered "near threatened," 16 (30.2%) as "least concern," 21 (39.6%) are classified as "data deficient" (IUCN 2020), and 12 species were not evaluated (Fig. 9.2). Further, information regarding population trends is only available for 52% of species. This means that there is limited knowledge regarding the true conservation status with only basic information available on population parameters for nearly half of the species. The conservation status, areas of occurrence, population trends, and major threats for species are listed in Table 9.1. We note that not all tuco-tuco species are included in this list.

The current state of knowledge of *Ctenomys* conservation is lacking, and the lack of data to guide conservation actions and species management is among the major roadblocks. In this context, the most recent compilations of the genus (Bidau 2015; Freitas 2016; Teta and D'Elia 2020) recognize that the number of species will increase with the combination of additional field collections and revisionary work. The need to investigate basic ecological information to provide adequate conservation measures is urgent for species that are not well described. Regardless, the conservation of soil and vegetation in regions of occurrence is certainly one of the best conservation strategies that can be implemented for these species.

Although it is broadly understood that subterranean rodents act as ecological engineers, obtaining ecological information about tuco-tucos remains challenging due to the primarily underground lifestyle. Since tuco-tuco populations vary spatially and temporally, the ecology of these animals is also expected to be strongly influenced by regional factors. Long-term ecological studies are needed to better understand the interactions between these rodents and their effects on inhabited ecosystems. Given the high degree of habitat specialization and endemicity of *Ctenomys*, regional scenarios of habitat transformation pose a serious threat to species viability, and the current lack of knowledge regarding tuco-tuco biology and ecology could negatively impact population management. Despite many studies that have utilized various approaches to estimate ecological data for these species, it remains difficult to corroborate those data with ecological parameters at broad scales.

Table 9.1 Species, distribution countries, IUCN conservation status, major threats, and population trends of extant *Ctenomys* species

Species	Countries[a]	IUCN conservation status[b]	Threats[c]	Population trends
C. andersoni	BOL	NE	–	–
C. argentinus	ARG	NT	AA	Decreasing
C. australis	ARG	EN	AA; RCD	Decreasing
C. bergi	ARG	EN	AA	Decreasing
C. bicolor	BRA	NE	–	–
C. bidaui	ARG	NE	–	–
C. boliviensis	ARG; BOL; BRA; PRY	LC	There are no major threats	Stable
C. bonettoi	ARG	EN	AA	Decreasing
C. brasiliensis	URY	DD	The threats to this species are unknown	Unknown
C. coludo	ARG	DD	There is no information available on the threats	Unknown
C. conoveri	BOL; PRY	LC	There appear to be no major threats	Stable
C. contrerasi	ARG	NE	–	–
C. dorbignyi	ARG	NT	BRU	Decreasing
C. dorsalis	PRY	DD	AA; RCD	Unknown
C. emilianus	ARG	LC	There are no imminent threats	Decreasing
C. erikacuellarae	BOL	NE	–	–
C. famosus	ARG	DD	There is no information available on the threats	Unknown
C. flamarioni	BRA	EN	EPM; HID; RCD	Decreasing
C. fochi	ARG	DD	There is no information available on the threats	Unknown
C. fodax	ARG	DD	There is no information available on the threats	Unknown
C. frater	BOL	LC	There do not appear to be any major threats to this species	Decreasing
C. fulvus	CHL	DD	There is no information available on the threats	Unknown
C. haigi	ARG	LC	There are no known threats to this species	Unknown
C. ibicuiensis	BRA	DD	AA, NSM	Unknown
C. johannis	ARG	DD	There is no information available on the threats	Unknown
C. juris	ARG	DD	There is no information available on any threats to this species	Unknown

(continued)

Table 9.1 (continued)

Species	Countries[a]	IUCN conservation status[b]	Threats[c]	Population trends
C. knighti	ARG	DD	AA	Unknown
C. lami	BRA	VU	AA; RCD	Decreasing
C. latro	ARG	EN	AA	Decreasing
C. lessai	BOL	NE	–	–
C. leucodon	BOL; PER	LC	BRU	Stable
C. lewisi	BOL	LC	There appear to be no major threats to this species	Stable
C. magellanicus	ARG; CHL	LC	AA	Decreasing
C. maulinus	ARG; CHL	LC	No major threats to this species are described	Unknown
C. mendocinus	ARG	LC	There are no major threats known to this species	Unknown
C. minutus	BRA	DD	AA; RCD	Stable
C. nattereri	BOL; BRA	NE	–	–
C. occultus	ARG	EN	AA	Decreasing
C. opimus	ARG; BOL; CHL; PER	LC	There appear to be no major threats to this species	Stable
C. osvaldoreigi	ARG	CR	AA; NSM	Decreasing
C. paraguayensis	PRY	NE	–	–
C. pearsoni	URY	NT	AA; RCD	Decreasing
C. perrensis	ARG	LC	There are no known threats to this species	Stable
C. peruanus	PER	LC	There appear to be no major threats to this species	Stable
C. pilarensis	PRY	EN	AA; BRU	Decreasing
C. pontifex	ARG	DD	There is no information available on any threats to this species	Unknown
C. pundit	ARG	EN	AA	Decreasing
C. rionegrensis	ARG; URY	EN	AA	Decreasing
C. roigi	ARG	CR	AA	Decreasing
C. rondoni	BRA	NE	–	–
C. rosendopascuali	ARG	NE	–	–
C. saltarius	ARG	DD	There is no information available on any threats to this species	Unknown
C. scagliai	ARG	DD	There is no information available on any threats to this species	Unknown

(continued)

Table 9.1 (continued)

Species	Countries[a]	IUCN conservation status[b]	Threats[c]	Population trends
C. sericeus	ARG	DD	There is no information available on any threats to this species	Unknown
C. sociabilis	ARG	CR	AA; BRU	Decreasing
C. steinbachi	BOL	LC	There appear to be no major threats to this species	Stable
C. talarum	ARG	LC	There are no known threats to this species	Unknown
C. thalesi	ARG	NE	–	–
C. torquatus	BRA; URY	LC	AA; EPM	Unknown
C. tuconax	ARG	DD	AA	Unknown
C. tucumanus	ARG	DD	AA	Unknown
C. tulduco	ARG	DD	The threats to this species are unknown	Unknown
C. validus	ARG	DD	There is no information available on threats to this species	Unknown
C. viperinus	ARG	DD	The threats to this species are unknown	Unknown
C. yatesi	BOL	NE	–	–

Data presented here are based on the Caraballo et al. (2020) and IUCN (2020)

aAcronyms (Countries): ARG, BOL, BRA, CHL, PER, PRY, and URY, correspond to Argentina, Bolivia, Brazil, Chile, Peru, Paraguay, and Uruguay, respectively

bAcronyms (IUCN conservation status): NE, DD, LC, NT, VU, EN, and CR, correspond to Not evaluated, Data deficient, Least concern, Near threatened, Vulnerable, Endangered, and Critically endangered, respectively

cAcronyms (Threats): AA, BRU, EPM, HID, NSM, and RCD, correspond to Agriculture and aquaculture; Biological resource use, Energy production and mining, Human intrusions and disturbance, Natural system modifications, and Residential and commercial development, respectively

Acknowledgments We thank the editors for the invitation to contribute to this collective book on tuco-tucos. We are grateful to all colleagues and students that we have discussed ideas on the ecology of tuco-tucos over the years. We thank Diego A. Caraballo for the invaluable help in sending us the distribution ranges of *Ctenomys*. BBK received financial support from Conselho Nacional de Desenvolvimento Científico e Tecnológico (CNPq; 158250/2018-4). Lastly, we would like to thank everyone who studies or has studied the ecology of these incredible animals, the tuco-tucos.

Literature Cited

Abaturov BD (1972) The role of burrowing animals in the transport of mineral substances in the soil. Pedobiologia 12:261–266

Albanese S, Rodríguez D, Dacar MA, Ojeda RA (2010) Use of resources by the subterranean rodent Ctenomys mendocinus (Rodentia, Ctenomyidae), in the lowland Monte desert, Argentina. J Arid Environ 74:458–463

Altuna CA, Francescoli G, Tassino B, Izquierdo G (1999) Ecoetologia y conservacion de mam-iferos subterraneos de distribucion restringida: el caso de Ctenomys pearsoni (Rodentia, Octodontidae) en el Uruguay. Etología 7:47–54

Andersen D (1987) Belowground herbivory in natural communities: a review emphasizing fosso-rial animals. Q Rev Biol 62:261–286

Antinuchi CD, Zenuto RR, Luna F, Cutrera AP, Perissinotti PP, Busch C (2007) Energy budget in subterranean rodents: insights from the tuco tuco Ctenomys tralarum (Rodentia: Ctenomyidae). In: Kelt DA, Lessa EP, Salazar-Bravo JA, Patton JL (eds) The quintessential naturalist: hon-oring the life and legacy of Oliver P. Pearson. The University of California Press, Berkeley, pp 111–140

Bennett NC, Jarvis JUM, Davies KC (1998) Daily and seasonal temperatures in the burrows of African rodent moles. S Afr J Zool 3:189–195

Bidau CJ (2015) Family Ctenomyidae (Lesson, 1842). In: Patton JL, Pardiñas UFJ, D'Elía G (eds) Mammals of South America, vol 2. The University of Chicago Press, Chicago, pp 818–877

Bongiovanni SB, Nordenstahl M, Borghi CE (2019) Resources and soil influencing habitat selec-tion by a subterranean rodent in a high cold desert. J Mammal 100:537–543

Borges LR, Maestri R, Kubiak BB, Galiano D, Fornel R, De Freitas TRO (2017) The role of soil features in shaping the bite force and related skull and mandible morphology in the subter-ranean rodents of genus Ctenomys (Hystricognathi: Ctenomyidae). J Zool 301(2):108–117

Borghi CE, Giannoni SM, Martínez-Rica JP (1990) Soil removed by voles ofthe genus Pitymys species in the Spanish Pyrenees. Pirineos 136:3–18

Burda H, Sumbera R, Begall S (2007) Microclimate in burrows of subterranean rodents – revis-ited. In: Begall S, Burda H, Schleich C (eds) Subterranean rodents: news from underground. Springer, Heidelberg, pp 21–23

Busch C, Malizia AI, Scaglia OA, Reig OA (1989) Spatial distribution and attributes of a popula-tion of Ctenomys talarum (Rodentia: Octodontidae). J Mammal 70:204–208

Cameron GN (2000) Community ecology of subterranean rodents. In: Lacey EA, Patton JL, Cameron GN (eds) Life underground. The biology of subterranean rodents. University of Chicago Press, Chicago, pp 227–256

Campos CM, Giannoni SM, Borghi CE (2001) Changes in Monte desert plant communities induced by a subterranean mammal. J Arid Environ 47:339–345

Caraballo DA, López SL, Carmarán AA, Rossi MS (2020) Conservation status, protected area cov-erage of Ctenomys (Rodentia, Ctenomyidae) species and molecular identification of a popula-tion in a national park. Mamm Biol 100:33–47

Comparatoe V, Agnusdei M, Busch C (1992) Habitat relations in sympatric populations of Ctenomys australis and Ctenomys talarum (Rodentia, Octodontidae) in a natural grassland. Z Säugetierkd 57:47–55

Comparatore VM, Cid MS, Busch C (1995) Dietary preferences of 2 sympatric subterranean rodent populations in argentina. Chilean Review of Natural History 68:197–206

Contreras LC, Gutiérrez JR (1991) Effect of the subterranean herbivorous rodent Spalacopus cya-nus on herbaceous vegetation in arid coastal Chile. Oecologia 87:106–109

Contreras J, Reig O (1965) Dados sobre la distribuición de género Ctenomys (Rodentia: Octodontidae) en la zona costera de la Provincia de Buenos Aires entre Neocochea y Bahía Blanca. Physis 25:169–186

Cox GW, Roig V (1986) Argentinian Mima mounds occupied by ctenomyid rodents. J Mammal 67:428–432

Cutrera AP, Antinuchi CD, Mora MS, Vassallo AI (2006) Home-range and activity patterns of the South American subterranean rodent Ctenomys Talarum. J Mammal 87:1183–1191

Cutrera AP, Mora MS, Antenucci CD, Vassallo AI (2010) Intra- and interspecific variation in home-range size in sympatric tuco-tucos, Ctenomys australis and C. talarum. J Mammal 91:1425–1434

De Freitas TRO (1995) Geographic distribution and conservation of four species of the genus ctenomys in southern Brazil. Stud Neotropical Fauna Environ 30:37–41

De Freitas TRO (2016) Family Ctenomyidae (Tuco-tucos). In: Wilson D, Lacher T, Mittermeier RA (eds) Handbook of the mammals of the world lagomorphs and rodents I. Lynx Edicions Publications, Barcelona, pp 498–534

De Freitas TRO, Fernandes FA, Fornel R, Roratto PA (2012) An endemic new species of tuco-tuco, genus Ctenomys (Rodentia: Ctenomyidae), with a restricted geographic distribution in southern Brazil. J Mammal 93:1355–1367

Del Valle JC, Lohfelt MI, Comparatore VM, Cid MS, Busch C (2001) Feeding selectivity and food preference of Ctenomys talarum (tuco-tuco). Z Säugetierkd 66:165–173

Estevan I, Lacey EA, Tassino B (2016) Daily patterns of activity in free-living rio nego tuco-tuco (Ctenomys rionegrensis). Neotrop Mastozool 23:71–80

Fernandes FA, Fernández-Stolz GP, Lopes CM, de Freitas TRO (2007) The conservation status of the tuco-tucos, genus Ctenomys (Rodentia: Ctenomyidae), in southern Brazil. Braz J Biol/Rev Bras Biol 67:839–847

Freygang CC, Marinho JR, De Freitas TRO (2004) New karyotypes and some considerations about the chromosomal diversication of. Genetica 121:125–132

Galiano D, Kubiak BB, Overbeck GE, de Freitas TRO (2014a) Effects of rodents on plant cover, soil hardness, and soil nutrient content: a case study on tuco-tucos (Ctenomys minutus). Acta Theriol 59:583–587

Galiano D, Bernardo-Silva J, De Freitas TRO (2014b) Genetic pool information reflects highly suitable areas: the case of two parapatric endangered species of tuco-tucos (Rodentia: Ctenomiydae). PLoS One 9(5):e97301. https://doi.org/10.1371/journal.pone.0097301

Galiano D, Kubiak BB, Menezes LS, Overbeck GE, de Freitas TRO (2016) Wet soils affect habitat selection of a solitary subterranean rodent (Ctenomys minutus) in a Neotropical region. J Mammal 97:1095–1101

Gallardo MH, Kühler N, Araneda C (1996) Loss of genetic variation in Ctenomys coyhaiquensis (Rodentia, Ctenomyidae) affected by vulcanism. Mastozool Neotropical 3:7–13

Garcias FM, Stolz JFB, Fernández GP, Kubiak BB, Bastazini VAG, De Freitas TRO (2018) Environmental predictors of demography in the tuco-tuco of the dunes (Ctenomys flamarioni). Mastozool Neotropical 25:293–304

Gardner SL, Salazar J, Cook JA (2014) New species of Ctenomys Blainville 1826 (Rodentia: Ctenomyidae) from the Lowlands and Central Valleys of Bolivia. Special Publications Museum of Texas Tech University, Number 62

Garshelis DL (2000) Delusions in habitat evaluation: measuring use, selection, and importance. In: Boitani L, Fuller TK (eds) Research techniques in animal ecology: controversies and consequences. Columbia University Press, New York, pp 111–164

Gava A, De Freitas TRO (2003) Inter and intra-specific hybridization in Tuco-Tucos (Ctenomys) from Brazilian Coastal Plains (Rodentia: Ctenomyidae). Genetica 119:11–17

Gómez Fernández MJ, Boston ESM, Gaggiotti OE, Kittlein MJ, Mirol PM (2016) Influence of environmental heterogeneity on the distribution and persistence of a subterranean rodent in a highly unstable landscape. Genetica 144:711–722

Grant WE, Mc Bryer JF (1981) Effects of mound formation by pocket gophers (Geomys bursarius) on old-field ecosystems. Pedobiologia 22:21–28

Hagenah N, Bennett NC (2013) Mole rats act as ecosystem engineers within a biodiversity hotspot, the Cape Fynbos. J Zool 289:19–26

Hole FD (1981) Effects of animals on soils. Geoderma 25:75–112

Huntly N, Reichman OJ (1994) Effects of subterranean mammalian herbivores on vegetation. J Mammal 75:852–859

Inouye RS, Huntly NJ, Tilman D, Tester JR (1987) Pocket gophers (Geomys bursarius), vegetation, and soil nitrogen along a successional sere in east Central Minnesota. Oecologia 72:178–184

IUCN (2020) The IUCN Red List of Threatened Species. Version 2020–1. https://www.iucnredlist. org. Downloaded on 19 March 2020

Jarvis JUM, Bennett NC, Spinks AC (1998) Food availability and foraging by wild colonies of Damaraland mole-rats (Cryptomys damarensis): implications for sociality. Oecologia 113:290–298

Johnson DL (1990) Biomantle evolution and the redistribution of earth materials and artifacts. Soil Sci 149:84–102

Kerley GIH, Whitford WG, Kay FR (2004) Effects of pocket gophers on desert soils and vegetation. J Arid Environ 58:155–166

Kubiak BB, Galiano D, De Freitas TRO (2015) Sharing the space: distribution, habitat segregation and delimitation of a new sympatric area of subterranean rodents. PLoS One 10(4):e0123220. https://doi.org/10.1371/journal.pone.0123220

Kubiak BB, Galiano D, de Freitas TRO (2017a) Can the environment influence species home-range size? A case study on Ctenomys minutus (Rodentia, Ctenomyidae). J Zool 302:171–177

Kubiak BB, Maestri R, Borges LR, Galiano D, De Freitas TRO (2017b) Interspecific interactions may not influence home range size in subterranean rodents: a case study of two tuco-tuco species (Rodentia: Ctenomyidae). J Mammal 98:1753–1759

Kubiak BB, Gutiérrez EE, Galiano D, Maestri R, De Freitas TROD (2017c) Can niche modeling and geometric morphometrics document competitive exclusion in a pair of subterranean rodents (Genus Ctenomys) with tiny parapatric distributions? Sci Rep 7:16283. https://doi.org/10.1038/s41598-017-16243-2

Kubiak BB et al (2018) Evolution in action: soil hardness influences morphology in a subterranean rodent (Rodentia: Ctenomyidae). Biol J Linn Soc 125:766–776

Kubiak BB et al (2020) Hybridization between subterranean tuco-tucos (Rodentia, Ctenomyidae) with contrasting phylogenetic positions. Sci Rep 10:1502. https://doi.org/10.1038/s41598-020-58433-5

Lacey EA, Wieczorek JR (2003) Ecology of sociality in rodents: a ctenomyid perspective. J Mammal 84:1198–1211

Lacey EA, Braude SH, Wieczorek JR (1997) Burrow Sharing by Colonial Tuco-Tucos (Ctenomys sociabilis). J Mammal 78:556–562

Lacey EA, Braude SH, Wieczorek JR (1998) Solitary burrow use by adult patagonian tuco-tucos (Ctenomys haigi). J Mammal 79:986

Lacey EA, Patton JL, Cameron GN (2000) Life underground the biology of subterranean rodents. The University of Chicago Press, Chicago

Lara N, Sassi P, Borghi CE (2007) Effect of herbivory and disturbances by tuco-tucos (Ctenomys mendocinus) on a plant community in the southern Puna Desert. Arct Antarct Alp Res 39:110–116

Leipnitz LT, Fornel R, Ribas LEJ, Kubiak BB, Galiano D, De Freitas TRO (2020) Lineages of Tuco-Tucos (Ctenomyidae: Rodentia) from Midwest and Northern Brazil: late irradiations of subterranean rodents towards the Amazon Forest. J Mamm Evol 27:161–176

Lessa EP, Wlasiuk G, Garza JC (2005) Dymanics of genetic differentiation in the Rio Negro tuco-tucos (Ctenomys rionegrensis) at the local and geographical scales. In: Lacey E, Myers P (eds) Mammalian diversification. From chromosomes to phylogeography, A celebration of the career of James L. Patton. University of California Publications in Zoology, Berkeley, pp 155–174

Londoño-Gaviria M, Teta P, Ríos SD, Patterson BD (2019) Redescription and phylogenetic position of Ctenomys dorsalis Thomas 1900, an enigmatic tuco tuco (Rodentia, Ctenomyidae) from the Paraguayan Chaco. Mammalia 83:227–236

Lopes CM et al (2015) DNA metabarcoding diet analysis for species with parapatric vs sympatric distribution: a case study on subterranean rodents. Heredity 114:1–12

Lopes CM, De Barba M, Boyer F, Mercier C, Galiano D, Kubiak BB, Maestri R, Filho PJ, Gielly L, Coissac E, Freitas TRO DE, Taberlet P (2020) Ecological specialization and niche overlap of subterranean rodents inferred from DNA metabarcoding diet analysis. Mol Ecol (in press)

Luna F, Antinuchi CD (2006) Cost of foraging in the subterranean rodent Ctenomys talarum: effect of soil hardness. Can J Zool/Rev Can Zool 84:661–667

Madoery L (1993) Composición botánica de la dieta del tuco-tuco (Ctenomys mendocinus) en el piedemonte precordillerano. Ecol Austral 3:49–55

Malizia AI, Busch C (1991) Reproductive parameters and growth in the fossorial rodent Ctenomys talarum (Rodentia: Octodontidae). Mammalia 55:293–305

Malizia AI, Busch C (1997) Breeding biology of the fossorial rodent Ctenomys talarum (Rodentia: Octodontidae). J Zool 242:463–471

Malizia AI, Vassallo AI, Busch C (1991) Population and habitat characteristics of 2 sympatric species of ctenomys (rodentia, octodontidae). Acta Theriol 36:87–94

Malizia AI, Kittlein MJ, Busch C (2000) Influence of the subterranean herbivorous rodent Ctenomys talarum on vegetation and soil. Z Säugetierkd Saugetierkd 65:172–182

Manning JA, Edge WD (2004) Small mammal survival and downed wood at multiple scales in managed forests. J Mammal 85:87–96

Mapelli FJ, Kittlein MJ (2009) Influence of patch and landscape characteristics on the distribution of the subterranean rodent Ctenomys porteousi. Landsc Ecol 24:723–733

Margules CR, Pressey RL (2000) Systematic conservation planning. Nature 405:253

Marinho JR, De Freitas TRO (2006) Population structure of Ctenomys minutus (Rodentia, Ctenomyidae) on the coastal plain of Rio Grande do Sul, Brazil. Acta Theriol 51:53–59

Massarini IA, De Freitas TRO (2005) Morphological and cytogenetics comparison in species of the Mendocinus-group (genus Ctenomys) with emphasis in C. Australis and C. Flamarioni (Rodentia-Ctenomyidae). Caryologia 58:21–27

Massoia E (1970) Mamíferos que contribuyen a deteriorar suelos y pasturas em la República Argetina. INTA 276:14–17

Mielke HW (1977) Mound building by pocket gophers (Geomyidae): their impact on soils and vegetation in North America. J Biogeogr 4:171–180

Miller RS (1964) Ecology and distribution of pocket gophers (Geomyidae) in Colorado. Ecology 45:256–272

Miranda V et al (2019) Subterranean desert rodents (Genus Ctenomys) create soil patches enriched in root endophytic fungal propagules. Microb Ecol 77:451–459

Morgan CC, Verzi DH, Olivares AI, Vieytes EC (2017) Craniodental and forelimb specializations for digging in the South American subterranean rodent Ctenomys (Hystricomorpha, Ctenomyidae). Mamm Biol 87:118–124

Nevo E (1979) Adaptive convergence and divergence of subterranean mammals. Annu Rev Ecol Syst 10:269–308

O'Brien SL, Tammone MN, Cuello PA, Lacey EA (2020) Facultative sociality in a subterranean rodent, the highland tuco-tuco (Ctenomys opimus). Biol J Linn Soc 129:918–930

Parada A, Elía GD, Bidau CJ, Lessa EP (2011) Species groups and the evolutionary diversification of tuco-tucos, genus Ctenomys Species groups and the evolutionary diversification of tuco-tucos, genus Ctenomys (Rodentia : Ctenomyidae). J Mammal 92:671–682

Pearson OP (1959) Biology of subterranean rodents, Ctenomys, in Peru. Mem Mus Nat "Javier Prado" 9:1–56

Pearson OP, Binsztein N, Boiry L (1968) Social structure, spatial distribution and composition by ages of a population of tuco-tucos (Ctenomys talarum). Investigaciones Zoologicas Chilenas 13:47–80

Reichman OJ (2007) The influence of pocket gophers on the biotic and abiotic environment. In: Begal S, Burda H, Schleich CE (eds) Subterranean rodents: news from underground. Springer, Berlin/Heidelberg, pp 271–286

Reichman OJ, Seabloom EW (2002) The role of pocket gophers as subterranean ecosystem engineers. Trends Ecol Evol 17:44–49

Reichman OJ, Smith S (1985) Impact of pocket gopher burrows on overlying vegetation. J Mammal 66:720–725

Reichman OJ, Smith SC (1990) Burrows and burrowing behavior by mammals. In: Genoways HH (ed) Current mammalogy, vol 2. Plenum Publishers, New York

Reig O, Bush C, Ortellis M, Contreras J (1990) An overview of evolution, systematica, population biology and molecular biology. In: Nevo E, Reig O (eds) Evolution of subterranean mammals at the organismal and molecular levels. Wiley-Liss, New York, pp 71–96

Rosi M, Puig S, Videla F, Madoery L, Roig V (1992) Estudio ecológico del roedor subterráneo Ctenomys mendocinus en la precordillera de Mendoza, Argentina: ciclo reproductivo y estructura etaria. Rev Chil Hist Nat 65:221–233

Rosi MI, Puig S, Videla F, Coina MI, Roig VG (1996) Ciclo reproductivo y estructura etaria de Ctenomys mendoncinus (Rodentia:Ctenomyidae) del Piedemonte de Mendoza, Argentina. Ecol Aust Austraç 6:87–93

Schauer J (1987) Remarks on the construction of burrows of Ellobius talpinus, Myospalax aspalax and Ochotona daurica in Mongolia and their effect on the soil. Folia Zool 36:319–326

Schleich CE, Zenuto R (2007) Use of vegetation chemical signals for digging orientation in the subterranean rodent Ctenomys talarum (Rodentia: Ctenomyidae). Ethology 113:573–578

Šklíba J, Šumbera R, Chitaukali WN, Burda H (2009) Home-range dynamics in a solitary subterranean rodent. Ethology 115:217–226

Spencer SR, Cameron GN, Eshelman BD, Cooper LC, Williams LR (1985) Influence of pocket gopher mound on a Texas coastal prairie. Oecologia 66:11–115

Tammone MN, Lacey EA, Relva MA (2012) Habitat use by colonial tuco-tucos (Ctenomys sociabilis): specialization, variation, and sociality. J Mammal 93:1409–1419

Tassino B, Estevan I, Garbero RP, Altesor P, Lacey EA (2011) Space use by Río Negro tuco-tucos (Ctenomys rionegrensis): excursions and spatial overlap. Mamm Biol 76:143–147

Taylor SA, Larson EL (2019) Insights from genomes into the evolutionary importance and prevalence of hybridization in nature. Nat Ecol Evol 3:170–177

Teta P, D'Elía G (2020) Uncovering the species diversity of subterranean rodents at the end of the world: three new species of Patagonian tuco-tucos (Rodentia, Hystricomorpha, Ctenomys). PeerJ 8:e9259

Tomasco IH, Sánchez L, Lessa EP, Lacey EA (2019) Genetic analyzes suggest burrow sharing by río negro tuco-tucos (Ctenomys rionegrensis). Mastozoologia Neotropical 26:430–439

Tort J, Campos CM, Borghi CE (2004) Herbivory by tuco-tucos (Ctenomys mendocinus) on shrubs in the upper limit of the Monte desert (Argentina). Mammalia 68:15–21

Vassallo AI (1993) Habitat shift after experimental removal of the bigger species in sympatric Ctenomys talarum and Ctenomys australis (Rodentia: Octodontidae). Behaviour 127:247–263

Vassallo AI (1998) Functional morphology, comparative behaviour, and adaptation in two sympatric subterranean rodents genus Ctenomys (Caviomorpha: Octodontidae). J Zool 244:415–427

Zenuto R (1999) Sistema de apareamiento en Ctenomys talaru. (Rodentia: Octodontidae). Universidad Nacional de Mar Del Plata, Argentina

Zenuto R, Busch C (1998) Population biology of the subterranean rodent Ctenomys australis (tuco-tuco) in a costal dunafield in Argentina. Z Säugetierkd 60:277–285

Zenuto RR, Antinuchi CD, Busch C (2002) Bioenergetics of reproduction and pup development in a subterranean rodent (*Ctenomys talarum*). Physiol Biochem Zool 75:469–478

Chapter 10
The Diet of Ctenomyids

Carla Martins Lopes

10.1 An Overview

What does this animal eat? This is one of the first questions that comes out when someone is introduced to a new species. Researchers are trying to answer this question and all other aspects related to the diet of ctenomyids at least since 1989 (Torres-Mura et al. 1989). Assessing the dietary composition of species in their natural environment is a clue to unravel central questions in ecology and evolution. Ecological processes such as intra- and interspecific competition, interactions between predator-prey and herbivore-plant, population fluctuations, and species geographical distributions can be strongly influenced by the availability of food resources in the environment and how species use these resources (Begon et al. 2006). All these aspects, ultimately, affect ecosystem functioning and evolutionary processes.

Ctenomyids are considered strictly herbivorous, generalist, and opportunistic species able to consume a wide range of plant groups. This behavior has been associated with adaptation to high costs of burrowing and low levels of energy available at the subterranean ecotope (del Valle et al. 2001). The availability of food in the natural environment is what ultimately drives species selectively (Emlen 1966). Despite ctenomyids can use a wide variety of plants as a food source, they often show a preference for the aerial vegetative part of grasses (Fig. 10.1) (Comparatore et al. 1995; Puig et al. 1999; del Valle et al. 2001; Rosi et al. 2003, 2009). This food preference has been associated to specific nutritional requirements, palatability, and lower harvesting and handling times required for feeding these items, which could minimize the time that individuals expend outside their burrows, diminishing their predation risk (Puig et al. 1999; Rosi et al. 2003).

C. M. Lopes (✉)
Departamento de Biodiversidade e Centro de Aquicultura, Instituto de Biociências,
Universidade Estadual Paulista (UNESP), Rio Claro, SP, Brazil

© Springer Nature Switzerland AG 2021
T. R. O. de. Freitas et al. (eds.), *Tuco-Tucos*,
https://doi.org/10.1007/978-3-030-61679-3_10

Fig. 10.1 Specimen of
Ctenomys flamarioni
eating the aerial vegetative
part of grass. (Photograph
by Jose F. B. Stolz)

The feeding behavior of ctenomyids was described as individuals squat on their haunches when handling their food, hold the plants with one or both forepaws, between the three central digits and the thenar pad (Fig. 10.1), and shake it up and down quickly for cleaning. Coprophagy is also a common behavior observed but animals examine feces before deciding to eat or discard them. The individuals seat on their hindquarters, bring their head closer to the anus, and take the feces using lips and incisors. There is almost no feces manipulation with the forepaws. Only on few occasions animals were observed eating feces collected from the floor. The ingestion of feces was associated with better assimilation of nutrients and retention of water (Altuna et al. 1998; Martino et al. 2007).

10.2   Methods for Diet Analysis

Several methodologies have been developed and applied in order to better understand the dietary composition of animals. Some of them provide information about the ingested or egested items, as observation of foraging behavior and examination of fecal and gut contents by means of visual inspections, microscopy, or DNA-based methods. However, not all items ingested are assimilated and incorporated into the tissue of the consumers. The analysis of stable isotopes, biomarkers, and the near-infrared reflectance spectroscopy, e.g., can estimate the components that are assimilated by the animal (Johnson et al. 1983; Symondson 2002; Valentini et al. 2009; Pompanon et al. 2012; Nielsen et al. 2018). Each method addresses different aspects related to the diet of species and must be applied depending on the question to be further explored. Most of the knowledge about dietary habits and preferences of ctenomyids is based on the analysis of ingested and egested food items. The information provided in this chapter are mainly based on these studies.

The direct observation of foraging is one of the simplest methods to unravel the feeding habits of animals. However, it can be very time-consuming or even impossible to apply for elusive or generalist species living in complex environments, like herbivorous (Valentini et al. 2009). Cafeteria tests consist of capturing animals in their natural environment, and keep them at laboratory conditions, where the food availability is controlled and their feeding behavior can be observed. This approach provided much knowledge about harvesting and feeding behavior of ctenomyids and some of their food preferences (Camin and Madoery 1994; Altuna et al. 1998; del Valle et al. 2001; Martino et al. 2007). However, the controlled conditions in a laboratory do not recreate the natural environment where these species live, which may result in major changes in their behavior (Symondson 2002).

The most used method until recent years for unraveling diet preferences of ctenomyids was the microhistological analysis of fecal and gut samples. The epidermal cells of plants have taxon-specific characteristics that can be used for species identification. The microhistological technique consists of comparing the leaf epiderm morphology of food particles recovered from fecal and gut samples with those obtained from plants recovered in the surrounding area of study (Johnson et al. 1983). This technique has provided much valuable knowledge about the diet of ctenomyids, reviewed below. However, it is a very time-consuming method, prone to unspecific and context-dependent results, depending on the training of the observer and the amount and quality of samples (Soininen et al. 2009; Pompanon et al. 2012).

DNA-based methods have been largely applied in recent years to determine the diet of species, mainly after the development of the next-generation sequencers. These methods are particularly useful to assess the diet of generalist species that can explore a large variety of food resources available in highly diverse environments, fluid feeders, elusive species, or species for which the other methods cannot be easily applied or provide reliable results. These methods consist in amplify the DNA recovered from gut or fecal samples using species-specific or universal primers (Pompanon et al. 2012; Nielsen et al. 2018). The most recent and promising of these methods is the DNA metabarcoding approach, which has been successfully used to describe the diet composition of a wide range of animals, including herbivorous species (Pompanon et al. 2012). Basically, the total DNA is extracted from the gut or fecal samples, a small fragment of DNA is amplified using universal primers for targeting a group of the ingested species of interest. The amplified DNA is sequenced using the next-generation technology. Finally, a sequence reference database is used to assign taxonomy to the sequences recovered from fecal and gut samples (Fig. 10.2). The DNA metabarcoding approach allows the analysis of several samples in one single experiment, even when these samples are composed by a high

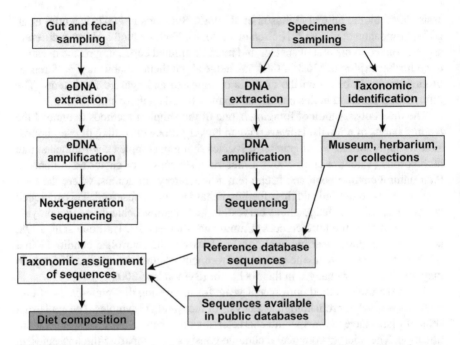

Fig. 10.2 Flowchart diagram showing the main steps for assessing the diet composition of species by means of the metabarcoding approach. The analysis of fecal and gut samples are represented in yellow. The sequence reference database assembling, for the ingested species, is represented in green

diversity of food items. The taxonomic resolution of this method depends on how informative is the fragment of DNA amplified and how complete is the sequence reference database (Pompanon et al. 2012; Taberlet et al. 2018). The dietary composition of ctenomyids was assessed only recently using this approach (Lopes et al. 2015, 2020), and is reviewed below.

10.3   What Do Ctenomyids Eat?

Despite the importance of knowing ctenomyids diet, just a few species of the genus *Ctenomys* were given appropriate attention in studies of diet composition and dietary habits. Feeding preferences of *C. mendocinus*, a species distributed in the Mendoza Province, Argentina, were better explored than in any other ctenomyid. The five studies performed to evaluate their diet were based on the microhistological technique (Torres-Mura et al. 1989; Madoery 1993; Puig et al. 1999; Rosi et al. 2003) or cafeteria tests (Camin and Madoery 1994). The results showed that *C. mendocinus* is generalist, preferring aboveground plant parts than roots (Torres-Mura et al. 1989; Camin and Madoery 1994; Rosi et al. 2003). They use about 65% of the plant genera present in their environment as food resource (Puig et al. 1999). However, a high preference for grasses was observed during the four seasons (70–94.5% of the diet content). The most eaten genera, depending on the population, were *Poa*, *Panicum*, *Stipa*, *Setaria*, *Aristida*, and *Elymus*. Some shrub, forb, and succulent forms were also recovered as part of their diet (Madoery 1993; Puig et al. 1999; Rosi et al. 2003). Males and females eat similar proportions of plants from different categories. However, males have a more varied diet during winter, and females showed a higher specialization on grasses during spring. These dietary differences in males and females have been associated with specific nutritional requirements during the reproductive season, pregnancy, and lactating periods (Puig et al. 1999). Camin and Madoery (1994) observed that not all plants harvested were consumed, suggesting that some part of plant material is used for storage and nesting.

Special attention was given to ecological aspects shaped by the food preferences of populations of *C. talarum* and *C. australis* distributed in sympatry. These species are allopatric distributed in the grassland dunes of the Province of Buenos Aires, Argentina, with some populations occurring in sympatry. In their sympatric zone, *C. talarum* occupy areas with dense vegetation, compact, and shallow soils, while *C. australis* occupy areas with sparse vegetation, sandy, and deep soils (Comparatore et al. 1995). Moreover, the latter species weights three times less than *C. talarum*. Comparatore et al. (1995) observed that populations of both species in the sympatric zone are generalist herbivorous, preferring to consume aerial than subterranean or reproductive plant parts, and grasses over forbs. del Valle et al. (2001) confirmed that the diet of *C. talarum* populations inhabiting Mar del Cobo, in Argentina, is predominantly composed of the aerial vegetative plant parts of perennial grasses. Their forage behavior changes seasonally, but they keep their preference for plant parts with a higher fiber/protein ratio. Despite *C. talarum* is able to feed selectively, and males are slightly more selective than females, this species consumes a high variety of plants present in the grasslands, which are an import source of nutrients.

Rosi et al. (2009) determined the effects of cattle grazed and ungrazed sites on the diet preferences of a *C. eremophylus* population inhabiting the arid plain of Mendoza, Argentina. The results of microhistological analyzes showed that the vegetative part of grasses (mainly *Panicum* and *Setaria* genera) was the dominant food source in their diet, followed by low shrubs. Tall shrubs and forbs were observed in minor proportions. In the grazed area, *C. eremophilus* showed higher dietary diversity, lower percentage of grasses, and higher consumption of low shrubs and reproductive plant parts than in the ungrazed area, reinforcing the expectation that cattle grazing force the species to change their dietary habits. During autumn-winter, the availability of their preferred plants becomes lower. As a response, these rodents shifted to a lower dietary selectivity, eating a higher variety of plants available closer to their burrows, as consequence, decreasing their risk of predation.

Most recently, the dietary composition of seven ctenomyids species inhabiting the southern and midwestern Brazil was assessed using the DNA metabarcoding approach (Lopes et al. 2015, 2020). One of the central questions in community ecology, the competitive exclusion principle, was addressed by Lopes et al. (2015) using

C. minutus and *C. flamarioni* as a model of study. Under this principle, two complete competitor species could not occupy the same habitat under limited resources, unless one species exclude the other or some kind of niche partition takes place (Pianka 2011). *Ctenomys minutus* and *C. flamarioni* are parapatrically distributed in the southern Brazilian coastal plain, with a narrow area where they occur in sympatry and can compete for food resources. The authors recovered 19 plant families as part of their diet, 13 for *C. minutus*, and 10 for *C. flamarioni*. Despite some differences in their diet composition, Poaceae, Araliceae, and Asteraceae were the most common plants identified in the feces of both species, together with Fabaceae for *C. minutus*. These plant families are highly frequent and have high species richness in the southern Brazilian coastal plain, reinforcing the previous knowledge about generalist and opportunistic feeding habits of ctenomyids. No remarkable diet overlapping was observed in the area where these species co-occur. *Ctenomys flamarioni* showed a more homogeneous and less variable diet in the sympatric region when compared to *C. minutus* or to other populations of *C. flamarioni*, suggesting that some level of dietary partitioning was developed to avoid competition, allowing these species to occur in sympatry.

Further investigations about the degree of diet specialization and diet overlap were assessed by Lopes et al. (2020) for ctenomyids inhabiting southern Brazil, using the DNA metabarcoding approach. Besides *C. flamarioni* and *C. minutus*, the authors evaluated the diet composition of *C. lami*, a species that inhabit more internalized dunes of the southern Brazilian coastal plain; *C. torquatus* and *C. ibicuiensis*, that inhabit lowlands of the State of Rio Grande do Sul; and, specimens sampled in the hybrid zone between *C. minutus* and *C. lami*. The results showed that these species consume more than 60% of the plant families recovered in soil samples collected around their burrows. Once again, grasses are the preferred food item by all ctenomyid species analyzed. Despite some particularities observed in the diet of each species, that are determined by the availability of food resources in the surrounding environment, these ctenomyids showed a high niche overlap in the plant families and Molecular Taxonomic Units consumed. Authors hypothesized that these species are ecologically similar and able to use the same range of resources when available in the environment, because they are allopatrically distributed and the interspecific competition is not a limiting factor. Lopes et al. (2020) also analyzed diet preferences of *C. bicolor* and *C.* sp., species that inhabit sandy soils in the States of Rondônia and Mato Grasso, respectively. They are distributed in a patched environment between fragments of the Amazonian forest and deforested areas occupied by crops. Both species are generalists, using grasses as the main food source. Plants from crops were absent or represent a small proportion of their diet, contrasting common knowledge that these species frequently use rubber trees and cassava as a food source (Lopes et al. 2020).

Literature Cited

Altuna CA, Bacigalupe LD, Corte S (1998) Food-handling and feces reingestion in *Ctenomys pearsoni* (Rodentia, Ctenomyidae). Acta Theriol 43:433–437

Begon M, Townsend CR, Harper JL (2006) Ecology from individuals to ecosystems, 4th edn. Wiley-Blackwell, Hoboken

Camin SR, Madoery LA (1994) Feeding behavior of the tuco-tuco (*Ctenomys mendocinus*): its modifications according to food availability and the changes in the harvest pattern and consumption. Rev Chil Hist Nat 67:257–263

Comparatore VM, Cid MS, Busch C (1995) Dietary preferences of two sympatric subterranean rodent populations in Argentina. Rev Chil Hist Nat 68:197–206

del Valle JC, Lohfelt MI, Comparatore VM, Cid MS, Busch C (2001) Feeding selectivity and food preference of *Ctenomys talarum* (tuco-tuco). Mamm Biol 66:165–173

Emlen JM (1966) The role of time and energy in food preference. Am Nat 100:611–617

Johnson MK, Wofford HH, Pearson HA (1983) Microhistological techniques for food habits analyses. United States Southern Forest Experiment Station USDA Forest Service, New Orleans, pp 1–45

Lopes CM et al (2015) DNA metabarcoding diet analysis for species with parapatric vs sympatric distribution: a case study on subterranean rodents. Heredity 114:525–536

Lopes CM et al (2020) Ecological specialization and niche overlap of subterranean rodents inferred from DNA metabarcoding diet analysis. Mol Ecol 29:3143–3153

Madoery L (1993) Composicion botánica de la dieta del tuco-tuco (*Ctenomys mendocinus*) en el piedemonte precordillerano. Ecol Austral 3:49–55

Martino NS, Zenuto RR, Busch C (2007) Nutritional responses to different diet quality in the subterranean rodent *Ctenomys talarum* (tuco-tucos). Comp Biochem Physiol A Mol Integr Physiol 147:974–982

Nielsen JM, Clare EL, Hayden B, Brett MT, Kratina P (2018) Diet tracing in ecology: method comparison and selection. Methods Ecol Evol 9:278–291

Pianka ER (2011) Competition. In: Pianka ER (ed) Evolutionary ecology, 7th edn. eBook

Pompanon F, Deagle BE, Symondson WOC, Brown DS, Jarman SN, Taberlet P (2012) Who is eating what: diet assessment using next generation sequencing. Mol Ecol 21:1931–1950

Puig S, Rosi MI, Cona MI, Roig VG, Monge SA (1999) Diet of Piedmont populations of *Ctenomys mendocinus* (Rodentia, Ctenomyidae): seasonal patterns and variations according sex and relative age. Acta Theriol 44:15–27

Rosi MI, Cona MI, Videla F, Puig S, Monge SA, Roig VG (2003) Diet selection by the fossorial rodent *Ctenomys mendocinus* inhabiting an environment with low food availability (Mendoza, Argentina). Stud Neotropical Fauna Environ 38:159–166

Rosi MI, Puig S, Cona MI, Videla F, Méndez E, Roig VG (2009) Diet of a fossorial rodent (Octodontidae), above-ground food availability, and changes related to cattle grazing in the Central Monte (Argentina). J Arid Environ 73:273–279

Soininen E et al (2009) Analysing diet of small herbivores: the efficiency of DNA barcoding coupled with high-throughput pyrosequencing for deciphering the composition of complex. Front Zool 6:1–9

Symondson WOC (2002) Molecular identification of prey in predator diets. Mol Ecol 11:627–641

Taberlet P, Bonin A, Zinger L, Coissac E (2018) Environmental DNA for biodiversity research and monitoring. Oxford University Press, Oxford

Torres-Mura JC, Lemus ML, Contreras LC (1989) Herbivorous specialization of the South American desert rodent *Tympanoctomys barrerae*. J Mammal 70:646–648

Valentini A et al (2009) New perspectives in diet analysis based on DNA barcoding and parallel pyrosequencing: the *trn*L approach. Mol Ecol 9:51–60

Chapter 11
Ecological Physiology and Behavior in the Genus *Ctenomys*

María Sol Fanjul, Ana Paula Cutrera, Facundo Luna, Cristian E. Schleich, Valentina Brachetta, C. Daniel Antenucci, and Roxana R. Zenuto

11.1 Introduction

Underground environments are considered among those that have most influenced the evolution of the organisms that inhabit them. Subterranean animals spend much of their lives in a moist, dark, poorly ventilated, hypoxic, and hypercapnic environment, also characterized by low primary productivity (Buffenstein 2000). Many of the sensory signals present aboveground are limited underground (Francescoli 2000), affecting vital activities such as food searching, reproduction, and territorial defense. Concomitantly, the underground habitat provides major advantages to its occupants: environmental stability (e.g., thermal insulation) and protection against predators (Nevo 1999). Therefore, similar adaptations are expected in underground mammals due to shared selective pressures. However, this expected evolutionary convergence is also accompanied by differences in biotic and abiotic factors that operate at different geographic scales, contributing to the adaptive divergence of underground mammals (Nevo 1999). Ecological physiology and behavior, as components of the biology of subterranean rodents, have been reviewed in several opportunities. In some of these reviews, the focus was on a particular species of subterranean rodents, as was the case with naked mole-rats (Sherman et al. 1991), South African mole-rats (Bennet and Faulkes 2000), and blind mole-rats (Nevo et al. 2001). Nevo and Reig (1990) put together the first and more comprehensive

Authors María Sol Fanjul, Ana Paula Cutrera and Facundo Luna have equally contributed to this chapter

M. S. Fanjul · A. P. Cutrera · F. Luna · C. E. Schleich · V. Brachetta · C. D. Antenucci
R. R. Zenuto (✉)
Grupo 'Ecología Fisiológica y del Comportamiento', Instituto de Investigaciones Marinas y Costeras, Universidad Nacional de Mar del Plata, Consejo Nacional de Investigaciones Científicas y Técnicas, Mar del Plata, Argentina
e-mail: msfanjul@mdp.edu.ar; acutrera@mdp.edu.ar; fluna@mdp.edu.ar; cschleic@mdp.edu.ar; vbrachetta@mdp.edu.ar; antinuch@mdp.edu.ar; rzenuto@mdp.edu.ar

© Springer Nature Switzerland AG 2021
T. R. O. de. Freitas et al. (eds.), *Tuco-Tucos*,
https://doi.org/10.1007/978-3-030-61679-3_11

review covering anatomical, physiological, ecological, and evolutionary studies in several groups of subterranean rodents. Subsequent efforts (Lacey et al. 2000; Begall et al. 2007) significantly expanded both the representation of species and the issues addressed. Studies on ecological physiology and behavior in *Ctenomys* were marginally included in the first review (Reig et al. 1990) but were more represented in later opportunities (Busch et al. 2000; Lacey 2000; Francescoli 2000; Lacey and Cutrera 2007; Schleich et al. 2007). The interest in understanding the multiplicity of physiological and behavioral responses associated with the challenges of underground life has grown since then. In this chapter, we aim to review the studies on physiological and behavioral responses to subterranean environmental conditions and food resources, as well as those that explore interactions with conspecifics (e.g., reproduction and territoriality) and heterospecifics (predators, parasites, and pathogens) in the genus *Ctenomys*. This genus is particularly interesting due to its high species diversity (63 species, Patton et al. 2015) in which solitary species are dominant while fewer are social (Lacey 2000; Tassino et al. 2011; Tomasco et al. 2019). These organisms conduct most of their activities within the burrow systems and hence are subject to many of the selection pressures typical of the underground environment. However, in many species, food collection is carried out aboveground (e.g., *Ctenomys peruanus,* Pearson 1959*; Ctenomys opimus,* Pearson 1959, *Ctenomys talarum,* Busch et al. 2000; *Ctenomys australis,* Busch et al. 2000; *Ctenomys sociabilis*, E.A. Lacey pers. comm.; *Ctenomys haigi*, E.A. Lacey pers. comm.; *Ctenomys fulvus*, Cortés et al. 2000; *Ctenomys mendocinus*, Puig et al. 1992) and in at least one of them (*C. opimus*, E.A. Lacey pers. comm.), vegetation is also consumed on the surface. This aboveground activity may lead to differences in the manner ctenomyids use sensory cues to assess their environment, how they cope with stressors both from above and below ground, how they thermoregulate and maintain energetic and water balance, how they reproduce and establish their territories, and how they face the challenges of parasite and predator exposure with respect to more strict subterranean rodents. These particular aspects of the ecological physiology and behavior of *Ctenomys* are reviewed in the following sections.

11.2 Sensory Biology

The sensory capabilities of individuals determine both the mode and extent to which animals sense changes in their environment and respond to them (Scott 2005). The sensory biology of the genus *Ctenomys* has been extensively studied. The first studies on this topic covered the olfactory and acoustic channels, two senses highly relevant for a genus that inhabits a dark and monotonous subterranean environment that limits the transmission of most signals and cues.

Up until now, most of the information regarding the use of olfactory cues in ctenomyids comes from studies conducted in *C. talarum*. Olfactory cues derived from urine, feces, or anogenital exudations are used by *C. talarum* to assess individual identity, reproductive condition, sex, and population of origin of conspecifics. Also, when the participation of the vomeronasal organ in the identification of odor cues is allowed, tuco-tucos can discriminate the reproductive condition of opposite-sex conspecifics (Zenuto and Fanjul 2002; Fanjul et al. 2003; Zenuto et al. 2004). Familiarization by odor cues reduces aggression between partners during courtship (Zenuto et al. 2007) and interacting males in a territorial context (Zenuto 2010, see below). In addition, the mating behavior of *C. talarum* seems to be linked to olfactory signals in mate evaluation and selection (Fanjul and Zenuto 2008a, 2012, 2013; Fanjul et al. 2018). Altogether, these findings reveal the key role of chemical cues in the territorial and reproductive biology of this rodent. Odor-based gender discrimination in males was also reported for *C. sociabilis* (Schwanz and Lacey 2003). Moreover, the use of odor cues is not restricted to conspecific interactions. Individuals of *C. talarum* use olfaction to orient their digging while foraging, both detecting substances released by the plants to the soil and discriminating different plant species (Schleich and Zenuto 2007, 2010). In addition, they also distinguish odors from predators, thus avoiding them, and choosing to feed in areas where these cues are not present (Brachetta et al. 2019a).

Vocalizations are an important means of communication in solitary and social members of *Ctenomys*. Acoustic signals emitted in territorial, mating, or aggressive encounters were described in several species (*C. peruanus*, Pearson 1959; *C. haigi*, Pearson and Christie 1985; *Ctenomys pearsoni*, Francescoli 1999, 2001, 2002, 2011, 2017; *C. talarum,* Schleich and Busch 2002; *C. mendocinus*, *C. sociabilis* Francescoli and Quirici 2010; Anillaco tuco-tuco, *Ctenomys* sp., Amaya et al. 2016). Most of the described vocalizations are within the mid- to low-frequency range, suggesting a convergent adaptation to the subterranean environment, where only low-frequency sounds can propagate over long distances. In addition, the acoustic characteristics of the vocalizations of subterranean rodents were also found to be relatively coincident with the hearing morphology – particularly the location and density of cochlear receptors – of the species studied so far (Mason 2004; Schleich and Busch 2004).

Although the subterranean ecotope is assumed to favor acoustic, olfactory, and tactile senses in detriment of vision, the only existing studies on the latter reveals a different perspective. Both solitary *C. talarum* and *Ctenomys magellanicus* present normally-developed eyes – in terms of size and functionality – with a significant proportion of two spectral cone types, indicating that photopic vision has a functional significance in these facultative subterranean rodents. Although active vision seems to be lost in both species studied, it is believed that the diurnal surface activity exhibited by these ctenomyids may be responsible for the maintenance of their reactive vision capabilities, that is, visual surveillance, escape reactions, and predator detection (Schleich et al. 2010; Vega-Zuniga et al. 2017).

Which sensorial cues are used by subterranean rodents to orient themselves in the complex and dark subterranean tunnels has deserved much attention. In *Ctenomys*, the orientation of the burrows of tuco-tucos (NNW-SSE) suggests that individuals may use the Earth's magnetic field as a common heading indicator (Malewski et al. 2018). However, until today, no further evidence that this genus relies on the geomagnetic field to orient underground has been obtained (Schleich and Antinuchi 2004), leaving the question of whether this genus has the sensory capability to use information from the Earth's magnetic field still unresolved.

11.2.1 Spatial Cognition

Spatial orientation, or the ability of individuals to learn to find their way through the environment without getting lost (Vorhess and Williams 2014), has been studied in different families of subterranean rodents due to the particularly complex structural characteristics of their habitats. In the genus *Ctenomys*, the species for which this type of information is available is *C. talarum*. Members of this species display a highly developed capacity to learn and memorize structurally complex labyrinths; individuals rapidly improved their spatial performance after the initial trials and were able to memorize a complex maze for a period of between 30 and 60 days after the learning process (Schleich and Antinuchi 2004; Mastrangelo et al. 2010). Interestingly, and regardless of sex differences in home-range size, males and females performed similarly in both tasks (Mastrangelo et al. 2010).

Spatial performance of an animal can influence its ability to perform crucial activities that depend on proper spatial orientation, such as food searching or mate localization. In *C. talarum*, animals exposed to predatory cues or an immune challenge, similar to what may be triggered by a parasite infection, showed a poorer navigation capacity (Mastrangelo et al. 2010; Brachetta et al. 2014; Schleich et al. 2015). This clearly indicates that life-threatening or energetically-challenging stimuli can negatively impact on spatial memory formation and recall, which play important roles in cognitive processes in tuco-tucos.

11.3 Stress Response and Individual Condition

Most organisms face stressful situations from different origins (physical, social, and/or psychological) on a daily basis. The consequent loss of homeostasis depends on the nature of the stress factor, as well as the magnitude of the response triggered (Armario 2006). The underground habitat is particularly interesting due to their distinctive physical and ecological characteristics that condition individual performance; e.g., poorly ventilated environments, with low primary productivity, and deprived of many sensory signals present aboveground (Buffenstein 2000; Francescoli 2000). To understand how tuco-tucos are affected by different ecological and environmental challenges, it is crucial to know about the physiology that underlies the response to a stressor. In vertebrates, the activity of Hypothalamic-Pituitary-Adrenal (HPA) axis is stimulated by several harmful stimuli, or stressors, triggering the secretion of glucocorticoids (GCs) from the adrenal glands. Therefore, both cortisol and corticosterone are commonly used to assess stress condition; basically, short-term high GCs levels are associated to the adaptive "fight or flight" response, while sustained high levels during long-lasting exposure to stressors are harmful to individuals (Sapolsky et al. 2000; Boonstra 2005). GCs are important hormones regulating energy balance since they are involved in the mobilization of energy reserves (Sapolsky et al. 2000) and high levels are considered indicative of high energetic demands (Boonstra 2005; Mac Ewen and Wingfield 2003; Vera et al. 2018; Mac Dougall-Shackleton et al. 2019). In rodents, corticosterone is the dominant GC, although there is variability within this group. In the genus *Ctenomys,* only two species were studied in relation to these hormones, showing clear differences between them with regard to the dominant GC found. GCs were assessed in *C. sociabilis* using blood and feces matrices (Woodruff et al. 2010, 2013), and higher baseline levels of corticosterone than cortisol were detected in both free-living and captive females. Individuals challenged with Adrenocorticotropic (ACTH) hormone showed an increase in the levels of corticosterone metabolites, confirming that corticosterone is responsive to stress. Furthermore, corticosterone levels were higher in free-living than captive females, and challenging conditions in the wild were proposed to explain such differences (Woodruff et al. 2010). Moreover, the physiological consequences of group living were explored in *C. sociabilis*, which live alone or in groups in nature (Woodruff et al. 2013). Data from captive and free-living individuals provide evidence that females living alone or in groups differ in their baseline GCs levels. Lone females showed higher levels of metabolite of GCs, so the achievement of basic activities such as territory defense and food provision may have more physiological consequences than social interactions with conspecifics. On the other hand, studies in *C. talarum* revealed that both cortisol and corticosterone circulate in the plasma, but females have higher levels of cortisol (Vera et al. 2012). These hormones differ in their seasonal and annual variation patterns in free-living individuals, suggesting differences in their endogenous regulation and affectation by environmental stimuli (Vera et al. 2011a, 2012, 2013). Cortisol is responsive to the factors that typically regulate GC concentrations (acute stressors and ACTH), but corticosterone is not (Vera et al. 2011a, 2012, 2013). Furthermore, only cortisol levels were affected by sustained low-quality diet and fasting (Vera et al. 2019). In contrast, angiotensine II – the main biologically-active hormone of the renin-angiotensin system – stimulated corticosterone, but not cortisol secretion, denoting its participation in mineral-water balance (Vera et al. 2019). Decreased levels of both GCs – although more so for corticosterone – and lower negative feedback efficacy for corticosterone in captivity, may account for differential responses to chronic stress conditions (Vera et al. 2019; Dickens and Romero 2013). Overall, these results reveal that cortisol and corticosterone are not interchangeable hormones and suggest different physiological roles in *C. talarum* (Vera et al. 2019).

Even though GCs are commonly considered as synonymous of "stress hormones," the activation of HPA is only one component of the complex stress response in vertebrates (MacDougall-Shackleton et al. 2019). In this sense, the idea of "stress profile" emerges as a more complex alternative, where a suite of biological parameters are considered instead of a single indicator for evaluating animal condition and dysregulation (Milot et al. 2014). The evaluation of changes in several indicators of animal condition, such as the neutrophils: lymphocytes ratio (N/L), body mass, hematocrit, glucose, triglycerides, accompanied GCs response (mainly cortisol) to challenging conditions in *C. talarum*. Decreased levels of glucose, triglycerides, body mass, and inflammatory response were related to food restriction (Merlo et al. 2016a). In some cases, increases in both cortisol and N/L followed long-lasting nutritional stress (Vera et al. 2019), while in others, no changes in cortisol levels were found, although a higher N/L was detected (Schleich et al. 2015; Merlo et al. 2016a). In contrast, increases in cortisol levels without changes in N/L were found under short-term exposure to stress conditions, such as a 24 h fasting or a brief immobilization (Vera et al. 2019; Brachetta et al. 2019b). Although linear relationships between GC levels and N/L are frequently proposed (Davis and Maney 2008, 2018), these parameters may be complementary in understanding the stress status of animals due to their different sensitivities in the responses according to different stress factors and duration of stimuli (Müller et al. 2011).

11.4 Energetics, Water Balance, and Thermoregulation

One of the most important aspects of the biology of species is the understanding of how different factors, internal (e.g., digestive capacity) and/or external (e.g., ambient temperature, water availability, or social interactions) impose limits on their energy budget (Karasov 1986; Weiner 1992; Withers et al. 2016). As animals can be understood as open systems to the flow of materials and the energy (Wiegert 1968), energy balance can be interpreted as the integration of intake, storage, and loss of energy (Fig. 11.1).

11.4.1 Energy Intake

Animals can increase efficiency in obtaining energy by modifying ingestive and digestive processes, as well as decreasing the cost of different internal mechanisms (Karasov 1986). For subterranean species, maintaining efficiency, hence a balanced energy budget, is expected to be more challenging due to the high cost of living underground, where food is scarce and/or quality of food is poor (Antenucci et al. 2007). Low energy intake challenges animals to display compensatory mechanisms. For instance, when *C. talarum* were fed a low-quality diet in captivity, they consumed more food and devoted more time to feeding activities (Martino et al. 2007; Perissinotti et al. 2009). In mammals, changes in the morphology of the gastrointestinal tract and the activity of digestive enzymes could represent other digestive strategies to address seasonal variation in food quality/availability. Such changes were also observed in *C. talarum* (del Valle and López Mañanes 2008, 2011). Increased efficiency can also rely on the re-ingestion of fecal items (coprophagy), as observed in *C. pearsoni* (Altuna et al. 1998) and *C. talarum* (Martino et al. 2007; Perissinotti et al. 2009; Fig. 11.1).

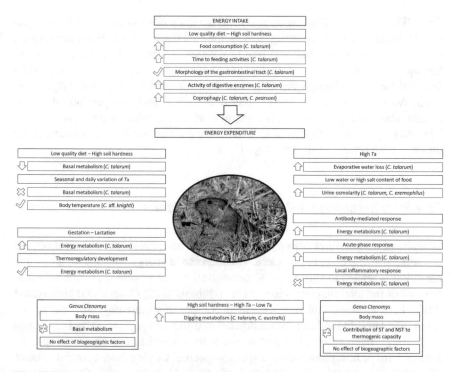

Fig. 11.1 Factors that affect energetic, thermoregulatory, and osmoregulatory variables in *Ctenomys* species. The effect – or the lack of effect – of different factors on physiological variables are indicated by different symbols. Upward and downward arrows indicate increment and decrease, respectively; x marks indicate no variation; tick marks indicate variation (both increments and decreases) in relation to developmental stages (growth, reproduction) and time (seasons, days) in the specific physiological variable; unequal marks indicate differences between species

11.4.2 Basal Energy Expenditure

A balanced energy budget depends on how energy intake is maximized and expenditure is minimized. Experimental estimation of the minimum energy expenditure needed by an individual to maintain homeostasis is the basal metabolic rate (BMR, Hulbert and Else 2004). In *Ctenomys*, basal metabolism was evaluated under different external conditions in different species. Depending on the species, different natural or experimentally induced changes in external factors might lead to different BMR responses (see Luna et al. 2017). Under a low quality diet or high soil hardness, *C. talarum* decreased their BMR (Perissinotti et al. 2009). However, no effect of controlled regimes of ambient temperature or seasonal changes in natural ambient temperatures was observed (Luna et al. 2012; Meroi et al. 2014). In *C.* aff. *Knighti* seasonal energy saving, based on BMR estimation, was detected (Tomotani et al. 2012; Tachinardi et al. 2017). Interestingly, when interspecific variability of BMR was evaluated, Luna et al. (2009) observed that body mass, but not biogeographic factors – such as latitude, ambient temperature, or precipitation – affect

BMR (Fig. 11.1). Similarly, body mass was the main determinant of BMR between *C. talarum* and *C. australis* living in the same area (Busch 1989). Basal metabolism can also be seen as the minimum heat production at thermoneutrality. Thus, maintaining a stable body temperature depends on the relationship between heat production and heat loss through the animal surface (thermal conductance; Naya et al. 2013; Withers et al. 2016). Even at thermoneutrality, body temperature of *C.* aff. *Knighti* shows daily variations (Tachinardi et al. 2014), which are synchronized to light-dark cycles by their rhythmic locomotory activity and excursions outside the burrow (Valentinuzzi et al. 2009; Fig. 11.1).

11.4.3 Thermoregulation and Water Balance

At ambient temperatures below thermoneutrality, an extra source of heat is needed to maintain stable body temperature. The increase in energy metabolism under this condition is related to shivering and non-shivering thermogenesis, which are components of the total thermogenic capacity (Hohtola 2002; Cannon and Nedergaard 2011). In *C. talarum,* shivering is the main component of thermogenic capacity and is not affected if individuals are acclimated to different temperature regimes (Luna et al. 2012). Interestingly, this pattern is not maintained across *Ctenomys* species. Mechanisms of heat production are species-specific, varying from a combination of shivering and non-shivering thermogenesis (e.g., *Ctenomys roigi, Ctenomys porteusi,* and *C. talarum*) to a complete use of shivering thermogenesis (e.g., *C. australis* and *Ctenomys tuconax*; Fig. 11.1). Similarly to basal metabolism, biogeographical factors did not affect the variability of total thermogenic capacity among *Ctenomys* species (Luna et al. 2019). At ambient temperatures above thermoneutrality, another problem arises: the need to lose heat. Individuals can lose heat by direct physical routes (i.e., conduction, convection, and radiation) and by evaporation. Evaporation is particularly effective at high temperatures since it does not rely on a thermal gradient (Withers et al. 2016). Evaporative water loss of *C. talarum* increases above thermoneutrality and remains low and stable within and below it (Baldo et al. 2015). The maximal value of evaporation does not change in individuals exposed to contrasting humidity acclimation (Baldo et al. 2016). Since the atmosphere of burrows can restrict evaporation – particularly at high ambient temperatures, due to the high water vapor content – non-evaporative routes of heat loss are also used (Cutrera and Antinuchi 2004; Luna and Antenucci 2007b). Evaporation cannot increase indefinitely because it is also a significant route of water loss. Water is an essential component of the body, and its balance must be regulated. One important non-evaporative venue for water loss is through urine excretion. Besides behavioral strategies (see Bozinovic and Gallardo 2006), mammals can minimize their urinary water loss by producing concentrated urine (Schmidt-Nielsen 1997). Interestingly, neither *Ctenomys eremophilus* (Diaz 2001 in Diaz et al. 2006) nor *C. talarum* (Baldo and Antenucci 2019) can concentrate urine. However, under environmental restrictions (i.e., low water or high salt content of food), *C. talarum* increased urine osmolarity

(Baldo and Antenucci 2019; Fig. 11.1). Further, as urine concentration capacity depends on the morphology of the kidney (McNab 2002), medullary thickness can be used as an index of renal performance. Medullary thickness is only reported for *C. fulvus*, *C. opimus*, and *C. eremophilus* showing values similar to those found in mesic surface-dwelling rodent species.

11.4.4 Non-basal Energy Expenditure

Measurement of non-basal metabolism in a wide range of conditions provides information about different components of energy budget at different timescales. In general, building new tunnels involves an extremely high cost of digging (see Vleck 1979). Soil hardness was proposed to be the main factor explaining the convergent morpho-physiological features found in phylogenetically unrelated subterranean species. In both *C. talarum* and *C. australis*, hard soils increased the digging metabolic rate (Luna et al. 2002; Luna and Antenucci 2006, 2007a). Also, in *C. talarum*, the increment in digging expenditure affects burrow architecture (Antinuchi and Busch 1992), associated with the energy restriction to construct tunnels with angles greater than 40° (Luna and Antenucci 2007b). At low ambient temperatures, *C. talarum* used the heat produced during digging to supply the energy required for thermoregulation, but at high temperatures, digging increases the risk of overheating, which is minimized by conduction to the soil (Luna and Antenucci 2007c; Fig. 11.1). Thus, digging represents a high physiological cost for *Ctenomys* species having a profound impact on the daily and annual energy budget (Antenucci et al. 2007). On the other hand, reproduction is considered the most energetically costly period of the life of a female (Tomasi and Horton 1992). Females of *C. talarum* increase their energy metabolism during periods of gestation and lactation (Zenuto et al. 2002a). After birth, pups show an increase in metabolic rate until day 10, when they start to eat solid food but still cannot maintain stable body temperature. During this critical period, contact with their mother and siblings is crucial for a normal thermoregulatory development (Baldo et al. 2014). After day 10, metabolic rate starts to decrease, reaching adult's BMR and body temperature values (Zenuto et al. 2002a; Cutrera et al. 2003; Fig. 11.1).

From an evolutionary perspective, both internal (e.g., digestive constraints) and external factors (e.g., ambient temperature) may have determined the balance between energy gains and losses in *Ctenomys*. The metabolic scope is defined as the difference between the lower and the upper limit of energy expenditure (i.e., basal and maximal metabolism, respectively), and represents the amount of energy available to the individual for activities over the lower limit of energy required to maintain basal physiological mechanisms. In *Ctenomys* species, the metabolic scope can be affected by the degree of underground commitment, since it appears to be lower than in surface-dwelling species (Luna et al. 2015). Thus, restricted metabolic scope can have a great impact on the species-specific energy budgets, and factors (e.g., low O_2 content within burrows) that constrain limits in energy expenditure can probably act independently on different energetic variables, such as basal metabolism or thermogenic capacity (Luna et al. 2009, 2017, 2019).

11.5 Intraspecific Interactions: Physiology and Behavior of Territory Defense and Reproduction

11.5.1 Social Behavior and Territoriality

Most *Ctenomys* species show solitary habits with one individual occupying each burrow system, only shared during mating and the maternal care of the young. Sources of evidence to account for solitary behavior are diverse: minimal or no home-range overlap, animal capture and visual verifications of no further activities in a given burrow, and anecdotal reports (Table 11.1). In most solitary species, both sexes show aggressive reactions toward conspecifics, but few studies addressed this topic from a behavioral perspective. In *C. talarum,* both sexes defend their burrow systems, but only males utter the typical "tuc-tuc" territorial vocalization warning potential intruders about the presence of the owner in its territory (Schleich and Busch 2002). Studies using seminatural enclosures in captivity showed that males engage in aggressive interactions with other males (Zenuto et al. 2002b). Such aggressive interactions result in dominance hierarchies, which are also linked to the access to females during the mating season. Moreover, as mentioned before, territorial aggression is modulated by odor familiarity (i.e., the "dear enemy phenomenon"; Temeles 1994), as a mechanism mediating territorial behavior in *C. talarum* (Zenuto 2010). Despite dominance not being a simple function of aggressiveness, the acquisition and maintenance of certain status often require different degrees of aggression. Androgenic steroid hormones, such as testosterone, mediate aggressive behavior. In the wild, male *C. talarum* testosterone levels peaked during the reproductive season but were also highly variable among individuals, ranging from barely detectable to extremely high concentrations in comparison to other mammals (Vera et al. 2011b, 2013). This could be related to their capacity to defend a territory and/or monopolize the access to females. On the other hand, the role of aggression in habitat segregation was identified by means of experimental interactions involving two species naturally living in sympatry, *C. talarum* and *C. australis* (Vassallo and Busch 1992).

Considering group living as part of a sociality continuum, few social species have been identified in the genus *Ctenomys* so far. The first reported social species was *C. peruanus* (Pearson 1959), followed by *C. sociabilis* (Pearson and Christie 1985). In *C. sociabilis,* the groups are stable and defined, and individuals are involved in cooperative tasks such as excavation of tunnels, nest sharing, and offspring attendance during the reproductive season (Lacey et al. 1997; Izquierdo and Lacey 2008). Later, radiotelemetry and genetic studies have found that burrow systems of *C. rionegrensis* are not strictly exclusive, and sporadic overlap occurs among adult residents (Tassino et al. 2011, Tomasco et al. 2019). Also, for *C.*

Table 11.1 Social and reproductive physiology and behavior in *Ctenomys* species

Species	Climate[a]	Social behavior	Territorial defense	Reproductive seasonality	Ovulation type (Postpartum estrum, PPE)	Mating system		Parental care	References
						Sexual dimorphism (sex ratio)	Sexual selection		
C. flamarioni	Humid subtropical			Sep to March[FCS] (births)		Male > female (1:2.21)			Fernández-Stolz et al. (2007) and Garcias et al. (2018)
C. minutus	Humid subtropical	Solitary[HR]		Oct to Dec[FCS]		Male > female (1:1.26)			Marinho and de Freitas (2006) and Kubiak et al. (2017)
C. lami	Humid subtropical	Solitary[FC]		June to Dec[FCS] (perforated and pregnant females)	(No PPE)	Polygyny? Male > female (1:1)			El Jundi and de Freitas (2004)
C. torquatus	Humid subtropical	Solitary?						Female[CO]	Talice and Laffitte de Mosera (1958)
C. pearsoni	Humid subtropical	Solitary[FC]	Yes?	June to Oct[FCS(?)]	Induced (No PPE)	Male > female			Altuna et al. (1991, 1998) and Francescoli (2011)
C. opimus	High Mountain	Solitary[FC] Facultatively social[HR]		Ago to Feb[FCS] (births)					Pearson (1959), O'Brien et al. (2020)
C. rionegrensis	Humid subtropical	Solitary/Social[HR, GE]		Jun to Dec[FCA]	Induced?	Male > female			Tassino and Passos (2010, Tassino et al. 2011) and Tomasco et al. (2019)

(continued)

Table 11.1 (continued)

Species	Climate[a]	Social behavior	Territorial defense	Reproductive seasonality	Ovulation type (Postpartum estrum, PPE)	Mating system Sexual dimorphism (sex ratio)	Sexual selection	Parental care	References
C. mendocinus	Cold arid or semiarid	Solitary	Yes [FC]	Ago to Feb [FCA]	(No PPE)	Male > female (1:1)		Female [CO]	Puig et al. (1992), Rosi et al. (1992, 1996, 2005) and Camín (1999, 2010)
C. talarum	Cool oceanic	Solitary[HR, CE]	Yes [CE]	Jun to Dec (pregnant females)[FCA, CO]	Induced [CE] (PPE[E])	Polygyny[CE,GE] Male > female (1:1.63)	Female sexual selection [CE, FC, GE]	Female exclusively [CO]	Busch et al. (1989), Zenuto et al. (2001, 2002a), Fanjul et al. (2006, 2018), Fanjul and Zenuto (2008a, 2012, 2017), Zenuto (1999, 2010) and Cutrera et al. (2012)
C. australis	Cool oceanic	Solitary [HR]	Yes [FC]	Absent [FCS] (pregnant females)	(PPE)	Polygyny? Male > female (1:2.12)			Zenuto and Busch (1998)
C. haigi	Mountain cold semiarid	Solitary [HR]				(1:1.2)			Lacey et al. (1998)
C. sociabilis	Mountain cold semiarid	Social [HR]		May to Dec [FCS]	(No PPE)	Polygyny?: Multi-female groups and single females		Communal female care[FC, HR]. Female care	Lacey et al. (1997), Lacey (2004) and Lacey and Wieczorek (2004)

Supraindexes indicate data source: HR home range and polygons estimated by telemetry, *FC* field capture (*A* annual survey, *S* seasonal survey), *CO* observational studies in captivity, *CE* experimental studies in captivity, *GE* genetic studies,*?*: no data provided. [a]Köppen–Geiger climate classification system of study sites *opimus* (previously reported as a solitary species; Pearson 1959), radiotelemetry studies revealed their facultative social behavior, as some individuals tend to aggregate during the night (O'Brien et al. 2020).

11.5.2 Seasonality and Regulation of Reproduction

Most ctenomyids are distributed within latitudes characterized by seasonal variation in climate and food availability (i.e., subtropical, temperate, arid, or mountain regions, Reig et al. 1990). Therefore, it is expected that reproduction occurs in a restricted period of time where food and climate conditions are optimum, thus maximizing reproductive output. Studies conducted suggest that the timing and length of reproductive activities are adjusted so that the most energetically-demanding activities (e.g., gestation and lactation, Zenuto et al. 2002a) are assured (Table 11.1). Specifically, the reproductive season of some *Ctenomys* species occurs when food is abundant (*C. talarum*, Fanjul et al. 2006), and temperatures are more benevolent (*C. pearsoni*, Tassino and Passos 2010). In addition, the length of the reproductive period – and the potential for a second litter – in *C. mendocinus* varies depending on the climatic rigorousness of the habitat (Rosi et al. 1992, 1996). Environmental cues regulate such timing, contributing to secure these favorable conditions. In this sense, studies in *C. talarum* in captivity showed some evidence of female reproductive responsiveness to photoperiodic cuing (Fanjul and Zenuto 2008b). However, the relevance of photoperiod in triggering reproduction in *C. talarum* could be more important in the wild, where food quantity (and possibly quality) is variable, particularly at the onset of the reproductive season (Fanjul et al. 2006).

11.5.3 Ovulation and Receptivity

Induced ovulation appears to be the rule among rodents that are solitary, occur at low population densities and/or inhabit highly seasonal environments (Zarrow and Clark 1968; Milligan 1982). So far, the only ctenomyid studied in this regard (*C. talarum*, Weir 1974) is an induced-ovulator, in which the induction depends on the amount of copulatory stimulation that the female receives, but not on the sole male presence or presence of its chemical cues (Fanjul and Zenuto 2008a). Penis morphology in this genus is characterized by the presence of spines and spikes which is consistent with the stimulatory requirement to induce ovulation (Balbontín et al. 1996; Rocha-Barbosa et al. 2013). Even though ovulation in *C. talarum* is triggered by copulation, female reproductive behavior varies with progesterone and oestradiol levels, with both hormone levels and vaginal cytology being affected by male presence (Fanjul and Zenuto 2012).

11.5.4   Courtship and Mating

In solitary species, courtship leads to lower levels of aggression and allows the assessment of potential partners in relation to their receptivity and mate quality (*C. pearsoni, C. mendocinus,* and *C. talarum*; Altuna et al. 1991; Camín 1999; Fanjul and Zenuto 2008a). As expected, courtship in ctenomyids involves chemical and vocal signals (Altuna et al. 1991; Camín 1999; Fanjul and Zenuto 2008a). The occurrence of multiple intromissions – and possibly multiple ejaculations – observed in *C. talarum* (Fanjul and Zenuto 2008a, 2012) and *C. pearsoni* (Altuna et al. 1991) during copulation supports the hypothesis of induced ovulation in these species.

11.5.5   Mating Preferences

In mammals, the internal development and lactation impose a differential cost of reproduction assumed by females. Then, females maximize their reproductive success by being more choosy in terms of mate preference (Andersson 1994). Female mate choice seems to be crucial for *C. talarum* as we found that is a prerequisite for mating in captivity (Zenuto et al. 2007). Female familiarization with male odors affects the outcome of a reproductive encounter (Zenuto et al. 2007), and females prefer novel than familiar males (Fanjul and Zenuto 2013). Moreover, females prefer dominant males (Fanjul and Zenuto 2017) and those that hold territories even after they were scent-marked by a competitor (Fanjul et al. 2018). Furthermore, the Major Histocompatibility Complex (MHC) genes play a key role in mate choice in *C. talarum*, since females mate preferentially with heterozygous males and those that carry specific MHC alleles (Cutrera et al. 2012), thus supporting the idea that females prefer males with "good genes," which would increase the chances of resisting infections for the offspring.

11.5.6   Mating System and Parental Care

Polygyny is the predominant mating system in mammals, which is characterized by the monopolization and defense of several females by a single male and exclusive maternal care of the young (Clutton-Brock 1989). The extent of polygyny has been associated to sexual size dimorphism (males of bigger size) and female-biased operative sex ratios (Mitani et al. 1996). Considering this indirect evidence, polygyny seems to be the most frequent mating system among solitary and social ctenomyids (Table 11.1). Confirmation of this hypothesis in *C. talarum* comes from molecular data (Zenuto et al. 1999). Parental care is conducted only by females (Zenuto et al. 2001) involving feeding of the young as well as thermal protection via direct contact (*C. mendocinus,* Camín 2010; *C. talarum,* Zenuto et al. 2001, 2002a; Cutrera et al. 2003). For *C. sociabilis,* telemetry data revealed communal nest attendance by several females (Izquierdo and Lacey 2008).

11.6 Interspecific Interactions: Physiology and Behavior of Predator Avoidance and Defense Against Parasites and Pathogens

11.6.1 Predator Avoidance

Predators affect prey both directly by killing them and indirectly by affecting their behavior, foraging patterns, reproduction, and stress physiology, thus affecting their fitness (Clinchy et al. 2013; Moll et al. 2017). The nonlethal impact of predators could be even of greater demographic magnitude than that produced by the death of prey, also involving trans-generational effects (Clinchy et al. 2013). So far, studies on the physiological and behavioral effects of predatory risk in the genus *Ctenomys* are limited to *C. talarum.* Although the subterranean environment provides protection against predators, individuals become vulnerable to aerial and terrestrial predators while dispersing or foraging aboveground (Busch et al. 2000). This species is often predated by owls, foxes, wildcats (Vassallo et al. 1994; Busch et al. 2000), dogs, and domestic cats (C.E. Schleich, pers. obs.). Experimental studies show that both acute and chronic exposure to direct cues indicating the presence of a predator (immobilization and cat urine) affected spatial performance in this species (Brachetta et al. 2014). Similarly, tuco-tucos exposed to predator odors generated an anxiety state and showed avoidance behaviors, even in juveniles prenatally exposed to predatory risk (Brachetta et al. 2015, 2016, 2018). The relationship between behavioral and physiological responses to stress by predation was experimentally proved. Moreover, the moderate magnitude found in both responses is proposed to be consistent with a predation pressure buffered by the use of the underground environment (Brachetta et al. 2019b).

11.6.2 Defense Against Parasites and Pathogens

The study of the interactions between host physiology (i.e., immune function) and disease ecology (i.e., pathogen prevalence) in a wide range of environments and animal species allows us to understand the extrinsic and intrinsic factors leading to immune function variation and disease susceptibility in natural populations (Demas and Nelson 2012). Variation in immune responsiveness among individuals and species may be the result of genetic factors, but also of the interplay among immunity, demography, and life-history traits in an ecological context (Schoenle et al. 2018). These factors have been explored in a few *Ctenomys* species, in an effort to understand the sources of variation in pathogen resistance in this group.

Previous studies in model species have shown that parasite resistance is under genetic control (see Charbonnel et al. 2006 for a review). More specifically, MHC genes code for glycoproteins involved in the recognition and binding of foreign antigens (Klein 1986). The high levels of polymorphism of MHC genes are the result of pathogen-mediated selection (Doherty and Zinkernagel 1975). This selective model has been explored in tuco-tucos in relation to variation in social habits between species (*C. haigi* and *C. sociabilis*, Hambuch and Lacey 2002), differences in demographic traits among populations (*C. talarum*, Cutrera and Lacey 2006), the impact of distinct demographic histories (*C. australis* and *C. talarum,* Cutrera et al. 2010a; Cutrera and Mora 2017), and patterns of evolution of MHC loci across 18 ctenomyid species (Cutrera and Lacey 2007). Further evidence of selection on MHC genes came from later studies that showed that (1) MHC allelic and genotypic variation are associated to parasite resistance and immunocompetence in *C. talarum* (Cutrera et al. 2011); (2) the strength of this association may vary among populations of this species (Cutrera et al. 2014a), and (3) as mentioned before in this review, female *C. talarum* choose their mates in relation to their MHC genotype, among other factors (Cutrera et al. 2012). Together, these studies suggest a role for parasite-driven selection and female mate choice in maintaining MHC variation, and hence parasite resistance, in natural populations of *C. talarum*.

Mounting an immune response is presumed to be costly, and these costs may mediate the trade-offs between immune function and other costly physiological processes. The different outcomes of these trade-offs may also explain the variation in immune responsiveness (Norris and Evans 2000) among species and individuals. Therefore, it becomes essential to estimate the magnitude of the immune response and the costs associated with its activation. One way of doing so is assessing the increase in oxygen consumption that may be associated with triggering an immune response (Demas et al. 2012). For *C. talarum*, the metabolic costs of mounting an antibody-mediated response (Cutrera et al. 2010b), a local inflammatory response (Merlo et al. 2014), and an acute-phase innate response (Cutrera, Luna, and Zenuto, unpublished data) to artificial antigens have been estimated directly using respirometry in captivity, suggesting that costs are variable among the different arms of immunity of tuco-tucos, with the antibody-mediated response being the most costly response to activate and maintain, followed by the acute-phase response. Surprisingly, the local inflammatory response was not associated with a significant increase in oxygen consumption (Merlo et al. 2014; Fig. 11.1). Another way of exploring the costs of immunity is to do so indirectly, by assessing if the magnitude of an immune response is negatively affected by other physiological processes occurring at the same time (i.e., growth, reproduction, see Sheldon and Verhulst 1996; Lochmiller and Deerenberg 2000). In *C. talarum*, possible trade-offs between mounting a local inflammatory response and several energetically-demanding processes, such as growth (Cutrera et al. 2014b), reproduction (Merlo et al. 2014), and mounting a simultaneous humoral response (Merlo et al. 2019) were assessed. Further, the effects of diet (Merlo et al. 2016a), parasitism (Merlo et al. 2016b), and body condition (Merlo et al. 2018) on the magnitude of the local inflammatory response were also explored in *C. talarum*, suggesting that additional costs, besides the energetic, maybe mediating the trade-offs between immune function and other energetically demanding activities of tuco-tucos. The studies conducted so far show a generally low immune responsiveness of the different arms of the immune system of *C. talarum* (innate and adaptive responses, both induced and constitutive) compared with other vertebrate species. This pattern of low immune responsiveness coupled with the low parasite richness found in *C. talarum* – which is presumed to be a consequence of their restricted mobility, high territoriality, and spatial isolation (Rossin and Malizia 2002) suggests an important role of the underground habitat in the evolution of immune strategies in this group of subterranean rodents.

11.7 Conclusions and Prospects

Early studies on physiology and behavior in members of the genus *Ctenomys* date from the 1970s with the contributions of Wise et al. (1972) and Weir (1974) reporting glucose regulation and reproductive behavior in captive *C. talarum*. Later, Busch (1987, 1989) performed her studies about physiological adaptations to underground conditions in *C. talarum* and *C. australis*, particularly assessing hematological traits that allow tuco-tucos to cope with the low oxygen and high CO_2 that characterize the burrow atmosphere, as well as their thermoregulatory capacity and basal metabolic rate. In addition, identifying dominance hierarchy between species as a possible cause of habitat segregation between sympatric species was an important first step for behavioral studies in *Ctenomys* (Vassallo and Busch 1992). Nowadays, there is a greater suite of studies on physiological and behavioral responses to diverse biotic and abiotic challenges faced by this group. While the number of species of tuco-tucos assessed is increasing, they still represent only a quarter of the diversity of the genus. In addition, some topics have only been addressed in one species – mostly *C. talarum* – or in a few species, making it difficult to identify general response patterns in an effort to synthesize our knowledge. Thus, to reach a better understanding of the multiple physiological and behavioral responses associated with living underground, we need to make a greater effort to include more *Ctenomys* species into the picture. Field and laboratory studies are needed for this commitment, even though the secretive habits of these organisms make it particularly difficult to obtain information in the wild and to develop adequate housing conditions for controlled experiments in captivity. Assessing glucocorticoid levels, together with a complete biochemical profile, will allow the evaluation of changes in energy demands and stress conditions associated with different challenges, both in nature and in captivity.

Despite these difficulties, it is now possible to identify some patterns of physiological and behavioral responses that are in accordance to what is expected for subterranean rodents. One example of this is the role of chemical and acoustic cues in intraspecific communication, especially in the contexts of territorial and reproductive behavior. The chemical channel also plays an important role in food searching as well as in predator avoidance. Further, spatial orientation appears as a critical skill in individuals living in structured and complex systems such as underground burrow systems. Energy balance is strongly modulated by restrictions on energy intake, as well as the reduction of costs associated with the maintenance of homeostasis and energy-demanding activities, such as digging. This energy limitation clearly affects reproductive seasonality. Finally, patterns of immune function variation and parasite resistance, at least in *C. talarum*, seem to have evolved according to the lower parasite exposure of the subterranean habitat in comparison to the challenges of the aboveground environment.

On the other hand, some studies suggest that the use of the aboveground environment during food collection impacts on other traits not closely related to living underground, such as the development of visual abilities, the relevance of photoperiod in the regulation of female reproductive activity, and the daily variation in body temperature synchronized to light periods by the rhythmic activity pattern and surface excursions. Moreover, detected differences in determinants of energy budget between *Ctenomys* species may account, at least in part, for different commitments to below and aboveground activities. Even though burrow systems protect individuals from aerial and terrestrial predators, response to predatory cues shows that individuals suffer predatory risk when exposed aboveground.

Beyond the adaptive convergence and divergence framework proposed by Nevo (1979), we suggest that future studies on ecological physiology of subterranean rodents will take into account the pace of life perspective. *Ctenomys* species may be considered "slow-living," given their relatively altricial development – at least for solitary species –, late acquisition of sexual maturity and longevity. In light of this, predictions can be made regarding the physiological responses of this group of rodents. For example, considering immune function, it is expected that tuco-tucos rely more strongly on the adaptive arm of immune defense, even when it is more energetically expensive than the innate arm, because the adaptive immunity confers memory against repeated infections that are more likely to occur in long-lived species (Lee 2006). Therefore, integrating the studies of the physiological ecology and behavior of ctenomyids into the pace of life syndrome (POLS) concept, which describes the covariation of life-history, physiological (particularly metabolic, hormonal and immunological), and behavioral traits or personality, under specific conditions (Stamps 2007), may offer insight into the relationship between the low levels of energy expenditure of ctenomyids and their slow pace of life.

Acknowledgments We wish to dedicate this chapter to Cristina Busch who initiated the study of ecology, physiology, and behavior of tuco-tucos in the Universidad Nacional de Mar del Plata. We are grateful to the editors for inviting us to contribute with this chapter. Our research was supported by grants from Agencia Nacional de Promoción Científica y Tecnológica, Consejo Nacional de Investigacion Científica y Tecnológica and Universidad Nacional de Mar del Plata.

Literature Cited

Altuna CA, Francescoli G, Izquierdo G (1991) Copulatory pattern of *Ctenomys pearsoni* (Rodentia, Octodontidae) from Balneario Solís, Uruguay. Mammalia 55:316–317

Altuna CA, Bacigalupe LD, Corte S (1998) Food-handling and feces reingestion in *Ctenomys pearsoni* (Rodentia, Ctenomyidae). Acta Theriol 43:433–437

Amaya JP, Areta JI, Valentinuzzi VS, Zufiaurre E (2016) Form and function of long-range vocalizations in a Neotropical fossorial rodent: the Anillaco Tuco-Tuco (*Ctenomys* sp). PeerJ 4:e2559

Andersson M (1994) Sexual selection. Princeton University Press, Princeton

Antenucci CD, Zenuto RR, Luna F, Cutrera AP, Perissinotti PP, Busch C (2007) Energy budget in subterranean rodents: insights from the Tuco-tuco *Ctenomys talarum* (Rodentia: Ctenomyidae). In: Kelt DA, Lessa E, Salazar-Bravo JA, Patton JL (eds) The quintessential naturalist: honoring the life and legacy of Oliver P. Pearson. University of California Publications in Zoology, Berkeley, pp 111–139

Antinuchi CD, Busch C (1992) Burrow structure in the subterranean rodent *Ctenomys talarum*. Z Säugetierkd 57:163–168

Armario A (2006) The hypothalamic–pituitary–adrenal axis: what can it tell us about stressors? CNS Neurol Disord 5:485–501

Balbontín J, Reig S, Moreno S (1996) Evolutionary relationships of *Ctenomys* (Rodentia: Octodontidae) from Argentina, based on penis morphology. Acta Theriol 41:237–253

Baldo MB, Luna F, Schleich CE, Antenucci CD (2014) Thermoregulatory development and behavior of *Ctenomys talarum* pups during brief repeated postnatal isolation. Comp Biochem Physiol A 173:35–41

Baldo MB, Antenucci CD, Luna F (2015) Effect of ambient temperature on evaporative water loss in the subterranean rodent *Ctenomys talarum*. J Therm Biol 53:113–118

Baldo MB, Luna F, Antenucci CD (2016) Does acclimation to contrasting atmospheric humidities affect evaporative water loss in the South American subterranean rodent *Ctenomys talarum*? J Mammal 97:1312–1320

Baldo MB, Antenucci CD (2019) Diet efect on osmoregulation in the subterranean rodent Ctenomys talarum. Comp Biochem Physiol A 235:148–158

Begall S, Lange S, Schleich CE, Burda H (2007) Acoustics, audition and auditory system. In: Begall S, Burda H, Schleich CE (eds) Subterranean rodents: news from underground. Springer, Heidelberg, pp 113–128

Bennett NC, Faulkes CG (2000) African mole-rats: ecology and eusociality. Cambridge University Press, Cambridge

Boonstra R (2005) Equipped for life: the adaptive role of the stress axis in male mammals. J Mammal 86:236–247

Bozinovic F, Gallardo P (2006) The water economy of South American desert rodents: from integrative to molecular physiological ecology. Comp Biochem Physiol 142:163–172

Brachetta V, Schleich CE, Zenuto RR (2014) Effects of acute and chronic exposure to predatory cues on spatial learning capabilities in the subterranean rodent *Ctenomys talarum* (Rodentia: Ctenomyidae). Ethology 120:563–576

Brachetta V, Schleich CE, Zenuto RR (2015) Short-term anxiety response of the subterranean rodent *Ctenomys talarum* to odors from a predator. Physiol Behav 151:596–603

Brachetta V, Schleich CE, Zenuto RR (2016) Source odor, intensity, and exposure pattern affect antipredatory responses in the subterranean rodent *Ctenomys talarum*. Ethology 122:923–936

Brachetta V, Schleich CE, Cutrera AP, Merlo JL, Kittlein MJ, Zenuto RR (2018) Prenatal predatory stress in a wild species of subterranean rodent: do ecological stressors always have a negative effect on the offspring? Dev Psychobiol 60:567–581

Brachetta V, Schleich CE, Zenuto RR (2019a) Feeding behavior under predatory risk in *Ctenomys talarum*: nutritional state and recent experience of a predatory event. Mamm Res 64:261–269

Brachetta V, Schleich CE, Zenuto RR (2019b) Differential antipredatory responses in the tuco-tuco (*Ctenomys talarum*) in relation to endogenous and exogenous changes in GCs. J Comp Physiol A 206(1):33–44. https://doi.org/10.1007/s00359-019-01384-8

Buffenstein RM (2000) Ecophysiological responses of subterranean rodents to an underground habitat. In: Lacey E, Patton J, Cameron G (eds) Life underground: biology of subterranean rodents. University of Chicago Press, Chicago, pp 62–109

Busch C (1987) Haematological correlates of burrowing in *Ctenomys*. Comp Biochem Physiol A 86:461–463

Busch C (1989) Metabolic rate and thermoregulation in two species of tuco-tuco, *Ctenomys talarum* and *Ctenomys australis* (Caviomorpha, Octodontidae). Comp Biochem Physiol A 93(2):345–347

Busch C, Malizia AI, Scaglia OA, Reig OA (1989) Spatial distribution and attributes of a population of Ctenomys talarum (Rodentia: Octodontidae). J Mammal 70:204–208

Busch C, Antinuchi D, Del Valle J, Kittlein M, Malizia A, Vassallo A, Zenuto R (2000) Population ecology of subterranean rodents. In: Lacey E, Patton J, Cameron G (eds) Life underground: the biology of subterranean rodents. University of Chicago Press, Chicago, pp 183–226

Camín S (1999) Mating behaviour of *Ctenomys mendocinus* (Rodentia, Ctenomyidae). Mamm Biol 64:230–238

Camín S (2010) Gestation, maternal behaviour, growth and development in the subterranean caviomorph rodent *Ctenomys mendocinus* (Rodentia, Hystricognathi, Ctenomyidae). Anim Biol 60:79–95

Cannon B, Nedergaard J (2011) Nonshivering thermogenesis and its adequate measurement in metabolic studies. J Exp Biol 214:242–253

Charbonnel N, de Bellocq JG, Morand S (2006) Immunogenetics of micromammal macroparasite interactions. In: Micromammals and macroparasites. Springer, Tokyo, pp 401–442

Clinchy M, Sheriff MJ, Zanette LY (2013) The ecology of stress: predator-induced stress and the ecology of fear. Funct Ecol 27:56–65

Cortés A, Rosenmann M, Bozinovic F (2000) Water economy in rodents: evaporative water loss and metabolic water production. Rev Chil Hist Nat 73:311–321

Cutrera AP, Antinuchi CD (2004) Cambios en el pelaje del roedor subterráneo *Ctenomys talarum*: posible mecanismo térmico compensatorio. Rev Chil Hist Nat 77:235–242

Cutrera AP, Lacey EA (2006) Major histocompatibility complex variation in talas tuco-tucos: the influence of demography on selection. J Mammal 87:706–716

Cutrera AP, Lacey EA (2007) Trans-species polymorphism and evidence of selection on class II MHC loci in tuco-tucos (Rodentia: Ctenomyidae). Immunogenetics 59:937–948

Cutrera AP, Mora MS (2017) Selection on MHC in a context of historical demographic change in 2 closely distributed species of tuco-tucos (*Ctenomys australis* and *C. talarum*). J Hered 108:628–639

Cutrera AP, Antinuchi CD, Busch C (2003) Thermoregulatory development in pups of the subterranean rodent *Ctenomys talarum*. Physiol Behav 79:321–330

Cutrera AP, Mora MS, Antenucci CD, Vassallo AI (2010a) Intra-and interspecific variation in home-range size in sympatric tuco-tucos, *Ctenomys australis* and *C talarum*. J Mammal 91:1425–1434

Cutrera AP, Zenuto RR, Luna F, Antenucci CD (2010b) Mounting a specific immune response increases energy expenditure of the subterranean rodent *Ctenomys talarum* (tuco-tuco): implications for intraspecific and interspecific variation in immunological traits. J Exp Biol 213:715–724

Cutrera AP, Zenuto RR, Lacey EA (2011) MHC variation, multiple simultaneous infections and physiological condition in the subterranean rodent *Ctenomys talarum*. Infect Genet Evol 11:1023–1036

Cutrera AP, Fanjul MS, Zenuto RR (2012) Females prefer good genes: MHC-associated mate choice in wild and captive tuco-tucos. Anim Behav 83:847–856

Cutrera AP, Zenuto RR, Lacey EA (2014a) Interpopulation differences in parasite load and variable selective pressures on MHC genes in *Ctenomys talarum*. J Mammal 95:679–695

Cutrera AP, Luna F, Merlo JL, Baldo MB, Zenuto RR (2014b) Assessing the energetic costs and trade-offs of a PHA-induced inflammation in the subterranean rodent *Ctenomys talarum*: immune response in growing tuco-tucos. Comp Biochem Physiol A 174:23–28

Clutton-Brock TH (1989) Mammalian mating systems. Proc R Soc B 236:339–372

Davis AK, Maney DL (2008) The use of glucocorticoid hormonesor leucocyte profiles to measure stress in vertebrates: what's the diference? Methods Ecol Evol 8:1556–1568

Davis AK, Maney DL (2018) The use of leukocyte profiles to measure stress in vertebrates: a review for ecologists. Funct Ecol 22:760–772

del Valle JC, López Mañanes AA (2008) Digestive strategies in the South American subterranean rodent *Ctenomys talarum*. Comp Biochem Physiol A 150:387–394

del Valle JC, López Mañanes AA (2011) Digestive flexibility in females of the subterranean rodent *Ctenomys talarum* in their natural habitat. J Exp Zool A 315A:141–148

Demas GE, Nelson RJ (2012) Introduction to ecoimmunology. In: Demas GE, Nelson RJ (eds) Ecoimmunology. Oxford University Press, New York, pp 3–6

Demas G, Greives T, Chester E, French S (2012) The energetics of immunity. In: Demas GE, Nelson RJ (eds) Ecoimmunology. Oxford University Press, New York, pp 259–296

Dickens MJ, Romero LM (2013) A consensus endocrine profile for chronically stressed wild animals does not exist. Gen Comp Endocrinol 191:177–189

Diaz GB (2001) Ecofisiología de Pequeños Mamíferos de Las Tierras Áridas de Argentina: Adaptaciones Renales. Universidad Nacional de Cuyo, Mendoza Argentina, Doctoral Thesis

Diaz GB, Ojeda RA, Rezende EL (2006) Renal morphology, phylogenetic history and desert adaptation of South American hystricognath rodents. Funct Ecol 20:609–620

Doherty PC, Zinkernagel RM (1975) Enhanced immunological surveillance in mice heterozygous at the H-2 gene complex. Nature 256:50

El Jundi TARJ, de Freitas TRO (2004) Genetic and demographic structure in a population of *Ctenomys lami* (Rodentia-Ctenomyidae). Hereditas 140:18–23

Fanjul MS, Zenuto RR (2008a) Copulatory pattern of the subterranean rodent *Ctenomys talarum*. Mammalia 72(2):102–108

Fanjul MS, Zenuto RR (2008b) Female reproductive responses to photoperiod and male odours in the subterranean rodent *Ctenomys talarum*. Acta Theriol 53:73–85

Fanjul MS, Zenuto RR (2012) Female reproductive behaviour, ovarian hormones and vaginal cytology of the induced ovulator, *Ctenomys talarum*. Acta Theriol 57:15–27

Fanjul MS, Zenuto RR (2013) When allowed, females prefer novel males in the polygynous subterranean rodent *Ctenomys talarum* (tuco-tuco). Behav Process 92:71–78

Fanjul MS, Zenuto RR (2017) Female choice, male dominance and condition-related traits in the polygynous subterranean rodent *Ctenomys talarum*. Behav Process 142:46–55

Fanjul MS, Zenuto RR, Busch C (2003) Use of olfaction for sexual recognition in the subterranean rodent *Ctenomys talarum*. Acta Theriol 48:35–46

Fanjul MS, Zenuto RR, Busch C (2006) Seasonality of breeding in wild tuco-tucos *Ctenomys talarum* in relation to climate and food availability. Acta Theriol 51:283–293

Fanjul MS, Varas MF, Zenuto RR (2018) Female preference for males that have exclusively marked or invaded territories depends on male presence and its identity in the subterranean rodent *Ctenomys talarum*. Ethology 124:579–590

Fernández-Stolz GP, Stolz JFB, de Freitas TRO (2007) Bottlenecks and dispersal in the Tuco-Tuco Das Dunas, *Ctenomys flamarioni* (Rodentia: Ctenomyidae), in Southern Brazil. J Mammal 88:935–945

Francescoli G (1999) A preliminary report on the acoustic communication in Uruguayan *Ctenomys* (Rodentia, Octodontidae): basic sounds types. Bioacoustics 103:203–218

Francescoli G (2000) Sensory capabilities and communication in subterranean rodents. In: Lacey E, Patton J, Cameron G (eds) Life underground: the biology of subterranean rodents. University of Chicago Press, Chicago, pp 111–144

Francescoli G (2001) Vocal signals from *Ctenomys pearsoni* pups. Acta Theriol 46:327–330

Francescoli G (2002) Geographic variation in vocal signals of *Ctenomys pearsoni*. Acta Theriol 47:35–44

Francescoli G (2011) Tuco-tucos' vocalization output varies seasonally (*Ctenomys pearsoni*; Rodentia, Ctenomyidae): implications for reproductive signaling. Acta Ethol 14:1–6

Francescoli G (2017) Environmental factors could constrain the use of long-range vocal signals in solitary tuco-tucos (Ctenomys; Rodentia, Ctenomyidae) reproduction. J Ecoacoust 1:R7YFP0

Francescoli G, Quirici V (2010) Two different vocalization patterns in *Ctenomys* (Rodentia, Octodontidae) territorial signals. Mastozool Neotropical 17:141–145

Garcias FM, Stolz JFB, Fernández GP, Kubiak BB, Bastazini VAG, de Freitas TRO (2018) Environmental predictors of demography in the tuco-tuco of the dunes (*Ctenomys flamarioni*). Mastozool Neotropical 25(2):293–305. https://doi.org/10.31687/saremMN.18.25.2.0.18

Hambuch TM, Lacey EA (2002) Enhanced selection for MHC diversity in social tuco-tucos. Evolution 56:841–845

Hohtola E (2002) Facultative and obligatory thermogenesis in young birds: a cautionary note. Comp Biochem Physiol A 131:733–739

Hulbert AJ, Else PL (2004) Basal metabolic rate: history, composition, regulation, and usefulness. Physiol Biochem Zool 77:869–876

Izquierdo G, Lacey EA (2008) Effects of group size on nest attendance in the communally breeding colonial tuco-tuco. Mamm Biol 73:438–443

Karasov WH (1986) Energetics, physiology and vertebrate ecology. Trends Ecol Evol 1:101–104

Klein J (1986) Natural history of the major histocompatibility complex. Wiley, New York

Kubiak BB, Galiano D, de Freitas TRO (2017) Can the environment influence species home-range size? A case study on *Ctenomys minutus* (Rodentia, Ctenomyidae). J Zool 302:171–177

Lacey EA (2000) Spatial and social systems of subterranean rodents. In: Lacey E, Patton J, Cameron G (eds) Life underground: the biology of subterranean rodents. University of Chicago Press, Chicago, pp 257–299

Lacey EA (2004) Sociality reduces individual direct fitness in a communally breeding rodent, the colonial tuco-tuco (*Ctenomys sociabilis*). Behav Ecol Sociobiol 56:449–457

Lacey EA, Wieczorek JR (2004) Kinship in colonial tuco-tucos: evidence from group composition and population structure. Behav Ecol 15:988–996

Lacey EA, Braude SH, Wieczorek JR (1997) Burrow sharing by colonial Tuco-Tucos (*Ctenomys sociabilis*). J Mammal 78:556–562

Lacey EA, Braude SH, Wieczorek JR (1998) Solitary burrow use by adult Patagonian Tuco-tucos (Ctenomys haigi). J Mammal 79:986–991

Lacey EA, Patton JL, Cameron GN (2000) Linking immuneLife underground: the biology of subterranean rodents. University of Chicago Press, Chicago

Lacey EA, Cutrera AP (2007) Behavior, demography, and Immunogenetic variation: new insights from subterranean rodents. In: Begall S, Burda H, Schleich CE (eds) Subterranean rodents: news from underground. Springer, Heidelberg

Lee KA (2006) Linking immune defences and life history at the levels of the individual and the species. Integr Comp Biol 46:1000–1015

Lochmiller RL, Deerenberg C (2000) Trade-offs in evolutionary immunology: just what is the cost of immunity? Oikos 88:87–98

Luna F, Antenucci CD (2006) Cost of foraging in the subterranean rodent *Ctenomys talarum* : effect of soil hardness. Can J Zool 84:661–667

Luna F, Antenucci CD (2007a) Effect of tunnel inclination on digging energetics in the tuco-tuco, *Ctenomys talarum* (Rodentia: Ctenomyidae). Naturwissenschaften 94:100–106

Luna F, Antenucci CD (2007b) Energetics and thermoregulation during digging in the rodent tuco-tuco (*Ctenomys talarum*). Comp Biochem Physiol A 146:559–564

Luna F, Antenucci CD (2007c) Energy and distribution in subterranean rodents: Sympatry between two species of the genus *Ctenomys*. Comp Biochem Physiol A 147:948–954

Luna F, Antenucci CD, Busch C (2002) Digging energetics in the south American rodent *Ctenomys talarum* (Rodentia, Ctenomyidae). Can J Zool 80:2144–2149

Luna F, Antenucci CD, Bozinovic F (2009) Comparative energetics of the subterranean *Ctenomys* rodents: breaking patterns. Physiol Biochem Zool 82:226–235

Luna F, Roca P, Oliver J, Antenucci CD (2012) Maximal thermogenic capacity and non-shivering thermogenesis in the South American subterranean rodent *Ctenomys talarum*. J Comp Physiol B 182:971–983

Luna F, Bozinovic F, Antenucci CD (2015) Macrophysiological patterns in the energetics of Caviomorph rodents: implications in a warming world. In: Vassallo AI, Antenucci CD (eds) The biology of Caviomorph rodents: diversity and evolution, SAREM Series A, pp 245–272

Luna F, Naya H, Naya DE (2017) Understanding evolutionary variation in basal metabolic rate: an analysis in subterranean rodents. Comp Biochem Physiol A 206:87–94

Luna F, Sastre-Serra J, Oliver J, Antenucci CD (2019) Thermogenic capacity in subterranean *Ctenomys*: species-specific role of thermogenic mechanisms. J Therm Biol 80:164–171

MacDougall-Shackleton SA, Bonier F, Romero LM, Moore IT (2019) Glucocorticoids an "stress" are not synonymous. Integrat Organ Biol 1:obz017. https://doi.org/10.1093/iob/obz017

Malewski S, Begall S, Schleich CE, Antenucci CD, Burda H (2018) Do subterranean mammals use the Earth's magnetic field as a heading indicator to dig straight tunnels? PeerJ 6:e5819

Marinho JR, de Freitas TRO (2006) Population structure of *Ctenomys minutus* (Rodentia, ctenomyidae) on the coastal plain of Rio Grande do Sul, Brazil. Acta Theriol 51:53–59

Martino NS, Zenuto RR, Busch C (2007) Nutritional responses to different diet quality in the subterranean rodent *Ctenomys talarum* (tuco-tucos). Comp Biochem Physiol A 147:974–982

Mason MJ (2004) The middle ear apparatus of the tuco-tuco *Ctenomys sociabilis* (Rodentia, Ctenomyidae). J Mammal 85:797–805

Mastrangelo ME, Schleich CE, Zenuto RR (2010) Spatial learning abilities in males and females of *Ctenomys talarum*. Ethol Ecol Evol 22:101–108

McEwen BS, Wingfield JC (2003) The concept of allostasis in biology and biomedicine. Horm Behav 43:2–15

McNab BK (2002) The physiological ecology of vertebrates: a view from energetics. In: Comstock publishing associates. Cornell University Press, Ithaca (New York)

Merlo JL, Cutrera AP, Luna F, Zenuto RR (2014) PHA-induced inflammation is not energetically costly in the subterranean rodent *Ctenomys talarum* (tuco-tucos). Comp Biochem Physiol A Mol Integr Physiol 175:90–95

Merlo J, Cutrera AP, Zenuto RR (2016a) Food restriction affects inflammatory response and nutritional state in Tuco-tucos (*Ctenomys talarum*). J Exp Zool A Ecol Genet Physiol 325:675–687

Merlo J, Cutrera AP, Zenuto RR (2016b) Parasite infection negatively affects PHA-triggered inflammation in the subterranean rodent *Ctenomys talarum*. J Exp Zool A Ecol Genet Physiol 325:132–141

Merlo JE, Cutrera AP, Kittlein MJ, Zenuto RR (2018) Individual condition and inflammatory response to PHA in the subterranean rodent Ctenomys talarum (Talas tuco-tuco): a multivariate approach. Mamm Biol 90:47–54

Merlo J, Cutrera AP, Zenuto RR (2019) Assessment of trade-offs between simultaneous immune challenges in a slow-living subterranean rodent. Physiol Biochem Zool 92:92–105

Meroi F, Luna F, Antenucci CD (2014) Variación estacional de la tasa metabólica de reposo en *Ctenomys talarum* (Rodentia, Ctenomyidae): Ausencia de efectos ambientales. Mastozool Neotropical 21:241–250

Milligan SR (1982) Induced ovulation in mammals. In: Finn CA (ed) Oxford reviews of reproductive biology. Clarendon Press, Oxford

Milot E, Cohen AA, Vézina F, Buehler DM, Matson KD, Piersma T (2014) A novel integrative method for measuring body condition in ecological studies based on physiological dysregulation. Methods Ecol Evol 5:146–155

Mitani JC, Gros-Louis J, Richards AF (1996) Sexual dimorphism, the operational sex ratio, and the intensity of male competition in polygynous primates. Am Nat 147:966–980

Moll RJ, Redilla KM, Mudumba T, Muneza AB, Gray SM, Abade L, Hayward MW, Millspaugh JJ, Montgomery RA (2017) The many faces of fear: a synthesis of the methodological variation in characterizing predation risk. J Anim Ecol 86:749–765

Müller C, Jenni-Eiermann S, Jenni L (2011) Heterophils/lymphocytesratio and circulating corticosterone do not indicate the same stressimposed on Eurasian kestrel nestlings. Funct Ecol 25:566–576

Naya DE, Spangenberg L, Naya H, Bozinovic F (2013) Thermal conductance and basal metabolic rate are part of a coordinated system for heat transfer regulation. Proc R Soc B Biol Sci 280:20131629

Nevo E (1999) Mosaic evolution of subterranean mammals: regression, progression, and global convergence. Oxford University Press, Oxford/New York

Nevo E, Reig OA (1990) Evolution of subterranean mammals at the organismal and molecular levels. Wiley-Liss, New York

Nevo E, Ivanitskaya E, Beiles A (2001) Adaptive radiation of blind subterranean mole rats. Backhuys, Leiden

Nevo E (1979) Adaptive convergence and divergence of subterranean mammals. Ann Rev Ecol Syst 10:269–308

Norris K, Evans MR (2000) Ecological immunology: life history trade-offs and immune defense in birds. Behav Ecol 11:19–26

O'Brien SL, Tammone MN, Cuello PA, Lacey EA (2020) Facultative sociality in a subterranean rodent, the highland tuco-tuco (Ctenomys opimus). Biol J Linn Soc 129:918–930

Patton JL, Pardiñas UFJ, D'Elía G (2015) Mammal fo Soth América, vol 2. The University of Chicago Press, Chicago, p 1336

Pearson OP (1959) Biology of the subterranean rodents, Ctenomys in Perú. Mem Mus Hist Nat Javier Prado 9:1–56

Pearson OP, Christie MI (1985) Los tuco-tucos (género Ctenomys) de los parques nacionales Lanín yNahuel Huapi, Argentina. Hist Nat 5:337–343

Perissinotti PP, Antenucci CD, Zenuto RR, Luna F (2009) Effect of diet quality and soil hardness on metabolic rate in the subterranean rodent Ctenomys talarum. Comp Biochem Physiol A 154:298–307

Puig S, Rosi MI, Videla F, Roig VG (1992) Estudio ecológico del roedor subterráneo Ctenomys mendocinus en la precordillera de Mendoza, Argentina: densidad poblacional y uso del espacio. Rev Chil Hist Nat 65:247–254

Reig OA, Busch C, Ortells MO, Contreras JR (1990) An overview of evolution, systematics, population biology, cytogenetics, molecular biology, and speciations in Ctenomys. In: Nevo E, Reig OA (eds) Evolution of subterranean mammals at the organismal and molecular levels. Wiley-Liss, New York

Rocha-Barbosa O, Bernardo JSL, Loguercio MFC, de Freitas TRO, Santos-Mallet JR, Bidau CJ (2013) Penial morphology in three species of Brazilian Tuco-tucos, Ctenomys torquatus, C. minutus, and C. flamarioni (Rodentia: Ctenomyidae). Braz J Biol 73:201–209

Rosi MI, Puig S, Videla F, Madoery L, Roig VG (1992) Estudio ecológico del roedor subterráneo Ctenomys mendocinus en la precordillera de Mendoza, Argentina: ciclo reproductivo -y estructura etaria. Rev Chil Hist Nat 65:221–223

Rosi MI, Puig S, Videla F, Cona MI, Roig VG (1996) Cielo reproductivo y estructura etaria de Ctenomys mendocinus (Rodentia, Ctenomyidae) del Piedemonte de Mendoza, Argentina. Ecol Aust 6:87–93

Rosi MI, Cona MI, Roig VG, Massarini AI, Verzi DH (2005) Ctenomys mendocinus. Mamm Species 777:1–6

Rossin A, Malizia AI (2002) Relationship between helminth parasites and demographic attributes of a population of the subterranean rodent Ctenomys talarum (Rodentia: Octodontidae). J Parasitol 88:1268–1270

Sapolsky RM, Romero LM, Munck AU (2000) How do glucocorticoids influence stress responses? Integrating permissive, suppressive, stimulatory and preparative actions. Endocr Rev 21:55–89

Schleich CE, Antenucci CD (2004) Testing magnetic orientation in a solitary subterranean rodent *Ctenomys talarum* (Rodentia: Octodontidae). Ethology 110:485–495

Schleich CE, Busch C (2002) Acoustic signals of a solitary subterranean rodent *Ctenomys talarum* (Rodentia: Ctenomyidae): physical characteristics and behavioural correlates. J Ethol 20:123–131

Schleich C, Busch C (2004) Energetic expenditure during vocalization in pups of the subterranean rodent Ctenomys talarum. Naturwissenschaften 91:548–551

Schleich CE, Zenuto RR (2007) Use of vegetation chemical signals for digging orientation in the subterranean rodent *Ctenomys talarum* (Rodentia: Ctenomyidae). Ethology 113:573–578

Schleich CE, Veitl S, Knotková E, Begall S (2007) Acoustic communication in subterranean rodents. In: Begall S, Burda H, Schleich CE (eds) Subterranean rodents. Springer, Berlin, Heidelberg. https://doi.org/10.1007/978-3-540-69276-8_10

Schleich CE, Zenuto RR (2010) Testing detection and discrimination of vegetation chemical signals in the subterranean rodent *Ctenomys talarum*. Ethol Ecol Evol 22:257–264

Schleich CE, Vielma A, Glösmann M, Palacios AG, Peichl L (2010) The retinal photoreceptors of two subterranean tuco-tuco species (Rodentia, *Ctenomys*): morphology, topography and spectral sensitivity. J Comp Neurol 518:400–4015

Schleich CE, Zenuto RR, Cutrera AP (2015) Immune challenge but not dietary restriction affects spatial learning in the wild subterranean rodent Ctenomys talarum. Physiol Behav 139:150–156

Schmidt-Nielsen K (1997) Animal physiology: adaptation and environment. Cambridge University Press

Schoenle LA, Downs CJ, Martin LB (2018) An introduction to ecoimmunology. In: Advances in comparative immunology. Springer, Cham, pp 901–932

Schwanz LE, Lacey EA (2003) Olfactory discrimination of gender by colonial tuco-tucos (*Ctenomys sociabilis*). Mamm Biol 68:53–60

Scott G (2005) Essential animal behavior. Blackwell, Hoboken

Sheldon BC, Verhulst S (1996) Ecological immunology: costly parasite defences and trade-offs in evolutionary ecology. Trends Ecol Evol 11:317–321

Sherman PW, Jarvis JUM, Alexander RD (1991) The biology of the naked mole-rat. Princeton University Press, Princeton

Stamps JA (2007) Growth-mortality tradeoffs and 'personality traits' in animals. Ecol Lett 10:355–363

Tachinardi P, Bicudo JEW, Oda GA, Valentinuzzi VS (2014) Rhythmic 24 h variation of core body temperature and locomotor activity in a subterranean rodent (*Ctenomys aff. knighti*), the tuco-tuco. PLoS One 9:e85674

Tachinardi P, Valentinuzzi VS, Oda GA, Buck CL (2017) The interplay of energy balance and daily timing of activity in a subterranean rodent: a laboratory and field approach. Physiol Biochem Zool 90:546–552

Talice RV, Laffitte de Mosera S (1958) Parto, comportamiento maternal y comportamiento filial en *Ctenomys torquatus* ("Tucu Tucu"). Revista de la Facultad de Humanidades y Ciencias, Universidad de la República. Montevideo 16:69–75

Tassino B, Passos CA (2010) Reproductive biology of Río Negro tuco-tuco, *Ctenomys rionegrensis* (Rodentia: Octodontidae). Mamm Biol 75:253–260

Tassino B, Estevan I, Garbero RP, Altesor P, Lacey EA (2011) Space use by Río Negro tuco-tucos (*Ctenomys rionegrensis*): excursions and spatial overlap. Mamm Biol 76:143–147

Temeles EJ (1994) The role of neighbours in territorial systems: when are they 'dear enemies'? Anim Behav 47:339–350

Tomasco IH, Sánchez L, Lessa EP, Lacey EA (2019) Genetic analyses suggest burrow sharing by Río Negro tuco-tucos (*Ctenomys rionegrensis*). Mastozool Neotropical 26:430–439

Tomasi TE, Horton TH (1992) Mammalian energetics: interdisciplinary views of metabolism and reproduction. Comstock Publishing Associates, Ithaca/New York

Tomotani BM, Flores DEFL, Tachinardi P, Paliza JD, Oda GA, Valentinuzzi VS (2012) Field and laboratory studies provide insights into the meaning of day-time activity in a subterranean rodent (*Ctenomys aff. knighti*), the Tuco-Tuco. PLoS One 7:e37918

Valentinuzzi VS, Oda GA, Araújo JF, Ralph MR (2009) Circadian pattern of wheel-running activity of a South American subterranean rodent (Ctenomys cf knightii). Chronobiol Int 26:14–27

Vassallo A, Busch C (1992) Interspecific agonism between two sympatric species of Ctenomys (Rodentia: Octodontidae) in captivity. Behaviour 120:40–50

Vassallo A, Kittlein M, Busch C (1994) Owl predation on two sympatric species of tuco tucos (Rodentia: Octodontidae). J Mammal 75:725–732

Vega-Zuniga T, Medina F, Marin G, Letelier JC, Palacios AG, Němec P, Schleich CE, Mpodozis J (2017) Selective binocular vision loss in two subterranean caviomorph rodents: *Spalacopus cyanus* and *Ctenomys talarum*. Sci Rep 7:41704. https://doi.org/10.1038/srep41704

Vera F, Antenucci CD, Zenuto RR (2011a) Cortisol and corticosterone exhibit different seasonal variation and responses to acute stress and captivity in tuco-tucos (*Ctenomys talarum*). Gen Comp Endocrinol 170:550–557

Vera F, Zenuto RR, Antenucci CD, Busso JM, Marín RH (2011b) Validation of a radioimmunoassay for measuring testosterone concentrations in plasma samples of the subterranean rodent *Ctenomys talarum*: outstandingly-elevated levels in the wild and the effect of captivity. J Exp Zool A 315:572–583

Vera F, Zenuto RR, Antenucci CD (2012) Differential responses of cortisol and corticosterone to adrenocorticotropic hormone (ACTH) in a subterranean rodent (*Ctenomys talarum*). J Exp Zool A 317:173–184

Vera F, Zenuto RR, Antenucci CD (2013) Seasonal variations in plasma cortisol, testosterone, progesterone and leukocyte profiles in a wild population of tuco-tucos. J Zool 289:111–118

Vera F, Antenucci CD, Zenuto RR (2018) Different regulation of cortisol and corticosterone in the subterranean rodent *Ctenomys talarum*: responses to dexamethasone, angiotensin II, potassium, and diet. Gen Comp Endocrinol 273:108–117. https://doi.org/10.1016/j.ygcen.2018.05.019

Vera F, Antenucci CD, Zenuto RR (2019) Different regulation of cortisol and corticosterone in the subterranean rodent Ctenomys talarum: responses to dexamethasone, angiotensin II, potassium, and diet. Gen Comp Endocrinol 273:108–117.

Vorhess CV, Williams MT (2014) Assessing spatial learning and memory in rodents. ILAR J 55:310–332. https://doi.org/10.1093/ilar/ilu013

Vleck D (1979) The energy cost of burrowing by the pocket gopher Thomomys bottae. Physiol Zool 52:122–136

Weiner J (1992) Physiological limits to sustainable energy budgets in birds and mammals: ecological implications. Trends Ecol Evol 7:384–388

Weir BJ (1974) Reproductive characteristics of Hystricomorph Rodents. Symp Zool Soc Lond 34:265–301

Wiegert RG (1968) Thermodynamic considerations in animal nutrition. Am Zool 8:71–81

Wise PH, Weir BJ, Hime JM, Forrest E (1972) The diabetic syndrome in the tuco-tuco (*Ctenomys talarum*). Diabetologia 8:165–172

Withers PC, Cooper CE, Maloney SK, Bozinovic F, Cruz-Neto A (2016) Ecological and environmental physiology of mammals. Oxford University Press, Oxford

Woodruff JA, Lacey EA, Bentley G (2010) Contrasting fecal corticosterone metabolite levels in captive and free-living colonial tuco-tucos (*Ctenomys sociabilis*). J Exp Zool 313A:498–507

Woodruff JA, Lacey EA, Bentley G, Kriegsfeld LJ (2013) Effects of social environment on baseline glucocorticoid levels in a communally breeding rodent, the colonial tuco-tuco (*Ctenomys sociabilis*). Horm Behav 64:566–572

Zarrow MX, Clark JH (1968) Ovulation following vaginal stimulation in a spontaneous ovulator and its implications. J Endocrinol 40:343–352

Zenuto RR (1999) Sexual size dimorphism, testes size and mating system in two populations of *Ctenomys talarum* (Rodentia: Octodontidae). J Nat Hist 33:305–314

Zenuto RR (2010) Dear enemy relationships in the subterranean rodent *Ctenomys talarum*: the role of memory of familiar odours. Anim Behav 79:1247–1255

Zenuto RR, Busch C (1998) Population biology of the subterranean rodent *Ctenomys australis* (Tuco-tuco) in a coastal dunefield in Argentina. Mamm Biol 63:357–367

Zenuto R, Fanjul MS (2002) Olfactory recognition of individual scents in the subterranean rodent *Ctenomys talarum* (Rodentia: Octodontidae). Ethology 108:629–641

Zenuto RR, Lacey EA, Busch C (1999) DNA fingerprinting reveals polygyny in the subterranean rodent *Ctenomys talarum*. Mol Ecol 8:1529–1532

Zenuto RR, Vassallo AI, Busch C (2001) A method for studying social and reproductive behaviour of subterranean rodents in captivity. Acta Theriol 46:161–170

Zenuto RR, Antinuchi CD, Busch C (2002a) Bioenergetics of reproduction and pup development in a subterranean rodent (*Ctenomys talarum*). Physiol Biochem Zool 75:469–478

Zenuto RR, Vassallo AI, Busch C (2002b) Comportamiento social y reproductivo del roedor subterráneo solitario *Ctenomys talarum* (Rodentia: Ctenomyidae) en condiciones de semicautiverio. Rev Chil Hist Nat 75:165–177

Zenuto R, Fanjul MS, Busch C (2004) Use of chemical communication by subterranean rodent *Ctenomys talarum* during the reproductive season. J Chem Ecol 30:2111–2126

Zenuto RR, Estavillo C, Fanjul MS (2007) Familiarity and mating behavior in the subterranean rodent *Ctenomys talarum* (tuco-tuco). Can J Zool 85:944–955

Chapter 12
Effects of Environmental Pollution on the Conservation of *Ctenomys*

Cristina A. Matzenbacher and Juliana da Silva

12.1 Introduction

All species on the planet have developed adaptations, in different ways, that promote enough aptitude to sustain different populations for long periods of time and environmental changes. This is the result of the interaction between the genome and the environment. The ability of species or population to adaptations from a polluted environment is called toxicology evolutionary (Bickham et al. 2000). Currently, the influence of epigenetic factors on adaptations has been discussed, alone and associated with mutations or natural selection or genetic drift. Epigenetics involves meiotically and mitotically stable alterations in gene expression that are not based on DNA sequence changes but involve processes that impact the packaging of DNA (chromatin structure) (Kalisz and Purugganan 2004). Although epigenetic variation can occur in the absence of genetic variation, genetic variation can influence epigenetic variation and the epimutation rate in several ways. Epigenetic variation can be a significant source of natural phenotypic variation; therefore, it has the potential to play a major role in adaptation to environmental change. Modern evolutionary theory is primarily based on the inheritance of random genetic variation, so it has been widely discussed whether evolutionary theory requires revision considering epigenetics (Jablonka and Raz 2009; Richards et al. 2010; Jablonka 2017).

Genomic instability and epigenetic changes induced by environmental pollutants have been related to different effects on populations, such as the reduction of population size due to loss of genetic variability and reproductive capacity. Pollution is

C. A. Matzenbacher (✉)
Programa de Pós-Graduação em Genética e Biologia Molecular, Universidade Federal do Rio Grande do Sul - UFRGS, Porto Alegre, Brazil

J. da Silva (✉)
Universidade LaSalle – UNILASALLE and Universidade Luterana do Brasil - ULBRA, Canoas, RS, Brazil
e-mail: juliana.silva@unilasalle.edu.br

© Springer Nature Switzerland AG 2021
T. R. O. de. Freitas et al. (eds.), *Tuco-Tucos*,
https://doi.org/10.1007/978-3-030-61679-3_12

any change in the environment that can be a natural or agrarian ecosystem, an urban system, or even a micro scale. It can cause changes in the proportions or character-istics of one of the elements that make up the environment itself, such as increased carbon dioxide concentration. It may be the result of the introduction of natural substances, although foreign to certain ecosystems, or artificial substances such as the deposition of pesticides in the soil (Maximillian et al. 2019). To assess complex chemical-biological interactions and predict the potential for chemical substances that cause harmful effects and impact ecological communities and populations, chemical and ecotoxicological methods are used to assess responses at both the organism and cellular levels (Schaeffer 1991). This kind of research integrates data from the field and laboratory biomarkers to understand and predict adverse effects; this way of study looks at the whole system. Adverse outcome pathways (AOPs) are conceptual frameworks that bring together what is known about chemical agents – with limited safety data – and their adverse effects on human health and ecosystems (Groh et al. 2015).

Contaminants transferred through the food chain play an important role in their action and persistence in the ecosystem and can slowly accumulate in the tissues of individuals over time. Many toxic substances such as metals and organic com-pounds can be transferred from individuals' tissues to their predators and reach a higher concentration at higher trophic levels. Pollutants can also act on DNA, lead-ing to teratogenic effects, mutations in germ cells, premature aging or induce neo-plasia in somatic cells. In addition, DNA damage (mutations) can affect population structure, in which genetic imbalance (change in genetic variability and allelic fre-quency) has a direct effect on biodiversity decline, increasing vulnerability to envi-ronmental stress and, consequently, possibly leading species to reduced survival and extinction (Hemminki et al. 1979; Groh et al. 2015).

Toxicity is the ability of a particular substance or even the complex mixture of elements present in the environment to affect a living organism. The effect will depend on chemical to chemical and possible interactions, with different effects on different types of cells and tissues. Environmental toxicology represents a complex triangular interaction between anthropogenic chemicals, the environment, and biota. A biomarker approach also involves the identification of molecules capable of responding to anthropogenic (xenobiotic) chemicals by positive or negative regula-tion, in order to use them to assess all environmental quality. In recent decades, interest in using biomarkers or bioindicators as monitoring tools to assess environ-mental pollution has increased considerably. This is because biomarkers are suitable not only for providing information on the health status of exposed organisms but also on the quality and/or quantity of the exposure situation. Thus, the biomarkers can be used as toxicity measurements or as fingerprints for exposure to chemicals. Depending on the character of the selected biomarkers, the evaluator will receive more data on the adverse effects on organisms or about the exposure situation (Sturla et al. 2014).

12.2   Biomarkers and Bioindicators

Different approaches are used to assess the effects and risks of exposure to chemicals, physical, and biological agents. Biomarker is a term for analyzing the interaction between a biological system and an environmental agent. In order to study the effects of exposure to environmental genotoxins in wildlife species, many biomarkers have been applied which have been derived from human cancer risk assessment studies, including markers of chemical DNA modifications such as DNA adducts, markers of cytogenetic effects such as chromosomal aberrations and micronucleate cells. Studies of markers of genotoxic effects in somatic or germ cells with wild animal populations can help in the analysis of exposure to environmental mutagens. It evaluates whether the agent is affecting biodiversity and the chance of survival, inducing changes or selecting critical environments for the survival of current environmental levels of pollution contributing to the assessment of ecological risks (Kleinjans and Van Schooten 2002).

Many studies suggest that exposure to pollutants is associated with genotoxicity biomarkers such as chromosomal aberrations, sister chromatic exchange (SCE), micronuclei (MN; See Fig. 12.1), and DNA damage observed by the comet assay (CA; See Fig. 12.2) (León-Mejía et al. 2011; Rohr et al. 2013; da Silva 2016; Espitia-Pérez et al. 2018). MN formation is widely used in molecular epidemiology as a biomarker of chromosomal damage, genome instability, and eventually cancer (Fenech 2002). The occurrence of MN represents an integrated response to phenotypes of chromosomal instability and altered cell viability caused by genetic defects and/or exogenous exposures to genotoxic agents. The MN test detects aneugens (numerical chromosomes aberrations) and clastogens (inducing breakages of chromosomes) in the cytoplasm of interphase cells that have undergone cell division during or after exposure to different agents (Fenech 2002, 2007; da Silva 2016).

Fig. 12.1 Images of the *Ctenomys torquatus* bone marrow cells representing micronuclei (MN). (Images from the author's archive)

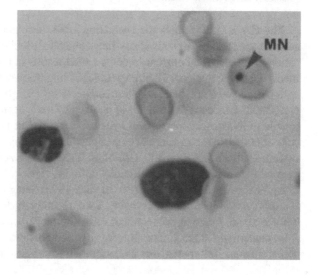

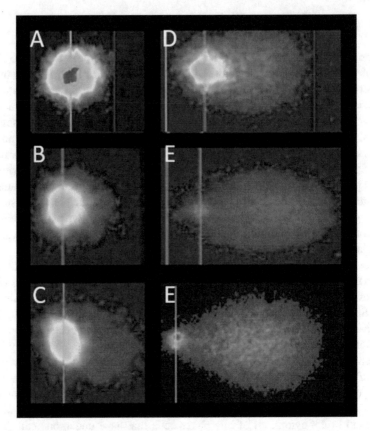

Fig. 12.2 Images of comets assay of *Ctenomys torquatus* in blood cells representing damage classes. (**a**) Class 0, undamaged; (**b**) class 1; (**c**) class 2; (**d**) class 3; (**e**) class 4, maximum damage. (Images from the author's archive)

The CA is a procedure for evaluating DNA lesions (i.e., strand breaks, DNA adducts, excision repair sites, cross-links, and alkali-labile sites) and involves application of an electrical current to cells, which results in the transport of DNA fragments out of the nucleus. The image of DNA migration obtained resembles a comet with a head and a tail, hence the term comet assay (Singh and Stephens 1998; Silva et al. 2000). Since the DNA damage induced by toxic agents is often tissue and cell-specific, CA is especially useful because it can detect DNA lesions in individual cells obtained under a variety of experimental conditions. The technique can also be used to evaluate DNA repair (Tice et al. 2000). The CA has already been used to detect DNA damage in several native animals, especially small mammal species, living near or in polluted areas (Petras et al. 1995; Salagovic et al. 1996).

Another biomarker that is being used recently is the telomere length (TL). Telomeres are found at both ends of each chromosome, which protect the genome from nucleolytic degradation, unnecessary recombination, repair, and interchromosomal fusion. Telomeres are nucleoprotein structures that protect the ends of eukaryotic chromosomes with a noncoding repeatable DNA sequence (TTAGGG). It has a single-strand 3'G adjacent nucleotide overhang linked by the shelterin hexameric protein complex (Fig. 12.3), which makes them sensitive to oxidative stress and single-stranded DNA breaks. Repeating units and shelter complexes harbor differences between species, but their homology is established in almost every respect. In normal somatic cells, telomeres decrease with each round of cell division due to limitations of the replication mechanism to replicate the ends of linear DNA molecules (De Lange 2005; Zakian 2012). When telomere length reaches a critical limit, the cell undergoes senescence and/or apoptosis. Telomere length may therefore serve as a biological clock to determine the lifespan of a cell and an organism, since telomere length decreases with age. Once telomeric DNA is less capable of repair, some agents associated with specific habits may expedite telomere shortening by inducing damage to DNA (Kahl and da Silva 2016).

It is known that DNA methylation has a function in telomere length variability (Blasco 2007; Yehezkel et al. 2008). DNA methylation patterns are formed by a family of DNA methyltransferases enzymes that transfer a methyl group from s-adenosyl-1-methionine (SAM) to the 5-position of cytosine forming 5-methyl-2-deoxycytidine (5-mdC), primarily found in CpG dinucleotides (Liou et al. 2017). Epigenetic mechanisms, such as DNA, RNA, histone modifications, and microRNAs have been shown to be potential links between the genetic and environmental exposure, which can be determinant to health life or reduction in life span. Environmental toxicants can alter epigenetic regulatory processes, and mediate specific mechanisms of toxicity and responses. Particularly, epigenetic modifications can alter genome expression and function under exogenous influence (Kahl et al. 2019). Growing evidence suggests that at least 15 environmental chemicals may lead directly to diseases via epigenetic mechanism-regulated gene expression

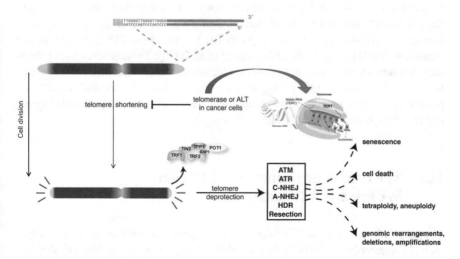

Fig. 12.3 Telomere with complex shelterin and telomerase. When TL reaches a critical limit, the cell undergoes senescence or apoptosis. (Modify from Jacobs (2013) and Alenalee (2008))

changes, such as hydrocarbons and inorganic elements (Santoyo et al. 2011; Liou et al. 2017). Other studies have brought evidence of the connection of global DNA methylation and mechanisms such as oxidative stress (Pavanello et al. 2010).

As important as knowing some biomarkers used to know and evaluate the effects of contaminants in free-living animals is knowing which are the best bioindicators for each assessment. Bioindicator is the organism that provides information about the environmental conditions of its habitat through its presence or absence or through its behavior (Van Gestel and Van Brummelen 1996). It is also defined as a species or group of species that readily reflect the abiotic or biotic state of an environment, representing the impact of environmental changes on a habitat, community or ecosystem. Either it is indicative of the diversity of a taxon subset or of all diversity within an area (Gerhardt 2000). According to Altenburger et al. (2003), bioindicators detect a biochemical aspect of toxic action (e.g., damage to the membrane, inhibition of enzymes, and damage to DNA), providing rapid and direct indications of the toxic impact on the environment. Bioindicators can reveal a lot about the mechanism of toxic action, which allows the extrapolation of related contaminants. However, bioindicators tend to be specific for toxic substances, as not all compounds inhibit the same biological processes. Thus, it is important to choose bioindicators relevant to the mechanisms of action to be known or evaluated.

The main anthropogenic activities studied using rodents refer to exhaustion of motor vehicles (Degrassi et al. 1999; Heuser et al. 2007), industrial emissions (Ieradi et al. 1998; Hazratian et al. 2017), polluted mining dump area (Andráš et al. 2006), and coal mining area (León et al. 2007). In general, contaminating compounds are complex mixtures and contain heavy metals and hydrocarbons. However, studies using subterranean rodents as bioindicators are rare. Subterranean rodents are interesting as bioindicators of soil pollution, mainly because they maintain the same territory for long periods and consume large amounts of vegetation. They play an important role in the dynamics of the ecosystem, due to their ability to modify the availability and dynamics of soil nutrients and resources for other species, because of their excavation system (Nevo 1979; Busch et al. 2000; Reichman and Seabloom 2002; Kerley et al. 2004; Dacar et al. 2010). Thus, studies with subterranean rodents could provide better information on contamination of the environment, the food chain, and, principally, on bioaccumulation, as well as data on potential exposures to wildlife and humans and on the effects of exposure to organic and metallic pollutants, mainly due to their ecology.

12.3 The Role of Environmental Pollution in *Ctenomys* Adaptation

The genus *Ctenomys*, popularly known as tuco-tucos, is composed of about 65 species of subterranean rodents (Teta and D'Elia 2020, Chap. 2 this volume) that live in burrows, but come to the surface to vocalize, clear the burrows, seek food, and have morphological characteristics adapted to this lifestyle, such as back paw bristles, reduced pinna, incisor teeth outside the mouth, as they use these teeth to aid excavation, not letting sand in their mouth (Parada et al. 2011). They share some common characteristics, including their solitary and territorial habits, small patchily distributed populations, and small effective population sizes, which are associated with low rates of adult dispersal and lead to a pattern of low genetic variation within populations and high genetic divergence among populations. Furthermore, the species commonly show high levels of karyotypic variation (Reig et al. 1990; Nowak 1999; Lacey et al. 2000).

The genus has a large geographic distribution extending from the extreme south of the Neotropical region to the south of Peru, spreading all over the Patagonic region with a great latitudinal variation, and recorded from the sea level up to 4000 m high in the Andean region (Reig et al. 1990). In Brazil, eight species of tuco-tucos are described. Three of them, *Ctenomys rondoni* Miranda Ribeiro, 1914, *C. bicolor* Miranda Ribeiro, 1914, and *C. nattereri* Wagner, 1848 are still poorly investigated. All the other five species of tuco-tucos, *Ctenomys torquatus* Lichtenstein, 1830, *C. minutus* Nehring, 1887, *C. flamarioni* Travi, 1981, *C. lami* Freitas, 2001, and *C. ibicuensis*, de Freitas et al., 2012, occur in southern Brazil, in the States of Rio Grande do Sul (RS) and Santa Catarina (SC).

C. torquatus is endemic to southern South America (Freitas and Lessa 1984). Despite the wide geographic range, there is a more common chromosomal form ($2n = 44$), and karyotype polymorphisms in the southern regions, restricted to the limits of the Atlantic Ocean ($2n = 46$), and Western ($2n = 40$ and 42) of its distribution (Freitas and Lessa 1984; Fernandes et al. 2009; Gonçalves et al. 2009). Its geographic distribution in RS almost coincides with the distribution of coal reserves (Freitas 1995). This coincidence took our research group to these regions to analyze possible genetic damage in the tuco-tucos that lived there. Because damage to DNA is not immediately recognized in organisms and has broad-ranging effects, this rodent could be an important system to monitor changes in environmental genotoxicity.

Since the late 1990s, our research group has been studying *C. torquatus* (Da Silva et al. 2000a, b; Silva et al. 2000a, b; Matzenbacher et al. 2019) and *C. minutus* (Heuser et al. 2002) as an environmental quality bioindicators in Rio Grande do Sul.

Silva et al. (2000a, b) conducted a 2-year study to detect the effects of coal, comparing the results with MN assay and CA to *C. torquatus* ($2n = 44$). At the end of 2 years, 240 rodents had been analyzed (capture-mark-recapture method). The locations studied covered three locations in RS: Candiota, a region about 2 km from the Presidente Médici coal power plant, Butiá, a region approximately 5 km from a strip coal mine, and Pelotas, a region without a coal mine and power plant. Biological hazards associated with Candiota coal field were investigated in a pilot study. The results showed that coal and derivatives induced DNA and chromosomal lesions in *C. torquatus* cells that were demonstrated by CA and MN test. The CA was more sensitive and showed a direct relationship between age and damage, and an inverse relationship between temperature and damage index. In addition to the authors demonstrated higher concentrations of heavy metals for soil samples from coal regions (Zn, Ni, Pb, Cd, V, and Cu), as well as hydrocarbons, and a relation of these concentrations and DNA damage. Other studies using free living rodents, yellow-necked mouse (*Apodemus flavicollis*), and bank vole (*Clethrionomys glareolus*), which were exposed to a coal mining area (Czech Republic) also demonstrated higher levels of DNA damage (Degrassi et al. 1999). Besides, León et al. (2007), using CA assay in wild rodents *Rattus rattus* and *Mus musculus*, exposed in a coal mining area (Cordoba, Colombia) show that mice and rats originating from the coal mining area exhibited a significantly higher extent of DNA damage as assessed by length of DNA migration, damage index, and percentage of damaged cells compared to animals from a control area.

Another study was conducted at the same sites of Silva et al. (2000a, b), with the same species of tuco-tucos, to evaluate the effect of exposure to coal and its derivatives and to examine the relationship of coal exposure with variations in TL, global DNA methylation, and genotoxicity (Matzenbacher et al. 2019). The study showed a significant reduction in the TL of the exposed tuco-tucos compared to the unexposed ones. Moreover, it demonstrated no association to factors such as sex and age with coal exposition. In this study, no relationship was found between global DNA methylation and exposure to coal, as well as no correlation between TL and DNA methylation, probably due to our small sample size. But a relation between more damaged cells in adults may be related with the reduction in the adults' number from Candiota. The reduction in TL is normal and expected, but this reduction is greater in exposed animals, which means that something is accelerating this loss in this region of the chromosome leading to a premature senescence. Our results demonstrated that *C. torquatus* suffer DNA damage/instability, as observed in the CA (DNA damage), and telomere shortening, likely as a consequence of the oxidative damage that results from their exposure to a complex mixture, including inorganic and organic elements. To prove the possibility of this mechanism, metal analyzes were carried out and Zn, Ni, Pb, Cd, V, and Cu were detected in soil. Martiniaková et al. (2010) studies determine the concentrations of heavy metal in the liver, kidney, and bone of yellow-necked mice (*A. flavicollis*) and bank voles (*C. glareolus*) trapped in a region with a chemical plant, coal power station, and coal mines in Nováky, Slovakia. Highest concentrations of Cd and Zn were found in the bone of both yellow-necked mice and bank voles. Cu and Fe accumulated mainly in the kidney and liver. There are many studies that report the relation between some coal exposure and the high concentration of chemical elemental and oxidative stress indexes such as the highest contents of S, Cl, Fe, Zn, and Br in frog *Hypsiboas faber* (Zocche et al. 2014); the reduced survival to metamorphosis in exposed larvae in grass shrimp (*Palaemonetes pugio*) (Kuzmick et al. 2007); a reduction in DNA repair capacity in *Mytilus edulis* with increasing duration of exposure to genotoxic agents (Steinert et al. 1998).

The other specie studied by our group was *C. minutus* which has a wide distribution, occurring from the south of "Farol de Santa Marta" (SC) to "São José do Norte" (RS). Its distribution is related to the formation of the coastal plain in southern Brazil and presents patterns of karyotype diversity that correspond to regions where there were paleochannels limiting the dispersion between parapatric locations (Lopes et al. 2013). They have seven parental karyotypes ($2n = 50a$, 46a, 48a,

42, 46b, 48b, and 50b) and current hybridization zones with intermediate karyotypes (Freygang et al. 2004; Marinho and De Freitas 2006; Lopes et al. 2013). These species have been used in genetic and population studies because they exhibit high karyotypic variability and deserves special attention in studies related to conservation.

Heuser et al. (2002) evaluated a possible genotoxic effect of vehicle emissions in *C. minutus* ($2n = 46$) on both sides of a highway on the coastal region (Amaral and Weber), and Maribo, a control area. Peripheral blood of *C. minutus* was used to perform MN test and CA, and the soil from their burrows to analyze the hydrocarbon concentration and the presence of some metals related to vehicle emission. In addition, concentration of NO_2 in the air also was measured. The study showed that the DNA damage rate was higher near the highway, as well as the average NO_2 concentration. Adult females showed greater DNA damage. The metals found in the soil of the highway with higher concentrations were Cr, Ni, Cu, and Zn, and the hydrocarbons were also shown higher in the two studied points in the highway in relation to the control region. This study provided chemical and biological data from areas exposed to automobile exhaust, indicating the association among environmental agents with levels of damaged cells observed in the wild rodent *C. minutus*. Similarly, Degrassi et al. (1999) found in the three parameters of measurement of the MN test (total number, average, and frequency) increase in the exposed area of Muro Torto, Rome (Italy) compared to the control area of the zoo under study with house mice (*Mus musculus domesticus*); however, this difference was not statistically significant. Vehicular emission also demonstrated affect the vegetation growth as shown in the study of Wagh et al. (2006) that evaluate pollution impact on the vegetation along the road in Jalgaon City, Maharashtra (India), and observed that vegetation at roadside with heavy traffic had less leaf area, total chlorophyll, and total proteins in leaves. Hazratian et al. (2017) to assess the potential use of the Norway rat, *Rattus norvegicus,* as a bioindicator for lead and cadmium accumulation in 10 urban zones in Tehran, Iran. The anthropogenic activities and vehicular emissions contribute to the entry of toxic metals to humans and other animal's food chains. The accumulation of heavy metals in some free-living rodents has been extensively studied (Šumbera et al. 2003; Guerrero-Castilla et al. 2014). Other studies also showed the relation between inorganic elements present in vehicles exhaustion and DNA damage in different species (Meireles et al. 2009; Brito et al. 2013).

Schleich et al. (2010) evaluated the concentrations of four heavy metals (Pb, Zn, Fe, and Cu) in muscle and liver from the subterranean rodent *Ctenomys talarum* from natural dunes, cultivated area, and military area of Buenos Aires Province, Argentina. The study revealed a higher concentration of metals in the livers than in the *C. talarum* muscles in the military (Fe and Cu) and cultivation areas (Pb and Cu). In soil samples, the highest concentrations of metals found were Fe in both military and cultivated areas. In the vegetation samples, low levels of metals were found, with Cu being most abundant in the military area, Pb in the cultivation area, and Zn in both cultivation and dunes areas.

In addition to the biomarkers that assess the effects of contamination on free-living animals, other markers of equal importance for the conservation and adaptation of these animals are interesting to be considered for a more comprehensive assessment of these effects. Knowledge of the genetic structures of these populations is an example. Lopes and de Freitas (2012) studied the effects of habitat reduction and fragmentation in four *C. lami* populations, due to human occupation, progressive urbanization, and expansion of agriculture and livestock. *Ctenomys* have a limited geographical distribution, as they are small populations distributed irregularly and low levels of adult dispersal (Reig et al. 1990; Nowak 1999; Lacey et al. 2000). In this study, it was demonstrated that the populations had no genetic structure associated with the distinct karyotype groups. However, molecular mitochondrial and nuclear markers demonstrated the existence of two populations instead of the four populations at the beginning of the study. These two populations are not completely isolated but are probably reinforced by a geographical barrier. The vulnerability of *C. lami* was greater than the authors previously assumed, then the author suggested the designation of evolutionarily significant units (ESU) to one population, based on their genetic differentiation for both molecular markers (nuclear and mitochondrial), and a Management Unit (MU) to another one, which considers statistically significant divergence in allele frequencies (nuclear or mitochondrial), no matter the phylogenetic differentiation of the alleles. In addition, they suggested the inclusion of the conservation status of this species as vulnerable (Lopes and de Freitas 2012).

Another interesting study involving conservation and habitat loss was carried out with Magellan tuco-tuco. Magellan tuco-tuco (*Ctenomys magellanicus*) is the southernmost Patagonian-Fueguian rodent, with a small distribution, which was categorized as vulnerable due to a strong population decline caused by overexploitation and loss and degradation of habitats produced by grazing sheep. In a study by Lazo-Cancino et al. (2020), the authors estimated the appropriate habitat distribution for *C. magellanicus* and predicted the appropriate habitat distribution and potential range changes under future climate change. According to the study, seven climatic variables, associated with the water regime and temperature variation between seasons in Patagonia, were the most important for the distribution of species. Under most future climate change scenarios, suitable habitat for *C. magellanicus* would decline mainly in its current continental distribution, with drastic loss and fragmentation of suitable habitats. This information is important for predicting a bottleneck, which results in a decrease in genetic diversity, which is extremely important for the viability of a population.

Animals such as tuco-tucos, which have reduced population size and still suffer from the impact on their environments, such as habitat fragmentation, are more susceptible to suffering DNA damage from exposure to pollutants and have a more catastrophic effect on your population structure. Figure 12.4 shows us a summary of the way that some chemical agents, in the form of a complex mixture, and fragmented habitats could be affecting small populations of mammals such as the tuco-tucos. These habitats are also a source of reduced genetic diversity, due to the degradation of the environment, and the emergence of new mutations that are leading them to an evolutionary adaptation to these effects (Matson et al. 2006; Pedrosa et al. 2017). Species have developed adaptations that promote enough aptitude to

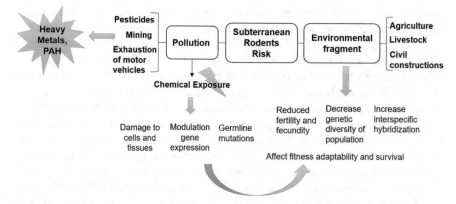

Fig. 12.4 Pressures that subterranean mammals suffer from anthropogenic actions

sustain different populations for long periods of time and environmental changes. The adaptations reflect responses to the selection imposed by toxins that characterized the environment (Brady et al. 2017).

12.4   Future

Little is really known about the physiology of subterranean mammals, and much less about the effects to which they are subjected in relation to the contamination that their habitats are exposed to every day. Although underground environments are considered more stable and simpler than those above ground, their characteristics are important ecologically and evolutionarily at the level of development of the species. The characteristics of underground niches have led to similar evolutionary pressures that have resulted in morphological, physiological, and behavioral adaptations converging on underground life, wherever they occur in the world. Subterranean rodents (e.g., *Geomys*, *Thomomys*, *Ctenomys*) play an important role in the dynamics of the ecosystem. They are considered ecosystem engineers because of their ability to modify the availability of resources directly or indirectly for other species. Its large excavation systems affect the texture, water holding capacity, soil nutrient dynamics, and vegetation composition and abundance (Nevo 1979; Jones et al. 1994; Busch et al. 2000; Reichman and Seabloom 2002; Kerley et al. 2004; Dacar et al. 2010).

Studies with conservation genetics or landscape genetics, conducted with populations of tuco-tucos, concluded that contemporary habitat fragmentation increases population differentiation. Genetic analysis of the landscape suggests that habitat quality and longitude were the most strongly associated environmental factors (Lopes and de Freitas 2012; Mora et al. 2017). It may be time by now, given that our knowledge of genetic toxicology has improved and that we also technically are better able to investigate DNA damage making use of modern molecular biological techniques, to start thinking on a new test strategy. Some examples are in silico methods, transcriptome approach, as well as next generation screening and sequencing tests (Perkins et al. 2003; Barzon et al. 2011). Transcriptomic profiles obtained from samples of wild animals considered to be environmental bioindicators may highlight candidate genes associated with pollution tolerance. Searching selection signatures using single nucleotide polymorphisms (SNP) genotyping offers a complementary route to explore adaptive evolution and has the gain of being able to provide information on selective pressures that affect all kinds of tissues (Hamilton et al. 2016). In silico toxicology can complement the predominant in vitro and in vivo toxicity tests, predicting toxicity, and prioritizing chemicals or drugs in order to minimize harmful effects (Parthasarathi and Dhawan 2018). New DNA sequencing techniques, known as "next generation sequencing" (NGS), offer high speed and throughput that can produce a huge volume of DNA sequences with many applications in research and diagnosis. The benefit is the determination of sequence data from single DNA fragments from a library that are secreted into chips, preventing the need for cloning into vectors before the acquisition of the sequence. At this time, NGS technologies are applied for complete genomic sequencing, research of genomic, metagenomic, epigenetic diversity, the discovery of noncoding RNAs, and protein-binding sites and gene expression profile by RNA sequencing (Barzon et al. 2011).

12.5   Conclusions

In conservation genetics, the analyses performed with *Ctenomys* described in this chapter represent a new approach to assess the anthropogenic effect on natural mammal populations. These studies are important due to the emphasis on the viability of genotoxic assays and epigenetic tools in conservation studies. We are currently watching a growing trend of cooperation between ecotoxicologists and conservationists, as evidenced by the rising number of studies with this approach, which can improve research in the areas of conservation and evolution. Genotoxicity and alteration in the methylation patterns in *Ctenomys* can result in environmentally induced phenotypic plasticity, which may be transgenerationally inherited. More studies, including modern molecular biological techniques, will always be needed in the field of ecogenotoxicity and epigenetics to continue to shape our understanding of the mechanisms of controlling and creating phenotypic variation and its implications for evolution.

References

Alenalee (2008) English wikibooks- transferred from en.wikibooks to commons. Public Domain
Altenburger R, Segner H, van der Oost R (2003) Biomarkers and PAHs – prospects for the assessment of exposure and effects in aquatic systems. In: PAHs: an ecotoxicological perspective. Wiley, Chichester, pp 297–328

ANDRÁŠ P, KRIŽÁNI I, STANKO M (2006) Free-living rodents as monitors of environmental contaminants at a polluted mining dump area. Carpth J Earth Environ Sci 1:51–62

Barzon L, Lavezzo E, Militello V, Toppo S, Palù G (2011) Applications of next-generation sequencing technologies to diagnostic virology. Int J Mol Sci 12:7861–7884

Bickham JW, Sandhu S, Hebert PDN, Chikhi L, Athwal R (2000) Effects of chemical contaminants on genetic diversity in natural populations: implications for biomonitoring and ecotoxicology. Rev Mutat Res 463:33–51

Blasco MA (2007) The epigenetic regulation of mammalian telomeres. Nat Rev Genet 8:299–309

Brady SP, Monosson E, Matson CW, Bickham JW (2017) Evolutionary toxicology: toward a unified understanding of life's response to toxic chemicals. Evol Appl 10:745–751

Busch C et al (2000) Population ecology of subterranean rodents. In: Life underground: the biology of subterranean rodents. University of Chicago Press, Chicago

da Silva J (2016) DNA damage induced by occupational and environmental exposure to miscellaneous chemicals. Mutat Res Rev Mutat Res 770:170–182

Da Silva J et al (2000a) Effects of chronic exposure to coal in wild rodents (*Ctenomys torquatus*) evaluated by multiple methods and tissues. Mutat Res Genet Toxicol Environ Mutagen 470:39–51

Da Silva J, De Freitas TRO, Heuser V, Marinho JR, Erdtmann B (2000b) Genotoxicity biomonitoring in coal regions using wild rodent *Ctenomys torquatus* by Comet assay and micronucleus test. Environ Mol Mutagen 35:270–278

Dacar MA, Ojeda RA, Albanese S, Rodrı D (2010) Use of resources by the subterranean rodent Ctenomys mendocinus (Rodentia, Ctenomyidae), in the lowland Monte desert, Argentina. J Arid Environ 74:458–463

de Brito KCT, De Lemos CT, Rocha JAV, Mielli AC, Matzenbacher C, Vargas VMF (2013) Comparative genotoxicity of airborne particulate matter (PM2.5) using Salmonella, plants and mammalian cells. Ecotoxicol Environ Saf 94:14–20

de Freitas TRO, Fernandes FA, Fornel R, Roratto PA (2012) An endemic new species of tuco-tuco, genus *Ctenomys* (Rodentia: Ctenomyidae), with a restricted geographic distribution in southern Brazil. J Mammal 93:1355–1367

De Lange T (2005) Shelterin: the protein complex that shapes and safeguards human telomeres. Genes Dev 19:2100–2110

Degrassi F et al (1999) CREST staining of micronuclei from free-living rodents to detect environmental contamination in situ. Mutagenesis 14:391–396

Espitia-Pérez L et al (2018) Genetic damage in environmentally exposed populations to open-pit coal mining residues: analysis of buccal micronucleus cytome (BMN-cyt) assay and alkaline, Endo III and FPG high-throughput comet assay. Mutat Res Genet Toxicol Environ Mutagen 836:1–12

Fenech M (2002) Chromosomal biomarkers of genomic instability relevant to cancer. Drug Discov Today 7:1128–1137

Fenech M (2007) Cytokinesis-block micronucleus cytome assay. Nat Protoc 2:1084–1104

Fernandes FA, Fornel R, Cordeiro-estrela P, Freitas TRO (2009) Intra- and interspecific skull variation in two sister species of the subterranean rodent genus Ctenomys (Rodentia, Ctenomyidae): coupling geometric morphometrics and chromosomal polymorphism. Zool J Linnean Soc 155:220–237

Freitas TRO (1995) Geographic distribution and conservation of four species of the genus ctenomys in Southern Brazil. Stud Neotropical Fauna Environ 30:53–59

Freitas TRO, Lessa EP (1984) Cytogenetics and morphology of Ctenomys torquatus (Rodentia: Octodontidae). J Mammal 65:637–642

Freygang CC, Marinho JR, de Freitas TRO (2004) New karyotypes and some considerations about the chromosomal diversification of Ctenomys minutus (Rodentia: Ctenomyidae) on the coastal plain of the Brazilian State of Rio Grande do Sul. Genetica 121:125–132

Gerhardt A (2000) Biomonitoring for the 21st century. In: Gerhardt A (ed) Biomonitoring of polluted water, vol 40. R. Trans Tech Publications Ltd, Zurich-Uetikon, pp 1–12

Gonçalves GL, de Freitas TRO, Freitas TRO (2009) Intraspecific variation and genetic differentiation of the collared tuco-tuco (Ctenomys Torquatus) in Southern Brazil. J Mammal 90:1020–1031

Groh KJ et al (2015) Development and application of the adverse outcome pathway framework for understanding and predicting chronic toxicity: II. A focus on growth impairment in fish. Chemosphere 120:778–792

Guerrero-Castilla A, Olivero-Verbel J, Marrugo-Negrete J (2014) Heavy metals in wild house mice from coal-mining areas of Colombia and expression of genes related to oxidative stress, DNA damage and exposure to metals. Mutat Res Genet Toxicol Environ Mutagen 762:24–29

Hamilton PB et al (2016) Population-level consequences for wild fish exposed to sublethal concentrations of chemicals – a critical review. Fish Fish 17:545–566

Hazratian L, Naderi M, Mollashahi M (2017) Norway rat, Rattus norvegicus in metropolitans, a bio-indicator for heavy metal pollution (Case study: Tehran, Iran). Casp J Environ Sci 15:85–92

Hemminki K, Sorsa M, Vainio H (1979) Genetic risks caused by occupational chemicals. Use of experimental methods and occupational risk group monitoring in the detection of environmental chemicals causing mutations, cancer and malformations. Scand J Work Environ Health 5:307–327

Heuser VD, Da Silva J, Moriske HJ, Dias JF, Yoneama ML, De Freitas TRO (2002) Genotoxicity biomonitoring in regions exposed to vehicle emissions using the comet assay and the micronucleus test in native rodent Ctenomys minutus. Environ Mol Mutagen 40:227–235

Heuser VD, Erdtmann B, Kvitko K, Rohr P, da Silva J (2007) Evaluation of genetic damage in Brazilian footwear-workers: biomarkers of exposure, effect, and susceptibility. Toxicology 232:235–247

Ieradi LA, Moreno S, Bolívar JP, Cappai A, Di Benedetto A, Cristaldi M (1998) Free-living rodents as bioindicators of genetic risk in natural protected areas. Environ Pollut 102:265–268

Jablonka E (2017) The evolutionary implications of epigenetic inheritance. Interface Focus 7:20160135

Jablonka EVA, Raz GAL (2009) Transgenerational epigenetic inheritance: prevalence, mechanisms, and implications for the study of heredity and evolution. Q Rev Biol 84:131–176

Jacobs JJL (2013) Senescence: back to telomeres. Nat Rev Mol Cell Biol 14:196

Jones CG, Lawton JH, Shachak M (1994) Organisms as ecosystem engineers. Oikos 69:373

Kahl VFS, da Silva J (2016) Telomere length and its relation to human health. In: Telomere – a complex end of a chromosome. InTech, Rijeka, pp 163–185

Kahl VS, Cappetta M, Da Silva J (2019) Epigenetic alterations: the relation between occupational exposure and biological effects in humans. In: Jurga S, Barciszewski J (eds) The DNA, RNA, and histone methylomes. Springer, Cham, pp 265–293

Kalisz S, Purugganan MD (2004) Epialleles via DNA methylation: consequences for plant evolution. Trends Ecol Evol 19:309–314

Kerley GIH, Whitford WG, Kay FR (2004) Effects of pocket gophers on desert soils and vegetation. J Arid Environ 58:155–166

Kleinjans JCS, Van Schooten FJ (2002) Ecogenotoxicology: the evolving field. Environ Toxicol Pharmacol 11:173–179

Kuzmick DM, Mitchelmore CL, Hopkins WA, Rowe CL (2007) Effects of coal combustion residues on survival, antioxidant potential, and genotoxicity resulting from full-lifecycle exposure of grass shrimp (Palaemonetes pugio Holthius). Sci Total Environ 373:420–430

Lacey EA, Patton JL, Cameron GN (2000) Life underground: the biology of subterranean rodents. University of Chicago Press, Chicago

Lazo-Cancino D, Rivera R, Paulsen-Cortez K, González-Berríos N, Rodríguez-Gutiérrez R, Rodríguez-Serrano E (2020) The impacts of climate change on the habitat distribution of the vulnerable Patagonian-Fueguian species Ctenomys magellanicus (Rodentia, Ctenomyidae). J Arid Environ 173:104016

León G, Pérez LE, Linares JC, Hartmann A, Quintana M (2007) Genotoxic effects in wild rodents (Rattus rattus and Mus musculus) in an open coal mining area. Mutat Res Genet Toxicol Environ Mutagen 630:42–49

León-Mejía G et al (2011) Assessment of DNA damage in coal open-cast mining workers using the cytokinesis-blocked micronucleus test and the comet assay. Sci Total Environ 409:686–691

Liou SH et al (2017) Global DNA methylation and oxidative stress biomarkers in workers exposed to metal oxide nanoparticles. J Hazard Mater 331:329–335

Lopes CM, de Freitas TRO (2012) Human impact in naturally patched small populations: genetic structure and conservation of the burrowing rodent, tuco-tuco (Ctenomys lami). J Hered 103:672–681

Lopes CM, Ximenes SSF, Gava A, De Freitas TRO (2013) The role of chromosomal rearrangements and geographical barriers in the divergence of lineages in a South American subterranean rodent (Rodentia: Ctenomyidae: Ctenomys minutus). Heredity 111:293–305

Marinho JR, De Freitas TRO (2006) Population structure of Ctenomys minutus (Rodentia, Ctenomyidae) on the coastal plain of Rio Grande do Sul, Brazil. Acta Theriol 51:53–59

Martiniaková M, Omelka R, Grosskopf B, Jančová A (2010) Yellow-necked mice (Apodemus flavicollis) and bank voles (Myodes glareolus) as zoomonitors of environmental contamination at a polluted area in Slovakia. Acta Vet Scand 52:1–5

Matson CW et al (2006) Evolutionary toxicology: population-level effects of chronic contaminant exposure on the marsh frogs (Rana ridibunda) of Azerbaijan. Environ Health Perspect 114:547–552

Matzenbacher CA, Da Silva J, Garcia ALH, Cappetta M, de Freitas TRO (2019) Anthropogenic effects on natural mammalian populations: correlation between telomere length and coal exposure. Sci Rep 9:6325

Maximillian J, Brusseau ML, Glenn EP, Matthias AD (2019) Pollution and environmental perturbations in the global system. In: Environmental and pollution science. Elsevier, Amsterdam, pp 457–476

Meireles J, Rocha R, Neto AC, Cerqueira E (2009) Genotoxic effects of vehicle traffic pollution as evaluated by micronuclei test in tradescantia (Trad-MCN). Mutat Res Genet Toxicol Environ Mutagen 675:46–50

Mora MS, Mapelli FJ, López A, Gómez Fernández MJ, Mirol PM, Kittlein MJ (2017) Landscape genetics in the subterranean rodent Ctenomys "chasiquensis" associated with highly disturbed habitats from the southeastern Pampas region, Argentina. Genetica 145:575–591

Nevo E (1979) Adaptive convergence and divergence of subterranean mammals. Annu Rev Ecol Syst 10:269–308

Nowak RM (1999) Walker's mammals of the world, 6th edn. Johns Hopkins University Press, Baltimore

Parada A, D'Elía G, Bidau CJ, Lessa EP (2011) Species groups and the evolutionary diversification of tuco-tucos, genus Ctenomys (Rodentia: Ctenomyidae). J Mammal 92:671–682

Parthasarathi R, Dhawan A (2018) Silico approaches for predictive toxicology. In: In vitro toxicology. Elsevier, London, pp 91–109

Pavanello S et al (2010) Shorter telomere length in peripheral blood lymphocytes of workers exposed to polycyclic aromatic hydrocarbons. Carcinogenesis 31:216–221

Pedrosa J et al (2017) Evolutionary consequences of historical metal contamination for natural populations of Chironomus riparius (Diptera: Chironomidae). Ecotoxicology 26:534–546

Perkins R, Fang H, Tong W, Welsh WJ (2003) Quantitative structure–activity relationship methods: perspectives on drug discovery and toxicology. Environ Toxicol Chem 22:1666

Petras M, Vrzoc M, Pandrangi R, Ralph S, Perry K (1995) Biological monitoring of environmental genotoxicity in southwestern Ontario. In: Butterworth J, Corkum BE, Guzmán-Rincón LD (eds) Biomonitors and biomarkers as indicators of environmental change. Plenum Press, New York, pp 115–137

Reichman OJ, Seabloom EW (2002) The role of pocket gophers as subterranean ecosystem engineers. Trends Ecol Evol 17:44–49

Reig OA, Busch C, Ortells MO, Contreras JL (1990) An overview of evolution, systematica, population biology and molecular biology in Ctenomys. In: Nevo OA, Reig E (eds) Biology of subterranean mammals at the organismal and molecular levels. Allan Liss, New York, p 442

Richards CL, Bossdorf O, Pigliucci M (2010) What role does heritable epigenetic variation play in phenotypic evolution? Bioscience 60:232–237

Rohr P et al (2013) Evaluation of genetic damage in open-cast coal mine workers using the buccal micronucleus cytome assay. Environ Mol Mutagen 54:65–71

Salagovic J, Gilles J, Verschaeve L, Kalina I (1996) The comet assay for the detection of genotoxic damage in the earthworms: a promising tool for assessing the biological hazards of polluted sites. Folia Biol 42:17–21

Santoyo MM, Flores CR, Torres AL, Wrobel K, Wrobel K (2011) Global DNA methylation in earthworms: a candidate biomarker of epigenetic risks related to the presence of metals/metalloids in terrestrial environments. Environ Pollut 159:2387–2392

Schaeffer DJ (1991) A toxicological perspective on ecosystem characteristics to track sustainable development. Ecotoxicol Environ Saf 22:225–239

Schleich CE, Beltrame MO, Antenucci CD (2010) Heavy metals accumulation in the subterranean rodent Ctenomys talarum (Rodentia: Ctenomyidae) from areas with different risk of contamination. Folia Zool 59:108–114

Silva J, De Freitas TRO, Marinho JR, S. G., and E. B. (2000) An alkaline single-cell gel electrophoresis (comet) assay for environmental biomonitoring with native rodents. Genet Mol Biol 23:241–245

Singh NP, Stephens RE (1998) Microgel electrophoresis: sensitivity, mechanisms, and DNA electrostretching. Mutat Res 175:184–191

Steinert SA, Streib-Montee R, Leather JM, Chadwick DB (1998) DNA damage in mussels at sites in San Diego Bay. Mutat Res Fundam Mol Mech Mutagen 399:65–85

Sturla SJ et al (2014) Systems toxicology: from basic research to risk assessment. Chem Res Toxicol 27:314–329

Šumbera R, Baruš V, Tenora F (2003) Heavy metals in the silvery mole-rat, Heliophobius argenteocinereus (Bathyergidae, Rodentia) from Malawi. Folia Zool 52:149–153

Teta P, D'Elía G (2020) Uncovering the species diversity of subterranean rodents at the end of the world: three new species of Patagonian tuco-tucos (Rodentia, Hystricomorpha, Ctenomys). PeerJ 8:e9259

Tice RR et al (2000) Single cell gel/comet assay: guidelines for in vitro and in vivo genetic toxicology testing. Environ Mol Mutagen 35:206–221

Van Gestel CAM, Van Brummelen TC (1996) Incorporation of the biomarker concept in ecotoxicology calls for a redefinition of terms. Ecotoxicology 5:217–225

Wagh ND, Shukla PV, Tambe SB, Ingle ST (2006) Biological monitoring of roadside plants exposed to vehicular pollution in Jalgaon city. J Environ Biol 27:419–421

Yehezkel S, Segev Y, Viegas-Péquignot E, Skorecki K, Selig S (2008) Hypomethylation of subtelomeric regions in ICF syndrome is associated with abnormally short telomeres and enhanced transcription from telomeric regions. Hum Mol Genet 17:2776–2789

Zakian VA (2012) Telomeres: the beginnings and ends of eukaryotic chromosomes. Exp Cell Res 318:1456–1460

Zocche JJ et al (2014) Heavy-metal content and oxidative damage in Hypsiboas faber: the impact of coal-mining pollutants on amphibians. Arch Environ Contam Toxicol 66:69–77

Index

© Springer Nature Switzerland AG 2021
T. R. O. de. Freitas et al. (eds.), *Tuco-Tucos*,
https://doi.org/10.1007/978-3-030-61679-3

Printed in the United States
by Baker & Taylor Publisher Services